Urban Groundwater Pollution

Editor

David N. Lerner
University of Sheffield, UK

A.A. BALKEMA PUBLISHERS LISSE / ABINGDON / EXTON (PA) / TOKYO

Library of Congress Cataloging-in-Publication Data

UNESCO and the International Association of Hydrogeologists (IAH) coordinated the preparation of this book as an IAH contribution to the 5th phase of the UNESCO International Hydrological Programme Project 3.4 "Groundwater contamination due to urban development".

Cover design: Studio Jan de Boer
Typesetting: Charon Tec Pvt. Ltd, Chennai, India
Printed in the Netherlands by Gorter, Steenwijk

Published by: A.A. Balkema Publishers, a member of Swets & Zeitlinger Publishers
www.balkema.nl and www.szp.swets.nl

ISBN 90 5809 629 7

Contents

LIST OF FIGURES

LIST OF TABLES

PREFACE

Most of the developed world's population lives in cities, but can cities become sustainable entities, or will they always be importers of resources and producers of pollution? Rapid urbanisation of the developing world is putting even greater pressure on their city environments as the pace of growth far outstrips the planning and control infrastructure. In particular, what of the urban water supply?

Groundwater is the main source of water for the world's population, because it is cheap, usually of excellent quality and is often directly underfoot at low capital cost. This ease of access places groundwater at severe risk of pollution in cities because of the density of polluting activities overhead – industrial plants, pipelines, sewers, landfills and the like. In Europe, many cities have abandoned their urban boreholes in favour of reservoirs and rural boreholes, and now suffer problems of rising water tables as too little water is withdrawn from below. In cities which are still developing, groundwater is often polluted but must be used anyway for public supply in the absence of the capital to built the expensive infrastructures to import sufficient water from rural areas.

The issue of urban groundwater encapsulates many of our modern problems – sustainability, rapid urbanisation and rising living standards, pollution and pollution prevention, inadequate data, and complex management issues of optimal use of resources. The story of each city is different, but some themes are common across climates, geologies, development status and continents. These include:
- industrial pollution, with attendant conflicts of wealth creation and environmental impacts;
- sewage disposal and associated health risks;
- protection of groundwater quality;
- appropriate development and sustainable use of resources;
- complexity of interpreting and managing groundwater;
- integrated management of surface and groundwater.

The complexity of urban groundwater issues makes them worthy of separate study, which was recognised by UNESCO and the International Association of Hydrogeologists. They formed a Working Group under UNESCO IHP-V project 3.4 to prepare a state-of-the-art report for Western and industrialising-country hydrogeologists, planners and engineers. The group comprised:

David Lerner (Chairman)
Cees van den Akker
Steve Appleyard
Kamal Hefny
Claus Otto

Oliver Sililo
Ken Howard

The inaugural meeting was held at the IAH 27th Congress "Groundwater in the urban environment" which took place in Nottingham, UK in September 1997. The outline of the book was agreed at this meeting, and other authors were invited to contribute. A further meeting took place at the IAH 30th Congress "Past achievements and future challenges" which took place in Cape Town in December 2000. Unfortunately Oliver Sililo died soon after the Congress in Cape Town. His hydrogeological expertise and infectious good cheer are missed in both the South African and international communities of hydrogeologists.

The book is in two parts. The first part (Chapters 1–3) is an introduction and overview. Chapter 1 gives a view of urban development from the groundwater perspective. Chapter 2 explains the disparate and conflicting values of groundwater for water supply, effluent disposal, amenity and ecology. Chapter 3 overviews pollutants, both by type (e.g. chlorinated solvents) and by source (e.g. landfill). The book makes no attempt to be a hydrogeology or chemistry textbook as many of these are readily available. It has the objectives of highlighting the important properties controlling pollutant behaviour and attenuation, as well as sources, prevention and cure.

Chapters 4–9 form the larger part of the book. They deal with case studies but, rather than describing individual cities in detail, they work with representative environments. Each chapter covers several cities to draw out the common threads across cultures and continents.

We hope the book will be of value to three groups. Professional hydrogeologists will find the environments chapters most interesting, where they will see how their colleagues have analysed and interpreted other cities. Urban planners and water supply engineers should find the whole book gives them a new perspective on the often-damaging effects that cities have on their hidden but valuable groundwater. Not least, students in hydrogeology, planning, water supply and environmental science will find this a useful source of understanding and case studies on the complex interactions in urban areas.

Just before you read the book, I would like to acknowledge the efforts and patience of the authors in helping to prepare this unique book on urban groundwater pollution. Most especially, I wish to acknowledge the tremendous efforts by Jenny Chambers in rewriting, editing, organising, and drafting figures. Without her the book would never have been finished. Thank you, Jenny!

David Lerner
Groundwater Protection and Restoration Group,
University of Sheffield

CONTRIBUTORS

Cees van den Akker
Delft University of Technology, P.O. Box 5048, 2600 GA Delft, The Netherlands

Ahmed Alderwish
University of Sana'a, Yemen

Steve Appleyard
Water and Rivers Commission, Perth, Australia

Mike H. Barrett
Robens Centre for Public and Environmental Health, University of Surrey, Guildford, Surrey, GU2 5XH, UK

Gino Bianchi-Mosquera
Geomatrix Consultants Inc, 330 W Bay Street, Suite 140, Costa Mesa, CA 92627, USA

Cors van den Brink
Royal Haskoning, P.O. Box 8520, 3009 AM Rotterdam, the Netherlands

Jenny Chambers
Groundwater Protection and Restoration Group, Civil and Structural Engineering, Mappin Street, University of Sheffield, Sheffield S1 3JD, UK

Aidan A. Cronin
Robens Centre for Public and Environmental Health, University of Surrey, Guildford, Surrey GU2 7XH, UK

Ken Howard
University of Toronto at Scarborough, 1265 Military Trail, Scarborough, Ontario, M1C 1A4, Canada

Kamal Hefny
National Water Research Centre, Cairo, Egypt

Bob Kent
Geomatrix Consultants Inc, 330 W Bay Street, Suite 140, Costa Mesa, CA 92627, USA

David N. Lerner
Groundwater Protection and Restoration Group, Civil and Structural Engineering, Mappin Street, University of Sheffield, Sheffield S1 3JD, UK

Claus Otto
CSIRO Land and Water, Australian Resources Research Centre, 26 Dick Perry Avenue, Kensington WA 6151, Australia

Veronica Ritchie
Centre for Groundwater Studies, CSIRO Land and Water, Private Bag, P.O. Wembley WA 6014, Australia

John M. Sharp Jr
The University of Texas, Department of Geological Sciences, Austin, TX 78712, USA

Oliver T.N. Sililo[+]
Cape Water Programme, Environmentek, CSIR, South Africa

Craig Stewart
Geomatrix Consultants Inc, 330 W Bay Street, Suite 140, Costa Mesa, CA 92627, USA

Richard G. Taylor
Department of Geography, University College London, 26 Bedford Way,
London, WC1H 0AP, UK

Callist Tindimugaya
Water Resources Management Department, Directorate of Water Development, P.O. Box 19, Entebbe, Uganda

Willem Jan Zaadnoordijk
Royal Haskoning, P.O. Box 8520, 3009 AM Rotterdam, the Netherlands

Sheng Zhang
Groundwater Protection and Restoration Group, Civil and Structural Engineering, Mappin Street, University of Sheffield, Sheffield S1 3JD, UK

[+] Deceased

CHAPTER 1

Values and functions of groundwater under cities

Willem Jan Zaadnoordijk, Cors van den Brink, Cees van den Akker and Jenny Chambers

1.1 INTRODUCTION

The development of a city may depend on favourable groundwater conditions or hydro-geology, but cities grow for all sorts of reasons. The functions of the city directly affect groundwater conditions, but often they are the last thing to be considered, and this has great implications for management policies, particularly in very rapidly growing areas. This chapter describes the relationship between city and groundwater, and explores how it may be managed.

Man is a social animal. He chooses to live in groups, perhaps because large breeding groups draw on diverse strains and thus introduce possibilities of genetic variation, ensuring survival. It is a human instinct to flock together, firstly in caves, then villages, then towns and cities. Early settlements provided shelter from the weather and were next to fertile areas where food could be found. If conditions were right the settlement gradually grew over time into a large city. Grouping people together leads to the creation of different tasks, which leads to further expansion, and the differentiated community with a variety of functions evolves. In general, early settlements had natural limitations such as the availability of land for food production, or technological difficulties in the construction of dwellings. Initially people had little means to change these limitations, but as technical capabilities advanced then problems began to be solved: the control of flooding, and provision of a reliable water supply are just two examples.

The growth of capitalism meant a big stimulus to urban expansion; firstly with merchants and financiers, then when mechanical invention and industrialisation kicked in. Trade and the market place became the driving force – towns meant lots of consumers. Economic factors were the dominant force in developing cities. Some towns grew quickly by allowing free trade (Antwerp), whilst others grew because of their location on trade routes, particularly sea ports (Liverpool). Once the Industrial Revolution got under way, urbanisation increased in direct proportion to industrialisation, and growth was so rapid that there were no provisions for water standards, fire protection, hospital care or education. Production was all-important, and by-products were just cast away. This process can be seen repeating itself in third world countries now. Gradually civic duty rose out of the chaos, and city fathers began to press for the standards we take as commonplace today. The symbiotic relationship between city and groundwater is now beginning to be recognised, and management policies for both urban planning and water conservation are being developed.

In this chapter we discuss the functions of the city through its land uses, and the demands on and consequences for the groundwater. Then, we look at the aspects of hydrogeology that are important in the formulation of groundwater management, by introducing the concepts of groundwater flow systems and contaminant transport. We then combine the hydrological knowledge with the demands and consequences for the groundwater to formulate integrated management policies, by describing real examples from cities around the world and concluding with an ideal but theoretical illustration.

1.2   FUNCTIONS, CONSEQUENCES AND DEMANDS

The city and its functions

A city is a living organism with many functions. Although little attention was historically paid to groundwater, water is necessary for most functions of the city. Whereas some have little effect, others have great effect and cause pollution, waste products and change in levels. Available water sources are usually surface water and groundwater, although groundwater has a more constant and generally better quality and is therefore a more reliable source. Some functions may coexist in particular areas, whilst others may use (ground) water one after another, such as irrigation with wastewater. This results in a wide variety of interactions with the groundwater in urban areas (Fig. 1.1).

The impact of the functions on the groundwater and subsurface system can be described as "pressures". Pressures are the agents that potentially stress the environment. They fall into three main categories (EEA, 1999): (1) emission of chemicals, waste and radiation,

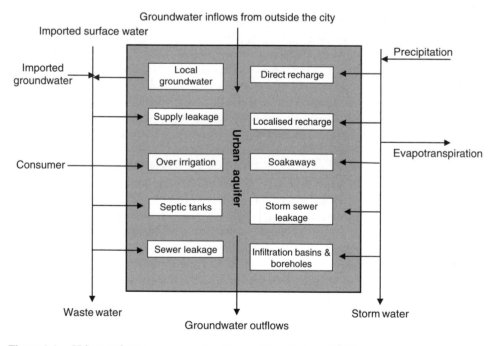

Figure 1.1.   Urban recharge sources and pathways (Barrett et al., 1999).

(2) use of environmental resources and (3) area of land used. We refer to the first category as the consequences and the latter two categories as the demands of the functions.

Functions of the city are linked to land use. Each function has consequences on the groundwater, regardless of the source of the water that is used. We distinguish between the following functions:

- Housing (drinking, washing, sanitation)
- Cultivation (allotments, gardens)
- Manufacturing
- Recreation
- Transport
- Commercial and social facilities

Their demands on soil and groundwater, and the consequences of these demands, are discussed in more detail below. They are also explored further in Chapter 2.

Housing

This is the primary function of a settlement. Within the houses, water is needed for drinking, washing, sanitation and for extinguishing fires. When there is a public water supply, it usually provides all water for households. Drinking water standards are normally specified in cities, in order to prevent disease and infection, and penalties are incurred where these are not met. However, this is not necessarily the case in the developing world.

The soil quality of at least the upper 1.5–2 m should meet high quality standards to prevent the city's inhabitants, particularly the more vulnerable members such as children and the elderly from the intake of soil-bound contaminants. Soil quality is also important with respect to the corrosion of underground infrastructure and storage tanks or deterioration of concrete.

The level of groundwater may affect the buildings in the city. Extraction of groundwater leads to lowering of groundwater levels, and levels that are too low may cause subsidence of the surface or rotting of wooden piles of foundations when they are no longer submerged below the water table. Levels that are too high may cause problems related with dampness or flooding of cellars and basements. In addition, there should be provision for storm water discharge and drainage to keep houses dry and to control the water level of the surface water. The buildings and pavements in the city reduce natural infiltration, which increases storm water flow and may create erosion problems.

Sewage disposal can affect the quality of groundwater, particularly in long-established settlements where the infrastructure may be aged and inefficient. Lerner (1986) showed that contamination of groundwater supplies occurs from leaking sewers. Septic tanks are not commonly found in developed cities, but are used in rapidly expanding settlements in the developing world and as such can provide a considerable source of pollution. When groundwater is used as a source of drinking water, it is vital that sanitation methods are of the highest efficiency to eliminate threats to human safety. This will be discussed further in Chapters 2, 3 and 6.

Cultivation

Originally, the inhabitants of a city would support themselves by farming. In today's city, garden plots and allotments provide food and greening, although in the fast-growing urban centres of the developing world agriculture is still an important function on small patches of land within the built-up areas (see Chapters 2 and 6). For this function, a particular soil water content or groundwater level is needed. Sometimes extra water (irrigation) or less water (drainage) is necessary, which has implications for the water levels. Control of the water level may cause lowering of groundwater levels or salinisation, and may vary during the year. The water used for irrigating agricultural areas does not have to meet drinking water quality. However, the salinity should not exceed certain limits to prevent salinisation of the soil in arid areas. The soil quality of at least the first 1.5–2 m should meet higher quality standards. In recent decades the nutrient condition of the soil has become of less importance because of the use of fertilisers. As a consequence, nitrate and metals from fertiliser and manure, together with persistent pesticides and metabolic products, leach to the groundwater.

The significance of non-agricultural sources of nitrogen for urban groundwater in Nottingham (UK) is discussed in Chapter 4. Fertiliser and agricultural practices are the main source of rural nitrates, but are insignificant in the city environment, where leaking sewers, landfills, and industrial spillages are much more important (Lerner et al., 1999). However, in many sub-Sahara cities small-scale agriculture within the city borders cause substantial loading of the shallow aquifer with nitrogen, and will be discussed in Chapter 6.

Manufacturing

Manufacturing industry is commonly the biggest source of nuisance in the city. Not only is it a big consumer of water for the various methods of production, which may include cleaning and cooling, but it also produces large quantities of waste, the disposal of which can have serious consequences for the functioning of the city. Many wastes contain potential groundwater contaminants: initially after the Industrial Revolution got under way it was mainly mining and foundry wastes that were predominant, but eventually they included tars, phenolic wastes, oils and more recently complex synthetic organic compounds. Accidental spills or leaks from tanks and pipelines carrying petroleum products have the potential to contaminate a very large volume of water to levels in excess of the low concentration acceptable in drinking water, and because of the slow rate of dissolution can remain as a source of pollution for decades.

Recreation

Playing fields and parkland usually represent a large area in terms of land use in the city, providing for the well-being of its citizens through recreation and on the other hand minimising the impact of the city on the ecosystem outside of the city. There is a trend in modern cities to use balancing lakes in parks as a recreational resource, as water sports are now popular recreational activities.

The surface water and groundwater leaving the city should be able to support the aquatic and terrestrial ecosystems outside the city.

Transport

We have seen that buildings and pavements in the city reduce natural infiltration and thus increase storm water flow. The major demands of road transport created by the huge numbers of cars owned by today's city dwellers means that large areas of the city are given over to roads and motorways, accentuating this process. The wide distribution of garages fuelling these cars and the extensive networks of pipelines carrying petroleum products to them are other potential hazards to the quality of groundwater.

Canals have historically been important for transport, both in bringing raw materials in and taking manufactured goods out, but increasingly now are used for recreation and stormwater drainage. Seepage from unlined canals can significantly lower the quality of groundwater.

Commercial and social facilities

Offices and most social facilities have the same kinds of demand as housing. Hospitals, however, are an exception, as they generate a certain amount of waste products which must be disposed of safely. There is further discussion relating to a case study in Chapter 5.

Consequences: when the relationship gets out of balance

What happens when the relationship between the city and its groundwater supply gets out of balance? To give a preliminary idea of the extent of urban groundwater pollution, the following paragraphs review some case studies of nitrogen, volatile organic compounds (VOCs) industrialisation, and landfills.

Kolpin et al. (1997) performed a systematic investigation of the occurrence of pesticides and VOCs in urban groundwater of nine population centres in the United States. Pesticides compounds were detected in 48% of the 208 urban wells sampled, with prometon by far the most frequently detected compound. VOCs were detected in 53% of the 208 urban wells sampled, with 36 compounds being found. Methyl *tert*-butyl ether (MTBE), a common fuel oxygenate, was the most frequently detected VOC.

Subba Rao and Subba Rao (1999) studied the impact of urban industrial growth on groundwater quality in Viskhapatnam, India. This port city has shown phenomenal urbanization and expansion since 1940. The population grew from 30 000 in 1900 to pass the one million mark in 1991. Major industries such as shipyards, petroleum refineries, zinc smelter plants, fertiliser units, ceramics and polymers have located in the city heartland. At the same time, 40% of the population lives in the industrial belt and the quality of groundwater has become seriously polluted. 15% of the industrial well waters were above $3\,000\,\mu S/m$, and major ions were above the "safe" limits at many places. Toxic elements including lead, chromium, arsenic, zinc and cobalt have been found in large amounts in the groundwater of the industrial zones. A zinc plant has been identified as the worst offender, followed by the port activities and polymer facility.

Mocanu and Mirca (1997) describe the groundwater contamination due to a landfill southeast of Bucharest, Rumania. This landfill, located in a former quarry on the south-eastern edge of the city in the Dambovita valley, receives 1 333 t/day of industrial and domestic waste from Bucharest. These wastes include dangerous substances from several hospitals and the chemical industry with components such as chromium, nickel and cyanide prevalent. The contaminated water on the floor of the quarry is in hydraulic connection with shallow aquifers discharging to the Dambovita river and to the deep aquifers which underlie and provide for the city. Some pollutants have already been identified in the deep aquifers, thus endangering the drinking water supply. Concentration ranges for chromium, nickel and cyanide in the upper part of this aquifer are 180, 1 900 and $20\,\mu g/L$ respectively.

Further examples to illustrate the interplay between pollution and groundwater behaviour are given in Box 1.1 for a case where induced infiltration of canal water is polluting groundwater, in Box 1.2 for a case where polluted shallow groundwater is being induced

Box 1.1.   Urban groundwater contamination by seepage: Hat Yai, Thailand (after Foster et al., 1998).

The city of Hat Yai in southern Thailand is situated on low-lying coastal alluvial deposits. The upper part of these deposits are of low permeability and have a shallow water table, which cause problems for wastewater and stormwater disposal. It is estimated that about 20% of wastewater disposal goes directly to the ground via unsewered sanitation units. The remainder discharges via drains to unlined drainage canals, which also receive storm-water runoff. As a result of heavy abstraction of groundwater within the urban area, the piezometric surface in the semi-confined aquifer has been significantly lowered.

Substantial leakage from the shallow water table to the semi-confined aquifer occurs, and canal seepage now represents the single most important component of groundwater recharge. Elevated concentrations of ammonium, chloride, and sulphate occur in the semi-confined aquifer beneath the city centre as a result of the poor quality of canal seepage. Where concentrations are highest, they represent mixing of some 60–80% canal seepage and 20–40% unpolluted groundwater.

Figure   Mixing of unpolluted regional groundwater flow and canal seepage in Hat Yai, Thailand. The most polluted urban groundwater has chloride concentrations indicating that they are largely derived from canal seepage, and occur where ground-water abstraction, and thus downward leakage, is greatest.

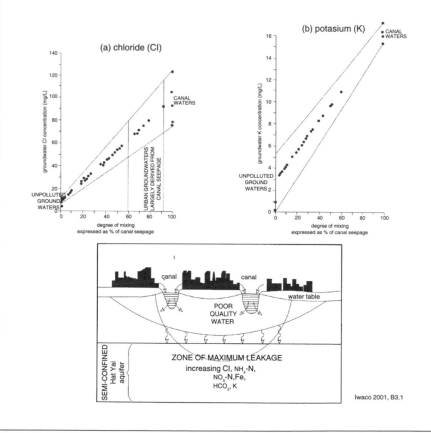

Iwaco 2001, B3.1

Box 1.2. Downward leakage of contamination induced by pumping: Santa Cruz, Bolivia (after Foster et al., 1998).

Santa Cruz, Bolivia, is a low-rise, relatively low-density, fast-growing city. The municipal water supply is derived entirely from wellfields within the city limits extracting from deep semi-unconfined alluvial aquifers. The supply is provided by co-operatives, and obtained from about 50 (90–350 m deep) boreholes, which provide 98 Ml/d (1994). There are also many private wells (some 550 in 1991), used for industrial, commercial, and some residential supplies. These wells are generally less than 90 m deep and draw principally from the shallow aquifer. The city has relatively good coverage of piped water supply, but until recently only the older central area had mains sewerage, and domestic/industrial effluent and pluvial drainage were mostly disposed to the ground. The main components of groundwater recharge (additional to natural infiltration of excess rainfall) are the on-site disposal of wastewater and leakage from the mains water supply. Seepage from the nearby Rio Piray is also believed to be significant, but is difficult to quantify precisely.

Groundwater in the deeper aquifer, below 100 m, is of excellent quality, similar to the shallow aquifer upgradient of the city, and this represents the natural condition. However, the uppermost aquifer above 45 m shows substantial deterioration with elevated nitrate and chloride concentrations beneath the more densely populated districts. These are derived from the disposal of effluent to the ground, mainly from on-site sanitation units. This urban recharge is then drawn downwards in response to pumping from the deeper semi-confined aquifers. Dissolved oxygen in the urban recharge is low, having been consumed as the carbon in the organic load is oxidised to carbon dioxide, which in turn reacts with carbonate minerals in the aquifer matrix to produce bicarbonate. The oxidation of the high organic load also mobilises naturally occurring manganese from the aquifer matrix, and some of the production boreholes in the main wellfields have started to show concentrations above 0.5 mg/L, leading to consumer taste and laundry problems.

Figure   Incipient contamination of public supply wells showing rises above background in those with screens starting above 90 m depth.

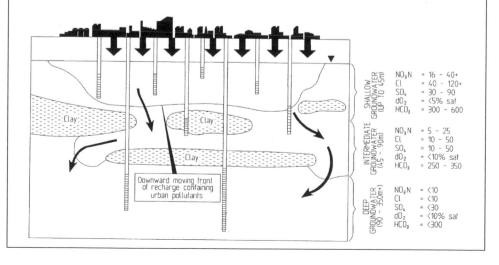

Box 1.3.   Groundwater contamination by saltwater intrusion: Barcelona, Spain (after Custodio, 1997).

Excessive urban and industrial water demand caused a strong drop of the piezometric water levels in Barcelona. As a result, saltwater intruded in some areas. In addition, the rivers that used to drain groundwater now infiltrate poor quality water into the aquifer. As a consequence of the decreasing quality and dropping water table, there was a progressive abandonment of wells and the recovery of the water table. Several underground car parks are waterlogged and a subway tunnel is now suffering from groundwater inflow. Some underground spaces have been abandoned due to the cost of pumping and the danger of creating geotechnical problems.

Because the aquifers in and around Barcelona are considered key sources for urban supply, management has been set up to preserve this role. This management is based on technical expertise, co-operation of the public and private sectors, and political will. Constant monitoring is needed.

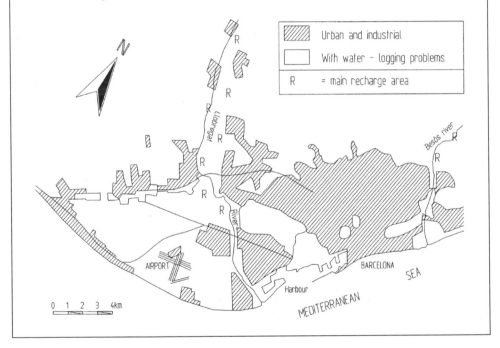

to migrate into deeper, cleaner aquifers, and in Box 1.3 for the common case where over-pumping has induced movement of saline groundwater into a freshwater aquifer.

1.3   GROUNDWATER FLOW AND FLOW SYSTEMS

Introduction

The flow of groundwater in the subsurface is determined by the permeability of the geologic layers in the underground and the interaction with the water on the surface. In general, groundwater is recharged by water from precipitation and discharges into local surface water or the sea. The subsurface flow is invisible, so it is not obvious to the general public what is happening. They usually have no notion how it is flowing, although they have found out that you can make a well and keep taking water. Decrease of the yield or deterioration of the water quality is a stimulus to find out what is happening underground.

The most common method used to investigate groundwater is to drill a hole in the ground. The water level in the borehole can then be measured, and this is referred to as the piezometric head. Water samples can also be taken and analysed for chemical and physical characteristics. The depth of groundwater below the surface is an important factor in its availability for plants and people, and thus for land use. As we have seen, the groundwater in an urban environment also affects the structural design of buildings, stability of the ground, and can be influenced by environmental risks such as spillages and releases of contaminants. The groundwater flow will determine how the depth of the groundwater, the piezometric head and the groundwater quality will change when water is extracted or other changes are made to the hydrogeological situation. Mathematical models were introduced long ago in the study of hydrogeology to make this easier. Nowadays, we cannot do without numerical models as a basis for groundwater management.

In the following sections an overview will be given of the aspects of groundwater that are important for the urban environment. More information can be found in textbooks such as Cherry and Freeze (1979), de Marsily (1986), Domenico and Schwarz (1990) on general hydrogeology, or Bear and Verruijt (1987), Anderson and Woessner (1992) on hydrogeological modelling, and Lerner et al. (1990) on recharge.

Principles of groundwater flow

Groundwater flows depend on the geology of the subsurface and the hydrological conditions at the surface (Fig. 1.2). The geology often allows us to distinguish layers in the subsoil that have distinctly different resistances to groundwater flow. These layers have a much larger lateral extent (kilometres to hundreds of kilometres) than vertical extent (metres to hundreds of metres). This means lateral flow takes place mostly in the layers with lower resistances, known as aquifers. Flow in the more resistant layers, known as aquitards, is almost vertical. An aquifer is called confined when an aquitard (a confining layer) line both top and bottom of the aquifer; a semi-confined aquifer has higher permeability in the associated aquitards so that some flow can take place in them.

An aquifer may contain a water table. Then, it is called a water table or phreatic aquifer. The part above the water table is called the unsaturated zone and the groundwater flow in this zone is referred to as unsaturated flow. It is multiphase flow, involving both air and water, and is much more complicated than saturated groundwater flow: here the resistance to flow of the soil is not constant but increases strongly when the amount of water in the soil decreases.

The resistance to flow of the layers determines the distribution of the groundwater flow. Aquitards are usually characterised by the total vertical resistance (with a physical dimension of time), since the flow is mainly vertical. For aquifers a reciprocal value is used: a conductance instead of a resistance. The hydraulic conductivity or permeability is the constant that relates the flow to the decrease of head over a unit distance. The total horizontal flow in an aquifer is governed by the transmissivity which is the product of the horizontal permeability and the saturated thickness.

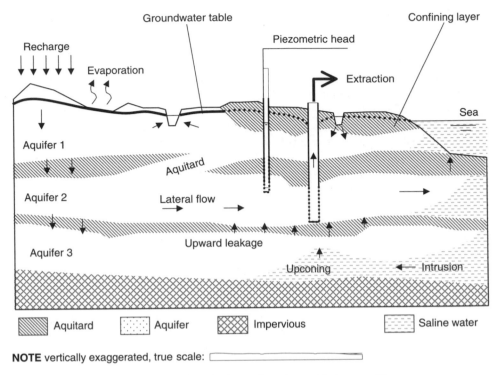

Figure 1.2.   A schematic cross-section of a layered geology with three aquifers.

Besides the hydrogeological properties of the subsoil, groundwater flow is determined by the conditions at the surface and features like pumping wells. Water from precipitation and snowmelt enter the ground to feed the groundwater. Pumping wells and evapotranspiration are sinks or discharges of groundwater. Lakes and rivers interact with the groundwater putting a control on the groundwater levels. They may either recharge to or discharge from the groundwater.

Groundwater flow continuously changes, since the hydrological conditions at the surface change. There are seasonal variations in precipitation, and there are one-directional trends such as putting pavement and buildings on the ground so that precipitation does not go into the ground any more. The ease in which extra water can be stored in the subsurface ("storativity") determines the reaction time to these changes. Changes are damped much more in an aquifer with a free water table than in an aquifer in between two confining layers. It can take as little as a day for a small confined aquifer to respond or as much as centuries for a thick phreatic aquifer. A determining parameter is the aquifer diffusivity, which is the ratio of the transmissivity and the storativity of the aquifer.

It is not always necessary to consider the transient character of the groundwater. In many practical cases, it is possible to consider an average situation over a period and use a steady state approximation of the transient flow.

Groundwater is recharged, mostly from precipitation (minus evaporation, minus drainage and capture on buildings and paved surfaces) and irrigation. Similarly, it discharges to low surface water levels in wetlands, rivers, sea and artificially drained areas. Groundwater can

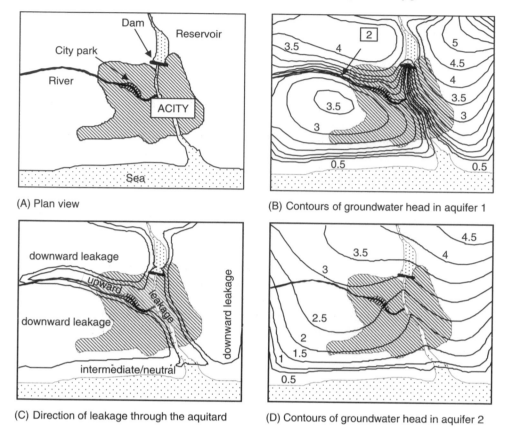

(A) Plan view

(B) Contours of groundwater head in aquifer 1

(C) Direction of leakage through the aquitard

(D) Contours of groundwater head in aquifer 2

Figure 1.3.   Heads and flows in a multilayer situation (see Fig. 1.2).

also be extracted by means of wells. Initially there is a local lowering of the groundwater heads after the pump has been switched on. Water is mainly drawn from storage. Gradually, a larger area is influenced, the regional inflow pattern is changed, and the influence of the well comes to a steady state. However, this steady state is only reached when the extraction is sustainable. The sustainable maximum yield is limited by available recharge minus required discharge, together with environmental and engineering constraints.

The mathematical description of groundwater flow extends the simple notion that water flows from high to low to a quantification of the amount of flow. For groundwater, "high" and "low" are defined in terms of the piezometric head measured with respect to a datum, which usually is sea level. The flux depends on the difference between the high and low head and the hydrogeological properties of the ground in between. The mathematical description forms the basis for simulation models. These models are computer codes that calculate piezometric heads and groundwater fluxes from information on geology and hydrological conditions.

Figure 1.3 shows plan views of the heads and flows in the aquifer system which is shown in Figure 1.2. The subsurface consists of two aquifers separated by an aquitard. The groundwater is recharged by precipitation, and is drained by rivers and the sea. The influence of the rivers and local recharge is much stronger in the shallow aquifer while the influence of the sea and the large-scale infiltration mainly determines the flow in the second aquifer. In most of the area the flow through the aquitard is downward due to the infiltration. The drainage of the rivers causes an upward flow toward the rivers.

Transport of chemicals in groundwater

So far we have explained the flow of groundwater and discussed piezometric heads, the water table, recharge and discharges. Groundwater quality is connected to the physical and chemical properties of the water, and is influenced strongly by the transport of dissolved substances and temperature of the groundwater. Transport by the flow of the water is the most important, although several other mechanisms can play a role (Cherry & Freeze, 1979; Domenico & Schwarz, 1990).

The transport velocity is expressed in the same units as groundwater fluxes such as recharge. However, the transport velocity is larger by a factor that is the inverse of the effective porosity. The total porosity is the fraction of the geological formation that is filled with water when it is fully saturated. The effective porosity, or the fraction that actually takes part in the groundwater flow, is less due to dead end pores and other effects. The difference is small for sand, large for clay, and variable for rock. We talk about dual porosity behaviour if the stagnant zones are important for transport. A good example of this is a rock formation that has large fractures in which almost all groundwater flow takes place, together with small pores that slowly accumulate contaminants when they enter the formation and then slowly release them when clean water flows again through the formation. In karst we find "preferential pathways", or shortcut routes for the transport. Chapter 7 gives an example from Kentucky, where a distance has been documented of 1 km travelled in one day. Cracks in clay due to drought can act in a similar way when the rain season follows. The initial precipitation may wash dirt from the surface and take it deep into the subsoil due through the cracks.

Groundwater flow systems

Water from precipitation or surface water infiltrates into the ground and feeds the groundwater in some areas, and water leaves the ground flowing to either wells or draining surface waters in other locations. Each drop of water follows some path through the subsoil. If we could follow an individual drop of water, we would determine a path line through the subsoil from the infiltration to the discharge point. The subsoil together with other influences on the groundwater flow uniquely determines this path line. We can also look back in history and determine where discharging water originally entered the ground.

Let us consider a pumping well and assume that we have enough knowledge of the flow situation to enable us to determine the origin of all water that is pumped out today. This knowledge would allow us to build a groundwater model in which we can trace path lines back in time. The results of the calculations show one or more areas where water that infiltrated at some specific time in the past flowed into our well today. Moreover, we would know through which part of the subsurface the water passed on its way to the well (Fig. 1.4).

The quality of the infiltrating water together with the geochemical composition of these parts of the subsoil determined the quality of the pumped water. Suppose that we could also trace back the water that was pumped half a year ago, and a year ago. We would find corresponding recharge areas to the water pumped at these times. The recharge areas for different times largely overlap if the flow does not change too much. This is usually the case when there are only seasonal variations but no strong trends in the groundwater and

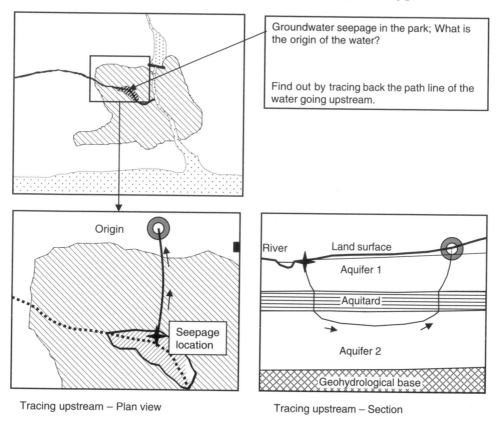

Figure 1.4. Tracing the origin of groundwater from discharge point upstream to recharge area.

is known as the capture zone of the well (e.g. Bair et al., 1991; Lerner, 1992; Bakker & Strack, 1996). Many countries have based their regulations for protection of drinking water wells on the delineation of capture zones. Usually, the government restricts land use within the capture zone to prevent contamination entering the soil and contaminating the well.

The capture zone together with the portion of the subsoil that the water flows through on its way to the well can be called the flow system of the well. Groundwater systems analysis stems from ideas put forward by Tóth (1963), and has been extended recently (e.g. Engelen & Kloosterman, 1996). It is closely related to the origin-path-use approach, also called source-pathway-target that is common in groundwater and soil remediation (Van den Brink et al., 2000).

As a complementary exercise, we can follow infiltrating water to the discharge location it will ultimately reach, which is the reverse of the tracing backwards exercise previously discussed (Fig. 1.4). The water particles are traced into the future. Now we can distinguish between systems based on the quality of the infiltration water. Land use is an important factor in this distinction. We can make subdivisions if the chemical composition of the subsoil varies significantly, to allow us to categorise groundwater systems with a typical groundwater quality. This type of groundwater system is determined by the future groundwater flow, but is useful only when the flow exhibits no strong temporal trends.

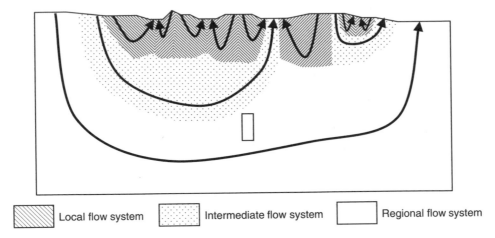

| Local flow system | Intermediate flow system | Regional flow system |

Figure 1.5.   Hierarchical groundwater flow systems (after Tóth, 1963).

Tóth (1963) introduced hierarchical flow systems in a cross section (Fig. 1.5). An important aspect is the fact that groundwater systems can be distinguished at different scales (Zijl, 1999). The distinction of the systems at different levels serves different purposes. The groundwater flow systems connected with large rivers play a role in (international) river basin management, for example as set up by the USA and Mexico for the Colorado river. Systems connected with single wells are the basis of the containment and remediation of groundwater contamination. Groundwater flow systems are determined by means of modelling supported by geochemical classification of groundwater samples (Bair & Roadcap, 1992; Bakker & Strack, 1996; Frapporti, 1993; Stuyfzand, 1999).

Groundwater flow systems are useful in the management of groundwater at all levels. The appropriate scale has to be chosen for the management tasks at hand, and the problems to be tackled must lead the choice of the recharge or discharge units selected to start the flow system analysis. Groundwater flow systems provide a means of linking the groundwater at a location to either the origin or the destination of that groundwater. This enables the consequences of the land use in an infiltration area to be translated into consequences for the entire downstream flow system. Similarly, the demands on the groundwater of a land use in an discharge area can be linked to restrictions in the corresponding infiltration area and underground activities in the subsoil part of the flow system.

For effective groundwater management in a city, the flow systems of all water that discharges in the city together with the water that infiltrates there have to be considered (Fig. 1.6), as well as the flow systems at the next lower level.

Spatial and time scales of systems

We have seen that recharge and drainage are the driving forces for groundwater flow. Both recharge and drainage change significantly when a rural or undeveloped area changes into an urban area, and also when the urban environment grows further. Groundwater systems change, and the quality of the discharging water does not correspond with the quality of the water infiltrating in the system. Groundwater system analysis can thus provide a tool

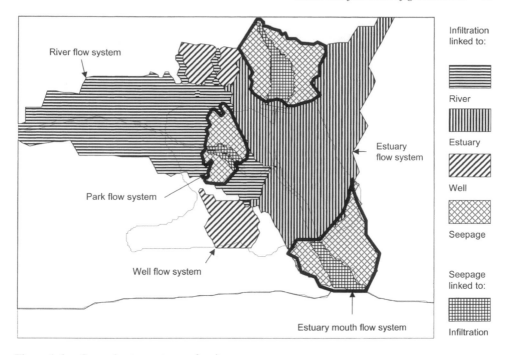

Figure 1.6.   Groundwater systems of a city.

for the assessment of the autonomous changes in the groundwater quality. These changes depend on changes in the boundary conditions, such as infiltration and drainage, and the response of the groundwater flow to these alterations. The response of the flow depends on the geo-hydrological situation. The time scale of adapting to current boundary conditions ranges from months to years or centuries. The determining factor is the aquifer diffusivity together with the size of the system. In a confined system with close boundaries, the flow may adapt fully within a month or even less, but it may take millennia in arid regions with a deep water table where the infiltration occurs over a large area.

The changes in a flow system can be clearly illustrated by changes in well extractions. The upper half of Figure 1.7 shows two plan views of the flow systems around a city. One shallow well field in the southern part of the city and two well fields to the north are active in the old situation on the left. In the new situation on the right, the southern well field has been shut down and the pumping in the northern ones has been increased four and six fold.

Comparison shows that not only the capture zones of the northern well fields have increased but also the infiltration zones connected with the discharge into the rivers have moved. The flow systems of the new situation seem to be valid after about fifty years from the change in pumping, when the piezometric heads and the groundwater flow have reached a new equilibrium. However, most travel times in these flow systems are a hundred years or more, as can be seen from the bottom part of Figure 1.7. This means that it will take many more years until the connection between the infiltration and discharge is established in terms of the water quality. In the meantime there will be water present in the flow systems that originated from outside the system and may have a different water quality. The southern well field had been closed down because of the contamination that increased in this part of the city.

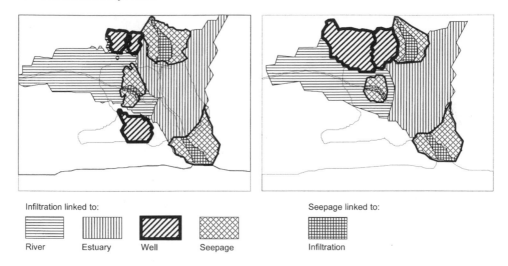

Infiltration linked to:

River   Estuary   Well   Seepage

Seepage linked to:

Infiltration

Figure 1.7.   Groundwater flow systems for different groundwater extractions.

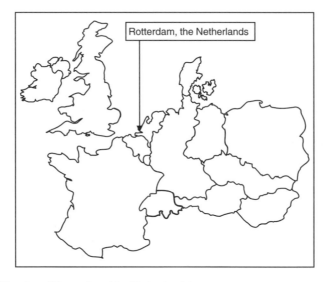

Rotterdam, the Netherlands

Figure 1.8.   The city of Rotterdam (the Netherlands).

After the closure, a portion of this water started heading toward the river, but it will not arrive there and reduce the water quality in the river until some centuries later.

Illustrations: Rotterdam, Cebu and Merida

What effects do these factors of change have on urban areas and their management? Three cities that contrast strongly in their situation, geology and management policies are described, illustrating how different elements influence change and development.

Rotterdam is a city in a delta area in The Netherlands (Fig. 1.8). Like most deltas the subsoil consists of sedimentary deposits, sands and clays and peat. This type of subsoil is often referred to as a porous medium, to set it apart from fractured rock and karst, which are distinctly different.

Rotterdam is the world's largest port, and many industries are situated here which provide the economic backbone of the entire region. The groundwater levels are relatively high, and in many places the groundwater is brackish or salt due to the open connection with the sea. Groundwater was abandoned as a source for drinking water long ago because of this, but is used for cooling water as illustrated by a large industrial abstraction in nearby Delft. The groundwater has recently attracted renewed attention for seasonal thermal storage. Plans for new infrastructure as the city continues to grow increasingly have to consider the subsurface.

The land-use and subsurface management is marginally limited by contamination problems. The relevant industries have mostly moved to a large area that has been established especially for industrial activities, which lies to the west of the city along the main waterway to the sea. Many historical contaminant plumes have now been remediated.

The ground surface in the new industrial area was elevated by some 5 m. A layer of sea sand was pumped on the existing land. This means the land surface is so high that there is no need to provide surface water or other drainage besides the main shipping canals in the area. There is no local seepage of groundwater, and therefore no direct health or environmental risks from contamination. This situation enables a different type of regulation by the planning authorities. Companies are required to monitor the subsurface and make sure that no contamination reaches the deep regional aquifer or leaves their plant laterally, but they are not required to maintain the shallow subsurface at a "naturally" clean level. This is possible because of the 10 m thick clay layer that was present below the original surface, and the moderate groundwater flow velocities. The clay layer protects the regional aquifer and the flow velocities are low enough that there is enough time to plan the necessary remediation when it has been concluded from the monitoring network that that is necessary.

Cebu City is the main city on Cebu Island, which is a part of the Philippine archipelago (Fig. 1.9). Away from the alluvial coastal plain, the main aquifer consists of fractured rock (Quiazon, 1972). The infiltration capacity of the ground is small. Most of the 1.6 m mean annual rainfall does not enter the ground to supply the groundwater but runs off over the surface. The surface runoff causes severe erosion that not only washes away the fertile soil deteriorating general conditions and lowering the infiltration capacity even more, but also causes sewerage problems. The soil is washed into the sewer system, severely reducing the capacity, and heavy isolated rain showers cause the sewer system to sporadically overflow.

The aquifer is overexploited by the unplanned urban expansion (Holman & Palmer, 1999). Groundwater extractions have already caused large drawdown due to the low porosity and relatively small transmissivity. The problem is aggravated by the diminishing recharge due to erosion and increased surface runoff. Because of the location near the sea, the decreased groundwater heads cause saline water to intrude the aquifer, posing a severe threat to groundwater quality.

It is difficult to work toward more sustainable groundwater management because of the rapid growth of the population, the lack of fresh water resources and the slowness of the system to reach equilibrium. It takes decades for the consequences of groundwater flow change to appear to their full extent. This is a general problem for all cities built on hard rock, as many African cities are (see Nairobi in Chapter 2, Kampala and Ilesha in Chapter 6). The transmissivities range from 0.01 to 100 m^2/d.

Figure 1.9.   Cebu city (Philippines).

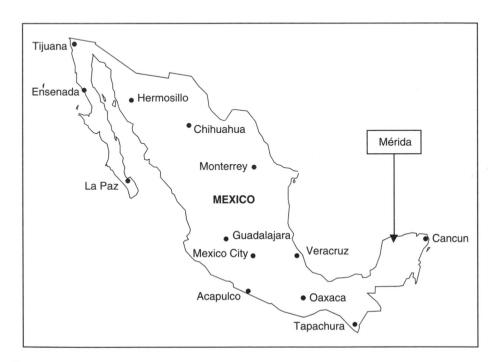

Figure 1.10.   The city of Mérida (Mexico).

The subsoil in the area of Merida in Mexico (Fig. 1.10) consists of karst. The rock matrix itself is quite impermeable, but the solution channels provide both high transmissivities and very fast pathways for the groundwater. Most of the wastewater is disposed into the subsoil without much treatment. The water reaches the groundwater very quickly, so that there is not even a reduction of the pathogenic bacteria and viruses, let alone any adsorption of contaminants. The drinking water extractions have therefore all been moved outside the city. The natural recharge of less than 0.1 m/y has increased to over 0.5 m/y, due to the high water usage of over $0.4\,m^3$ per capita. However, this large increase has not caused any major rise of the water table. Another illustration of the high transmissivities is the fact that the piezometric head hardly exhibits a gradient (2 cm/km). The subsoil has some inherent instability as the groundwater continues to dissolve the karst. Acidity accelerates this process is very much, so industries that handle acids are surrounded by extra measures.

Chapter 7 is entirely devoted to cities built on karst: Brasilia, Merida, Boshan, and Mount Gambier. Auckland is also discussed because, although the city is built on volcanic subsoil, the properties are similar.

1.4   MANAGEMENT

Philosophy

Over the last ten years, there has been increasing concern over groundwater resources and pollution problems. For example, the participants at the European Ministerial Seminar on groundwater in The Hague (Netherlands, 26/27 November 1991) recognised that groundwater is a natural resource with both ecological and economical value which is of vital importance for sustaining life, health, agriculture and the integrity of ecosystems. Groundwater resources are limited and should therefore be managed sustainably to ensure that future use is not constrained (EEA, 1999). Within an urban environment, it is not feasible to strive for a full restoration of ecological diversity, nor to protect completely the intrinsic value of groundwater and subsoil. Groundwater management should aim at the protection of the use and function of groundwater within cities and the protection of the intrinsic values and the related ecological diversity outside the cities. Due to the continuing increase of claims on land and groundwater, the main point of groundwater management will be the tuning and integration of these related uses. The starting point must be to increase the quality in terms of the physical and or the socio-economical environment.

Management objectives and strategies

Groundwater problems evolve over long time periods, so an important requirement for effective management is the detection of problems in an incipient stage. We have seen earlier in this chapter how cities and groundwater have a strong relationship with each other, and as the city grows the pressure upon groundwater as a resource base increases. The objective of sustainable groundwater management should be based upon an integrated approach, where the different land-use functions and groundwater demands are set within a framework of physical and socio-economic systems. As discussed by Burke et al. (1999), Notenboom (1999) and Ramesh (1999), institutional and political factors are the primary obstacles to sustainable management of groundwater resources, but what is clear is that the effective management of urban groundwater will always need to consider the constraint of groundwater abstraction and the control of contaminant loads (Foster et al., 1998; Burke et al., 1999).

Foster and Burke identify two main urban groundwater resource objectives:
- To improve the sustainability of resource exploitation in and around cities by avoiding irreversible degradation of aquifer systems;
- To use available resources more efficiently, avoiding anarchy in their exploitation and in land contaminant discharge.

These objectives can only be acceptable for the local stakeholders if they are the result of a combined regional and local approach in which priorities are set and the interests of all stakeholders are taken into account. The targets necessary to achieve these strategic goals are shown in Table 1.1, and are identified below:
- Constrain groundwater levels in aquifers underlying urban areas within a tolerable range by controlling groundwater abstraction;

Table 1.1.   Urban groundwater supply management: objectives, problems and mitigative measures (after Foster et al., 1998).

Objective	Problems experienced if objective not met	Targets	Management measures
Maintain groundwater supplies	Decline in well yields due to falling water table	Constrain groundwater levels	Redistribute/reduce abstraction Increase urban recharge
	Excessive treatment costs		Restrict density of residential development in vulnerable areas
	Secondary quality nuisance effects		Selective control of industrial activities/effluents Zone land for different uses Control landfill location and design Separate waste disposal from groundwater supply spatially
	Increase salinity from seawater intrusion	Constrain groundwater levels	Redistribute and/or reduce abstraction
	Induced contamination		Use scavenger boreholes Modify depth of water-supply boreholes
	Contaminants mobilised from contaminated land by rising water table	Constrain groundwater levels	Increase abstraction of shallow polluted groundwater for non-sensitive uses Reduce urban recharge

- Moderate subsurface contaminant load to acceptable levels, given local aquifer pollution vulnerability, by land-use planning to reduce potential pollution sources and selective controls over effluent discharges and other existing pollution controls.

A strong institutional framework is necessary in order to achieve such targets. The primary challenge for groundwater institutions and agencies lies in balancing formal regulations and delegated responsibilities at the local level. The regulations serve as boundary conditions in which all interests are represented. The delegation of responsibility to the local level is necessary to encourage sustainable use of groundwater by the users and stakeholders. To establish this fragile balance, groundwater agencies or institutions should: (a) increase public and political awareness of groundwater problems, (b) tune the control of ground- and surface water, environmental policy and regional physical planning, (c) communicate with a diverse array of groundwater users to encourage the sustainable use of groundwater, and (d) development a framework in which technical solutions can be found. Such a framework would set the objectives at a regional scale whilst allowing the tuning of the functions on a local scale.

The constraint of groundwater levels

In some countries, the regulation of groundwater exploitation is mainly achieved through control of borehole drilling, whilst other countries rely more on licensing their pumping; although both elements need to be present in a balanced policy. The most rational approach to defining annual groundwater charges is that a weighing factor should be applied to the charge per unit volume. Whatever measures are taken should aim at increasing the incentives to optimise the use of scarce high-quality groundwater resources in the vicinity of the urban areas.

For overexploited aquifers, abstraction control will often need to include measures to prohibit the construction of new wells and to reduce abstraction in existing wells. The most efficient way to restore groundwater levels is to declare special areas for the protection of groundwater resources, on the basis that priority is given to the common interest (protecting the resource) with rather than the individual interest (abstracting groundwater for any individual use). In practice, these strategic decisions are only technically and economically feasible when the main stakeholders, such as the groundwater agencies and the groundwater users group, are well organised. This facilitates the introduction of more realistic abstraction charges and may make possible the use of more sophisticated economic instruments, such as:

- The encouragement of non-sensitive groundwater users to switch from the exploitation of high-quality aquifers to poorer quality groundwater with major reduction of their abstraction charges;
- The restriction or withdrawal of abstraction rights from industrial companies that have not installed water-efficient technologies;
- The trading of treated wastewater for groundwater abstraction rights with agricultural irrigators in fringe urban areas;
- The provision of subsidies for improving the efficiency of irrigation water use on the fringes of urban areas in exchange for groundwater abstraction rights.

In general, constraining groundwater abstraction is simpler in cities where the bulk groundwater exploitation is by a few major water-supply utility and industrial boreholes or were the groundwater users are well organised. Such a problem is difficult to address where there are many individual small and privately operated domestic, commercial and industrial wells.

Controlling subsurface contaminant load

A well-balanced plan is necessary for effective control of the subsurface contaminant load. Normally a trade-off has to be made between the different interests, both public and private, within the boundaries set. Within these boundaries, the stakeholders will decide on common goals. As a result, groundwater protection strategies are defined which, while they constrain land-use activity, accept trade-offs between competing interests. The design and implementation of these groundwater protection measures has to be based on a strong institutional framework, well-organised stakeholders and an understanding of the geohydrological system. Relevant features of the groundwater system are necessary for management information, such as:

- Vulnerability map, based on distribution of travel times to abstraction wells, type of top soil, type of sub soil, and land-use;
- Natural attenuation map, based on the macro-chemical parameters, redox parameters and type of sub soil.

Such maps, together with maps showing defined priorities and the current situation, provide the basis for the groundwater protection strategy. In this way, the most vulnerable and threatened areas that require the most rigorous controls can be clearly shown. One type of measure can be the zoning of land surface so that land-uses that threaten the groundwater quality are restricted or forced to move. In addition, the remediation strategy of polluted sites can be changed from a site-specific approach towards a regional approach. This means that although the remediation goals are set on a regional level, the remediation goals of the individual sites become of minor importance (Van den Brink et al., 2000). Another measure can be a pro-active attitude of the planning authorities to stimulate the use of poor quality soil by non-sensitive land-uses, provided that this type of land-use is acceptable according to the vulnerability map. Information on the relation between different land-uses and the minimum soil quality needed for that type of land use can be found in Van Hesteren et al. (1999) and Van de Leemkule (1999).

It has to be recognised that shallow groundwater in urban areas is often likely to be contaminated, especially in the absence of a comprehensive mains sewerage system. However, to avoid excessive loads of persistent pollutants which may be transferred to deeper and less vulnerable aquifers in the longer term, there are ways in which the contaminant loading on vulnerable aquifers can be restricted (Morris et al., 1997):

- Prioritising areas of high groundwater vulnerability for main sewerage extension.
- Restricting the density of residential development served by in situ sanitation.
- Directing the location of landfill facilities to areas of negligible or low groundwater pollution vulnerability.
- Restricting the disposal of industrial effluents to the ground in vulnerable areas, through introduction of discharge permits and appropriate charging to encourage recycling and waste reduction.

- Introducing special measures for the handling of chemicals and effluents at any industrial sites located in vulnerable areas.
- Improving the location and quality of wastewater discharge from main sewerage systems with respect to the potential impacts on suburban and downstream municipal wellfields.

Illustrations

Institutional and political factors will mean that city authorities will react differently to both the problems identified and the way in which they formulate policies to be carried out. In Barcelona (Custodio, 1997), over-pumping has led to both quality problems and rising groundwater levels. Excessive urban and industrial water demand caused a strong drop of the piezometric water levels. This led to a drop below sea level in the water table in some areas, resulting in salt-water intrusion. In addition, the rivers began to infiltrate poor quality water into the aquifer. As a consequence of the decreasing quality and dropping water table, there was a progressive abandonment of wells and the recovery of the water table. Several underground car parks are waterlogged and a subway tunnel is now suffering from groundwater inflow. Some underground spaces have been abandoned because of the cost of pumping and the awkward geotechnical problems. Because the aquifers in and around Barcelona are recognised as key sources for urban supply, a positive management policy has been established through an action plan, based on a combination of technical expertise, co-operation of both public and private sectors, and political will.

In contrast, in Santa Cruz, Bolivia, management has been much more on a laissez-faire basis. The city is totally dependant on groundwater, and is served by a mixture of private and public supply. Many of the private wells that supply water for industry, small businesses and homes draw from no deeper than 90 m, but most water used for public supply comes from boreholes that tap the deep aquifer at a depth of between 90 and 350 m. Abstraction has induced downward contaminant movement from the shallow water table, but this does not appear to have penetrated beyond the 90 m mark, despite the heavy pumping of the deep public supply wells. It seems that the abstraction from the shallow aquifer for private water supply is effectively providing a degree of protection to deeper municipal well fields by intercepting, abstracting and recycling part of the polluted water. This is a fortuitously good management practice that has come about by accident, providing of course that none of the supplies from the shallowest wells is destined for sensitive use.

1.5   MANAGEMENT IN PRACTICE: ACITY, ANYWHERE

In Acity, there are some old problems, some new challenges and current groundwater situation that are all dealt with by the Acity Agency for Groundwater.

Acity developed autonomously at the junction of two waterways close to the sea. Like most cities, Acity grew from a small settlement (Fig. 1.11), and its economic activities were mainly based on the existence of the port: fishing and trade. Agriculture was important on the outskirts.

The demand for water was met by the use of both surface water and groundwater. The groundwater was taken from shallow urban wells, and was the preferred source for its quality, although surface water was easier to obtain and was taken from the river branch flowing

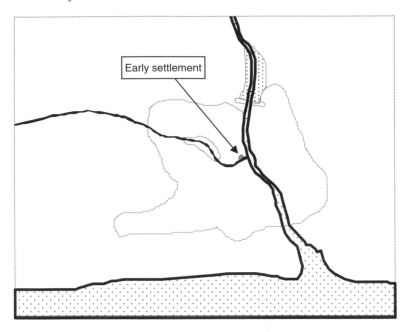

Figure 1.11.   Development stage: Acity as early settlement.

from west to east. The main river flowed south into the sea and caused yearly floods. The wastewater was not collected by any sewage system, but discharged directly to the ground. Only the city centre had a drainage system for rain, which was discharged into the river.

The expansion of Acity was initiated by the exploration of oil. The economic activities shifted to industrial and transhipment activities, resulting in petro-chemical industry, ship-yards and oil storage plants. Energy demand increased and was met firstly by the use of oil and later by a hydro power plant upstream. Both public and industrial water demand increased strongly, and was met by groundwater for its reliability and constant quality. This ground-water was initially obtained from shallow underlying aquifers, but the resource was limited. In addition, pollution caused a deterioration of the quality of the shallow groundwater. The extra water resources required were found by abstracting groundwater from deeper aquifers and surface water from the reservoir that was created to control the floods in the main river.

New districts developed, some on the other side of the river, and others linked to the existing town (Fig. 1.12). The districts developed for the housing of the workforce were built close to the factories, sometimes even on waste disposal sites, and had no water mains and sewage systems. In contrast, the residential area for wealthy inhabitants was built around a park. The shallow groundwater of Acity became more and more polluted by both industrial activity and the disposal of wastewater to the ground.

As a result, the water table beneath Acity dropped, which resulted in an increase in the pumping cost of the groundwater, the forced infiltration of polluted shallow groundwater and saltwater intrusion. Together with the increase in groundwater demand, the direct pre-cipitation was reduced due to an increase in the paved surface of the city.

The physical expansion of Acity in combination with the industrial activities meant the environmental conditions worsened. All formal waste disposal sites had to be abandoned and new waste disposal sites were established on the edges of the city. Spillage of all kinds

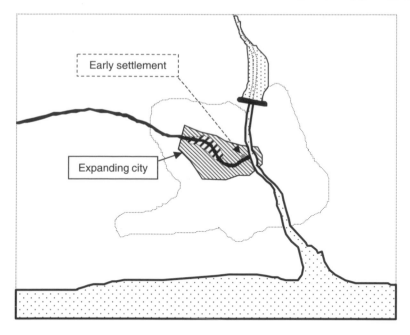

Figure 1.12.   Development stage: Acity as city.

of chemicals and seepage of oil products from fuel storage tanks occurred in large areas. In addition, chlorinated hydrocarbons, aromatic compounds and oil products polluted the groundwater, mainly at smaller industrial sites situated in the centre of the city. The groundwater throughout the entire city was diffusely polluted by leaking sewage systems and septic tanks, while the soil was polluted by PAH's and heavy metals.

Seawater intrusion and contamination led to progressive abandonment of wells and the recovery of the water table. As a result, extra water resources for the municipal water supply were required, found in the city's hinterland by constructing suburban wellfields. The marginal costs increased with the increase in distance. In addition, these water supplies had a competing prior use, notably agriculture, and a serious conflict resulted. The private groundwater abstractors were not able to move their abstraction. They were forced to treat the groundwater or, especially in the suburban residential areas, accept the risk of the contamination.

Today, Acity is still expanding (Fig. 1.13). The pressure to use both the soil and the subsoil is increasing. Due to environmental and economical considerations, the heavy industry is now concentrated in a large area in the south east of Acity, close to the port. The main economic activities are still industrial and transhipment activities, but the importance of the commercial centre is increasing. Investigation of existing pollution is undertaken before development of new districts goes ahead, and remediation measures are put in place if necessary to eliminate any risk to the inhabitants. In addition, waste disposal is concentrated at a limited number of controlled sites. In the commercial centre of Acity, office buildings use the subsoil for the storage of heat and cold. Most of the increasing public and industrial water demand is met by using bank filtered water.

The deterioration of the groundwater system of Acity was caused by the fact that the prevention of pollution was not high on the list of political objectives. In addition, there

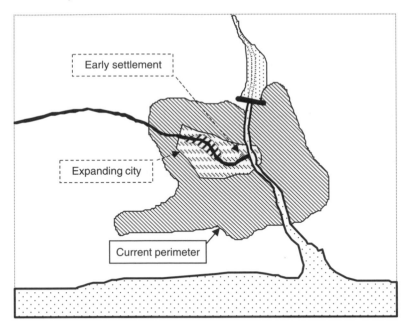

Figure 1.13.   Development stage: Acity end 20th century.

was a lack of public and political awareness of groundwater problems. Finally, the increased intensity of land-use caused an increased pressure on the groundwater system. The quality of the aquifer remained poor as a result of the large number of sites with polluted soil and groundwater. In addition, the recovery of the water table created waterlogging problems which threatened to mobilise both contaminants diffusely stored in the subsoil and contaminants stored in pits. The intense extraction of groundwater caused a decline of the water level in the city periphery.

The Acity Agency for Groundwater was established to deal with these problems and to improve the situation with respect to the groundwater system. The Agency initiated communication with both regional and city authorities to determine ambitious objectives for the groundwater system in Acity, and the issues, needs and management solutions can be seen in Table 1.2. The principal objective is the restoration and protection of the aquifers in and around Acity, because these are the key sources for urban and emergency water supply, and this can be done by working toward the twin targets of constraining groundwater levels and reducing subsurface contamination load.

Areas were identified where the restoration and protection of the aquifer had the first priority, using hydrogeological information. The following groundwater quality and quantity objectives were identified.

1.   The most vulnerable part of the aquifer must be protected as the key source for urban supply, therefore all activities have to be integrated for the restoration and protection of this aquifer. Measures include:
    - Pro-active and selective control of industrial activities and effluents
    - Restriction of contaminant loading by identified sources through the development of a regional remediation strategy

Table 1.2. Acity management issues.

Water quality and quantity issues	Management solutions
Wellfields unable to cope with increased demand and threatened by outward growth of city	Design vulnerability maps based on the travel times of the groundwater towards the wellfields, together with the type of subsoil and overlying strata, to obtain insight to the most important threats (areas with high pollution loads)
Recovery of water table causing problems of flooding, wastewater disposal etc.	Land zoning: designate high priority aquifers which must be protected and/or restored, and areas for city expansion and industry/landfills on less vulnerable (clay-covered) areas
Competing interests between public and private groundwater abstractions	Provide greater disincentives for non-sensitive industrial users to abstract high-quality groundwater more suited for portable supplies
Pollution load of the shallow groundwater likely to last many years because of existing (historical) pollution load in the subsoil and many active sources	Design remediation strategy to prevent further contaminant loading of the subsoil by identified sources

- Separation of waste disposal from groundwater supply
- Increase in abstraction of shallow polluted groundwater for non-sensitive uses
- Increase of urban recharge with unpolluted or treated water
- Increase of use of lightly polluted soil by non-sensitive uses

2. The soil groundwater quality in areas of recreational and environmental value should meet established ecological standards, and the groundwater level and fluctuations should not exceed certain limits. Measures include:
- Increase in the recharge of unpolluted water
- Restrictions in the use of fertilisers and pesticides

3. Quality standards for residential areas must be high to eliminate all risk to healthy living. Measures include:
- Restriction of density in vulnerable areas
- Control of demand from industrial and other uses to decrease abstraction
- Extension of mains sewerage and improvement of sewage treatment
- Increase in urban recharge with unpolluted or treated water

4. In the commercial centre and the industrial areas, the acceptable levels of contaminants should be based on the prevention of contamination of unpolluted groundwater. Measures include:
- Restriction of contaminant loading by identified sources through the development of a regional remediation strategy
- Separation of waste disposal from groundwater supply
- Increase in abstraction of shallow polluted groundwater for non-sensitive uses
- Increase in urban recharge with unpolluted or treated water
- Increase in use of lightly polluted soil by non-sensitive uses

5. In the remaining areas, soil and groundwater quality must be related to the potential human health risks. Measures include:
    - Increase of abstraction of shallow polluted groundwater for non-sensitive uses
    - Increase in urban recharge with unpolluted or treated water
    - Increase in use of lightly polluted soil by non-sensitive uses
    - Control landfill location and design
    - Selective control of industrial activities
    - Modification of depth of industrial water-supply boreholes
    - Control of water demand to decrease groundwater abstraction

A process was started within this framework with both the stakeholders and users of the soil and groundwater to encourage sustainable use of the groundwater. This process was financed from both private and public sources, because both public and private users are dependent on the protection and remediation of the groundwater system. As consultations proceed, several scenarios for the tuning and co-existence of functions will be defined. A mathematical model supplies the information needed. This model quantifies the wide variety of interactions with the groundwater in urban areas (Fig. 1.1) within a regional hydrological context. It is able to quantify the effects of the tuning and co-existence of functions with respect to the objectives: protection and remediation of the groundwater system to provide a healthy environment for the citizens and have no negative ecological impact outside the city.

1.6 SUMMARY

Groundwater is a hidden asset, and yet for many communities throughout the world it is the only source of water supply. The continued expansion of urban development means increased pressures from a variety of land-uses, threatening both quality and quantity of the water, and it is important to establish good management policies to ensure its sustainability. The science of hydrogeology through modelling and monitoring may not be sufficient on its own, and the co-operation of a variety of interests is vital in order to implement these management policies through a raft of measures.

CHAPTER 2

Characteristics of urban groundwater

Mike H. Barrett

2.1 THE DEVELOPMENT OF CITIES

Foster et al. (1998) highlight the provision of water supply, sanitation and drainage as a key requirement of the urbanisation process, with the subsurface playing an important role in all three of these elements, and in the disposal of industrial effluent and solid waste. This is illustrated by Figure 2.1. The subsurface is therefore a key consideration, and the presence of permeable subsoil and shallow groundwater are critical factors. As groundwater is generally of high natural quality, and a more constant source than seasonally affected surface waters, it is often the first resource to be exploited by overlying cities (Foster et al., 1998), and is shown by Figure 2.2.

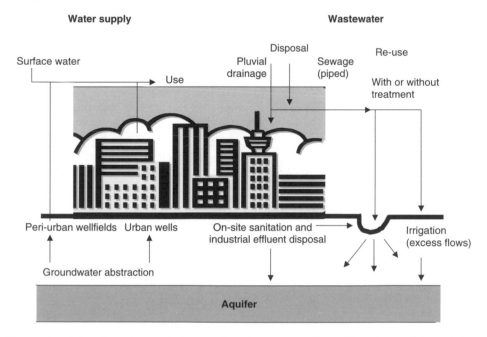

Figure 2.1.   Interaction of groundwater supply and wastewater disposal in a city overlying a shallow aquifer (after Foster et al., 1998).

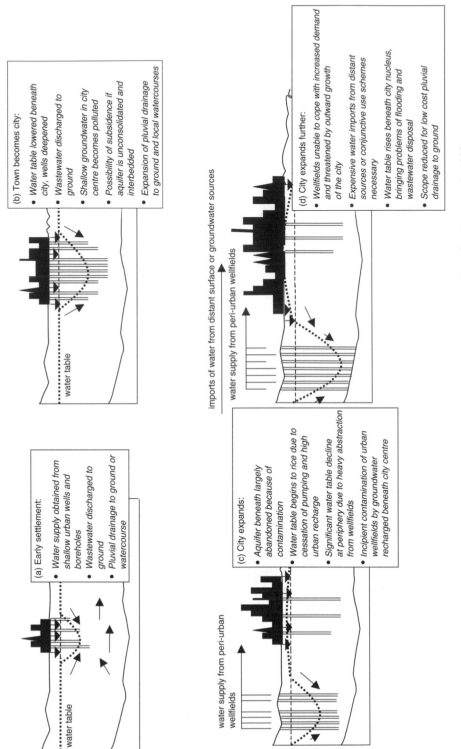

Figure 2.2.   Evolution of water supply and wastewater disposal in a city overlying a shallow aquifer (after Foster et al., 1998).

The subsurface also plays a key role in urban wastewater disposal in the absence of unsewered sanitation, and is found to be impacted even in the presence of sewers, described by Barrett et al. (1999). Foster et al. highlight an important point; that is, mains sewerage installation often lags behind population growth and mains water supply due to its high cost. This, unfortunately, may have a significant impact on the underlying groundwater which is often the source of the mains supply, as seen throughout urban sub-Saharan Africa and discussed in Chapters 5 and 6.

Urbanisation may lead to impermeabilisation of land surfaces (less so in sub-Saharan Africa than European, American or Asian cities), depending on land use changes. This has been shown to have a modifying effect on groundwater recharge, rather than an inevitable reduction (Yang et al., 1996).

The subsurface may receive industrial effluents either directly from disposal to the ground, or indirectly through seepage from lagoons, surface watercourses or drainage ditches. There can be spillage and leakage of hydrocarbons/petroleum products stored in tanks (industry, filling stations) or directly onto the road from vehicles.

Urban areas produce solid waste, and municipal authorities and/or private companies are responsible for waste collection. Disposal is to landfill or open dumps, which may produce leachate that penetrates the subsurface and threaten groundwater, or may represent a direct health hazard themselves.

Foster et al. also point out that cities are intimately associated with their surrounding areas. As they grow, water supplies that were originally obtained from shallow underlying aquifers may no longer be sufficient, or may be unacceptably contaminated. Alternatively, the city may grow to incorporate its own peri-urban wellfield. Extra water may be drawn from aquifers or surface waters in the city's hinterland, as is the case for Nottingham, discussed below. These water supplies may have a competing prior use, such as agriculture, and may be an important part of the rural ecology by supplying flow to rivers and wetlands, and serious conflicts can arise as a result of their exploitation.

The UK has some of the world's oldest continuously urbanised areas. These have developed through a number of stages including the Industrial Revolution, massive subsequent population growth, suburbanisation, industrial manufacturing decline and modern urban regeneration. These stages of development have been responsible for changes in the role of the urban environment in influencing underlying groundwater quality and quantity. In other parts of the world, urban development has been far more recent and dramatic. Many of the stages seen in the UK can be observed in these cities at different stages of development. In some cases, the stages that took centuries in the UK have occurred in a few decades. "Megacities" in Asia have undergone industrialisation, population growth, and a change in emphasis towards banking and commerce, but slum areas which were largely cleared from the UK by the 1930s co-exist with the modern metropolis.

Urban areas in sub-Saharan Africa are still at the early stages of industrialisation but nonetheless are seeing massive population growth, although not as yet a corresponding increase in wealth. This leads to problems where a massive population is unsupported by basic urban infrastructural development, and provides a contrast with the UK and other zones of mature urbanisation where population growth, wealth increase and infrastructural development were largely concurrent. In the Asian "megacities", population growth has been matched by an increase in urban wealth, but the distribution of this wealth is extremely uneven, allowing massive rich/poor divides to open up. This may be a temporary feature associated with the rapidity of urbanisation. As with the mature cities, the different stages of development including the land use, population density and infrastructure, may be seen to directly influence underlying groundwater quality and quantity of cities elsewhere (both "megacities" and sub-Saharan urbanisation).

Nottingham, UK: a mature city

The city of Nottingham, in the East Midlands of the UK, currently occupies an area of some $400\,km^2$ (Charsley et al., 1990). The current city is thought to have its origin as an important Anglo-Saxon settlement in the 6th century, but its beginnings may go back to Roman or pre-Roman times. It was settled continuously throughout the Viking and Norman periods and became a major trading centre in the Middle Ages. The siting of Nottingham relates to its important strategic position on high dry sandstone ground close to suitable bridging or fording locations on the River Trent, a major navigable river. Expansion occurred in the 19th century with the development of coal mining in the area and associated development of major industries in the city. The rapid urban growth of Nottingham commenced in the 1870s. The period since the 1960s has seen further rapid development of dormitory housing, transforming surrounding villages into populous suburban centres. Many of the city's older industries have declined and closed down over the past few decades, and redevelopment of the inner city has commenced during this time. The city is currently known mainly for the manufacture of pharmaceuticals, telecommunications equipment, bicycles, cigarettes and the knitting and textile industries. Surrounding the city are coalmines (open cast), sand and gravel extractions and gypsum mining, and arable farming.

Until expansion in the 19th century, Nottingham was wholly built on the Triassic Bunter Pebble Beds, now known as the Nottingham Castle Sandstone (Beckett, 1997). This forms an area of high ground, truncated by cliffs to the south that drop down to the plain of the River Trent. The cliffs overlook a shallow river crossing, now used by Trent Bridge. The topography of the area results from faulting and glacial and post-glacial erosion. The properties of the sandstone influenced the town's development and environment. It forms an excellent aquifer due to the high permeability and porosity, which guaranteed a supply of drinking water from sufficiently deep wells. The permeability also ensured excellent drainage for the town. The sandstone is sufficiently competent (below the upper weathered $2\,m$) that by 1840 it had more man-made underground structures, which included rock houses, malt kilns, public house cellars, storage vaults and tanneries, than any other town in Britain (Beckett, 1997). Subsequent urban expansion has seen building to the west on the Mercia mudstone, overlying the Nottingham Castle Sandstone, and on the gravels and sands of the Trent flood plain, thought to be in hydraulic continuity with the Sandstone.

The Trent has played a critical role in the siting and development of Nottingham. The town was the last navigable point from the coast on the river. The Trent flows into the Humber, connecting Nottingham with the other important towns of York and Hull as well as the North Sea.

A major development during the improvement of the town's infrastructure was an attempt in the 1690s to provide piped water, using mechanical abstraction from the Leen and pumping the water through wooden pipes laid under the streets. Water had been traditionally derived from both public and private wells sunk into the Sandstone, and an estimated 300 were in use in 1700. These became inadequate, and additional water was brought into the town from the River Leen, a tributary of the Trent located to the west of the town, by hand or in water carts. The advent of a piped water system led to a reduction in used wells from 300 to 200 (Beckett, 1997).

In the 17th century, the flood plain of the Trent immediately south of the town known as the Meadows area was used for agricultural activity. The soil was kept fertile through application of human manure taken from urban cesspits. Until this time, Nottingham's basic economic role was to serve as a market centre for the county for the distribution and exchange of the basic commodities of food and drink. During the 17th century, the traditional industries such as clothmaking, tanning and ironworking declined, while malting and hosiery boomed. This led to an increase in wealth and migration into the town resulting in a rapid increase in population. It was the widespread development of hosiery that gave the town a major manufacturing base, particularly after 1730. The population almost doubled in the first half of the 18th century (Beckett, 1997), and increased from 11 000 in 1750 to 260 000 in 1911 as labour from the surrounding countryside was attracted by the economic growth.

As industrial expansion increased during the 19th century, Nottingham did not develop the pattern of social and industrial zoning and commuting characteristic of other major towns such as London. An attempt to create an industrial zone in the Meadows area failed. Consequently, individual factories (largely lace and hosiery) were scattered around the town, intermixed with residential areas. Pockets of discrete industrial activity did not emerge until around 1900. By the 1880s a range of new interests was beginning to broaden the basis of Nottingham's economy including bicycle making, pharmaceuticals, tobacco, coal mining and service industries. However, it remained principally a textile town. Development of the hosiery industry included improved methods of bleaching and dyeing to enhance appearance, originally using vegetable dyes. Back in the 1790s, Robert Hall's mill in Basford had been the first to use chlorine for bleaching (Beckett, 1997).

Nottingham's prosperity in the 20th century derived largely from the success of three major firms; Boots (pharmaceuticals), Players (cigarettes) and Raleigh (bicycles). All of these had private boreholes for groundwater abstraction. However, the textile industry including bleaching and dyeing remained the largest industrial sector within Nottingham in 1995. Again, many of these factories continue to take the water for their industrial processes from private boreholes.

The result of rapid population growth in Nottingham in the 19th century has been suburbanisation. Since 1911 growth has been modest, peaking at 311 899 in 1961 with a net increase of 1.4% from 1911 to 1991, and this is characteristic of most large British towns. More significant has been the fall in household size, from 4.33 persons in 1911 to 2.37 in 1991 (Beckett, 1997), which has resulted in an increasing demand for properties. These trends stem from an increased standard of living and improvements in housing and public health. In addition, there has been an ageing of the population, with an increase in the over 60 age-group and a decrease in the under 15 group. Consequently, the number of households has almost doubled. The residential area of Nottingham was 4 times larger in 1994 than in 1916 (Beckett, 1997). In 1919, an application to extend the town boundaries was turned down by the government because sewage disposal arrangements in the existing urban area were inadequate. A massive modernisation programme was undertaken, and by 1926 around 30 000 pail (bucket) closets had been converted to water (flush) closets, and by 1928 a new main drainage system had been approved.

A distinctive feature of mature cities is that they have moved into a new phase of post-industrial development. Mature cities are increasingly shifting from industrial economies that are mass producing commodities to service-led systems characterised by more flexible forms of production individualised consumption (Crewe & Beaverstock, 1998). Crewe and

Beaverstock suggest that places are increasingly distinguished from each other by their "consumptional identities" rather than by what they produce. Crewe and Beaverstock also suggest that the chief impetus for current urban restructuring is the rise of the new middle class. In Nottingham this is demonstrated by the redevelopment of the Lace Market.

The Lace Market lies on the eastern edge of the main commercial centre and was historically the centre of the once world leading mechanised lace industry. It degenerated through the post-war period into a part derelict city space, which was home to various printing, clothing and warehousing businesses interspersed with dilapidated and empty buildings. Redevelopment occurred in 1969 with the designation of the area as a Conservation Area. This was followed by the development of the Town Scheme Plan in 1974, which aimed to actively promote renovation, and the declaration of the Lace Market as an Industrial Improvement Area. By 1982, more than 100 buildings were renovated. Development accelerated in 1989 with the creation of the Lace Market Development Company, a public-private partnership between the local authorities and private investors and property companies aimed at regenerating the Lace Market over a 5 year period. The Lace Market is now populated by new, small, entrepreneurial firms (240 of which were non-textile based and half of which had fewer than 5 employees by 1990). By 1996 there were 450 firms of which over 80% were involved in cultural production and/or consumption. These included fashion, retail services, visual communications (e.g. advertising and marketing), leisure and entertainment, arts and media, banking and finance, computer services and charities (Crewe & Beaverstock, 1998). Crewe and Beaverstock conclude that cultural production and consumption play a key part in the revitalisation and regeneration of contemporary cities. This shift in land use has critical impacts on water consumption (e.g. private boreholes are no longer required to supply large volumes of water to mass producing industries), and is also likely to considerably change the risks to underlying groundwater.

Public water supply to the Nottingham area is currently the responsibility of Severn Trent Water plc (STW). The mains supply system is currently supplied by a number of reservoirs in and around the city of Nottingham. Water from these reservoirs supplies a large number of gravity driven water supply zones. Quality within each of these zones is regularly monitored by STW at a number of sampling points. The reservoirs are largely filled by pumping from a number of public water supply (PWS) boreholes in the rural area to the north and east of the city, with augmentation from the River Derwent in Derbyshire. Chlorination is carried out at the borehole sites prior to pumping to the reservoir. Only one public water supply borehole is located within the city itself, which is the Basford borehole at the regional STW offices at Hucknall Road. This is currently only used in drought situations due to concerns regarding quality. Limited blending of sources is carried out prior to filling some reservoirs (Papplewick [Papplewick PWS with River Derwent] and Halam [Farnsfield PWS with Ompton PWS]) to ensure nitrate drinking water limits are not exceeded. In addition to the STW mains supply, a number of industrial sites and hospitals have private boreholes for on site supply.

Edwards (1966) gives a useful publication describing the historical water of Nottingham, and a summary of this publication follows:

The first recorded public water supplies in Nottingham were given in 1696. Prior to that, the inhabitants were dependent upon the Rivers Trent and Leen, private wells, springs and rainwater storage in barrels. In 1696 the Company of Proprietors of Nottingham Waterworks was formed and erected an engine house, water wheel and pumps on the River Leen and used this water to fill a reservoir which supplied the city by gravitation through a pipe system. This sufficed until the 19th century when quality problems and scarcity resulted in the company constructing further works in the Basford area to use nearby springs and the Leen at a point where the river was free of contamination from the town's sewers.

The waterworks remained a wholly private company until the 1770s. Water pipes were laid to various streets, but only around 150 households were connected and water was only available on set days of the week. By 1782, 39 streets in the town had customers attached to the water supply. Pipes had to be paid for by the consumer, putting them beyond the reach of many people. Water quality was also an issue; one pump (in Sheep Lane) was removed in around 1804 as human sewage from on-site sanitation had "penetrated the rock as to ooze into the well". The River Leen was the common sewer of the town; this had serious health impacts in low-lying areas of housing during floods. In the 1790s the streets were described as being covered by sludge "of the blackest kind". Refuse was often allowed to accumulate in the courtyards of poorer housing until such time that it had acquired (through "putrefaction") value as manure, at which point it was carted away by "muck majors" or manure collectors (Beckett, 1997).

Towards the end of the 18th century a private water supply (Zion Hill Waterworks) was established raising water from a well approximately 60 m deep by a steam engine. Supply was by a combination of pipes and water carriers. In 1824 and 1826 two more companies were established. The Nottingham New Waterworks Company pumped water from a well into a large cistern to supply the north-east of the town and the Nottingham Trent Waterworks Company supplied the south and south-east of the town by filtering the Trent through brick tunnels in the sand and gravel on the north side of the river and pumping by a steam engine to a reservoir. By 1844, around 4 000 houses in addition to breweries, dye-houses and steam engines were supplied. The Nottingham Water Act of 1845 amalgamated the three companies into the Nottingham Waterworks Company, founding the present system of water supply to the city and its immediate environment.

By the middle of the 19th century all the major rivers in the proximity of Nottingham were too polluted for public supply, yet demand for water was rising rapidly due to population growth and industrial development. Since 1850 the public water supply of virtually the whole region has been derived from the Sherwood Sandstone aquifer and from the River Derwent in Derbyshire.

The first of the existing pumping stations abstracting water from the Sherwood Sandstone was constructed at Basford (Hucknall Road) in 1857, using steam powered pumps. Two wells were constructed to about 35 m and connected by an adit of 30 m. From 1880 the yield fell from 2.75 to 1.96 gallons per day and to 1.2 in 1966, reflecting long term over pumping. Further stations were constructed at Bestwood in 1871 and Papplewick in 1883 (both sites being as close as practicable to the populated area). At Bestwood the two wells were about 60 m deep and connected by an adit extending a considerable distance from the wells. The construction at Papplewick was similar. Both stations used rotative steam engines to drive ram pumps which were still in use in 1966.

The development of wells and boreholes in this southern part of the area for industrial as well as public supply resulted in a lowering of the water table. In response the Nottingham authority sought sources more remote from the city. Three wells connected by adits were sunk at Boughton, 19 miles from the city centre. In 1898 a new type of station was developed at Burton Joyce, the water being pumped from one borehole of 14 inches diameter and about 140 m deep, which was initially artesian. This was the first borehole in the confined aquifer, confined by approximately 45 m of Mercia Mudstone. Three further boreholes were sunk at this site in 1908, and electrically operated pumps were installed in

1928. The station had ceased to be artesian before 1966. In 1945 an all electric borehole pumping station was completed at Rufford and in the years following the second world war, boreholes were constructed at Lambley (completed 1957), Halam and Markham Clinton, all on the confined aquifer at considerable distance from the city. In 1966 the Ompton source was developed. The post war sources were developed using a pattern exemplified by Markham Clinton, three boreholes dispersed as widely as possible on the available site with the nominal output of the station to be provided by two of the bores, the third providing standby in the event of failure or overhaul of either of the other two. At Markham Clinton, the boreholes were sunk to the base of the Sherwood Sandstone, a depth of approximately 220 m, and the top parts passing through the Mercia Mudstone were lined with steel tubes grouted in with concrete. The pumps at each station were located at sufficient depth to be "well drowned" when pumping at maximum capacity. The pumps discharged to a sedimentation tank to remove particulates and booster pumps then pumped the water from this tank to the delivery main towards a terminal reservoir. Chlorination was applied at the inlet to the sedimentation tank.

The sources constructed since 1850 as described are mostly still in use, subject to modifications. Additional supply is taken from the River Derwent at a point near Church Wilne.

Jakarta, Indonesia: an Asian "Megacity"

The formation of "global cities" or "megacities" (large cities resulting from global economic restructuring) in Asia, is characterised by interlinkages of those cities in a functional system built around telecommunications, transportation, services and finances (Firman, 1998). Such cities include Tokyo, Seoul, Taipei, Hong Kong, Manila, Bangkok, Kuala Lumpur, Singapore and Jakarta. Within the global economy they function as centres of financial services, information flows and commodity transactions. This process of globalisation was spurred by the development of communication and transportation technology. Urban restructuring in Asian cities is characterised by the following features (Firman, 1998):

- Development of economic activities at a world scale
- Spatial division of labour between the core and periphery of the city
- Shifting from single core to multicores of metropolitan area
- Land use change in the urban centre and agricultural land conversion in the periphery
- Development of large scale urban infrastructure
- Substantial increase in space production
- High growth of commuters and increasing commuting time

Modern Jakarta's origins may be traced back to the early part of the 17th century when a European style town, known as Batavia, was constructed by the Dutch East India Company. However, human settlement of the area may be traced back to prehistoric times (Abeyasekere, 1987). Its origins as a port may be traced back to the 12th century, when it was a harbour for the Hindu-Javanese kingdom. The Dutch transformed the settlement into their headquarters in the Indonesian archipelago; a site to repair ships, a store for goods, and a military and administrative centre. The town declined during the 18th century, as the Dutch East India Company went bankrupt, and the town suffered from severe drainage problems. Additionally, by this time the shallow aquifer in the town was so contaminated as to be undrinkable, with most people taking their water from the river. To the south of the town, men collected buckets of water, rowed them along the canals constructed by the Dutch, and offered the water for sale. Its impurity caused outbreaks of dysentery and typhoid. Ships that had previously stopped at Jakarta to take on provisions and water now avoided the town. In 1770 Captain Cook was forced to stop for repairs to his ship and saw his crew fall sick during his stay. A number subsequently died (Abeyasekere, 1987). Mortality in general amongst European settlers declined during the 19th century. This resulted in a movement away from the malaria ridden lower areas to zones less subject to flooding and with access to better water supplies. Wells in the upper town reached safer water than in the lower areas, and in the late 19th century, the administration sank deeper artesian wells in the European part of town.

Migration to cities from their hinterland depended on good transport and this only came to Jakarta in the late 19th century with road and rail links. From 1900 to 1940 Jakarta trebled its population to 435 000 (Abeyasekere, 1987). The building of European-style residential suburbs formed part of this expansion. Roads, formerly of clay, were asphalted enabling an increase in the use of motorised transport. The Indonesian, Chinese and Arab population was largely restricted to neglected, low-income "kampung" areas. These areas were scattered around the city in close proximity to European places of work and residence. The kampungs did not receive comprehensive piped water supply, drainage or sewerage and solid waste removal. The Dutch colonial era was ended by the Japanese invasion of 1942. The Dutch temporarily resumed control at the end of the war, but Indonesia achieved independence in 1949, with Jakarta assuming the role of capital. At this time, Chinese and European businessmen still controlled manufacturing, banking, export/import firms and the main urban utilities. A policy of "Indonesianisation" was implemented by the government, when only Indonesians were granted import licences, but this was abandoned in 1957 having promoted corruption rather than developing Indonesian run business. The late 1950s saw a process of nationalisation of land, industry and utilities. However, further corruption led to the formation a wealthy minority and poverty for the majority. This eventually led to the coup of 1965 when Suharto effectively took control of the country. This marked the start of Jakarta's rapid growth to become one of the new Asian "megacities", with the encouragement of foreign investment in manufacturing which utilised local cheap labour. The subsequent economic boom provided a higher tax base for Jakarta's municipal authority to undertake urban restructuring, alongside the many privately funded construction projects.

Jakarta is the government, industrial, commercial and transport hub of Indonesia (Hadiwinoto & Leitmann, 1994). The modern metropolitan area of Jakarta (Jabotabek) consists of an agglomeration of the cities of Jakarta, Bogor, Tanggareng and Bekasi. In 1990 the population was estimated at 17 million, at 20 million in 1995, and was expected to grow to 26 million by 2010 (Hadiwinoto & Leitmann, 1994). The population in the Jabotabek area grew at 3.8% annually between 1980 and 1990 and 3.3% from 1990–1995. Two thirds of the growth was attributed to natural increase and the remainder to rural-urban migration (Hadiwinoto & Leitmann, 1994; Firman, 1998).

It is the socio-economic impact of Indonesia's growing integration with the world economy that according to Leaf (1994) has provided the impetus for Jakarta's urban transformation into a city of international standards. He states that the most significant factor affecting these changes is the movement of the Indonesian economy away from a purely colonial role within the world economy to that of a diversifying economy with growing levels of indigenous ownership, the growth of indigenous middle management positions, and a growing service economy as the country industrialises. The most visible impact is the growth of Jakarta's middle class (or "consumer class"). This has resulted in growing demand for suburban homes. Leaf summarises the essential components (now in place) for growing suburbanisation of Jakarta:

- Demand stimulated by rising incomes
- Supply being created through industrial development
- Policies fostering low-density suburban development through subsidised housing finance and government sponsored provision of trunk infrastructure and a land use plane promoting separation of residential and commercial functions in the city.

DKI Jakarta is the central city with a population density of 12 400 people per sq km, compared to more than 30 000 in high-density, low-income areas of Kampala (Howard, 1999 *[pers. com.]*). According to Hadiwinoto and Leitmann (1994), the area consists of:

1. Low-lying coastal strip with poor drainage (subject to flooding)
2. Low-lying plains (also subject to flooding)
3. Higher land rising from the coastal plains with good drainage
4. Steeper slopes
5. Mountainous area with rapid run-off, subject to erosion

DKI Jakarta is located in zones 1–3 and is primarily residential (45% of the area), recreational or open (42%) with the remainder being industrial and commercial. Jabotabek as a whole uses 72% of its land for agriculture. Jabotabek accounts for 12% of GDP, 61% of all banking/financial activities and 17% of industrial output in Indonesia. There remain 1.4 million people below the poverty line in DKI Jakarta (those with household income less than $38 per month). These people are concentrated in unplanned slums with high population densities and inadequate infrastructure; just 6% of the lowest income quintile have in-house piped water (37% for the highest income quintile) and 64% share toilet facilities (2%).

In the periphery of Jakarta there has been extensive foreign investment in manufacturing industries such as metal products, textiles, chemical industries (US $4.3 billion from 1990–1994). In DKI Jakarta, the services sector which include hotels and restaurants stand out, demonstrating the shift of DKI Jakarta from a manufacturing centre to a centre of service activities with migration of manufacturing the periphery of the city (Firman, 1998). These changes are characterised by both rapid changes of land-use in the core and conversion of agricultural land to urban land-uses in the periphery. New town development at the fringe of Jakarta is extensive. These are built mostly for middle and upper income groups, although the government is also promoting low-cost housing projects for low-income households. The increasing scarcity of available urban land for development has resulted in some slum clearance, with conversion to residential areas, shopping centres and office spaces.

Urbanisation has been identified as affecting key natural resources, principally related to the water cycle. Groundwater is being degraded by saline intrusion due to overexploitation, and pollution linked to infiltration from sewers. All surface waters within Jakarta are polluted by grey water from households and commercial buildings as well as discharges from industries, pesticides and fertilisers, solid waste and sewage (Hadiwinoto & Leitmann, 1994).

Water resources in the region are ample, but storage and distribution to Jakarta and the surrounding fast-growth urban areas is reaching a critical state. There is an urgent need for demand management, loss reduction, new storage and long distance transmission of raw water for urban uses. The water company supplies less than half of households and businesses, and the system is subject to 50% losses and other technical problems. Remaining consumers rely primarily on groundwater supplies, mainly from shallow wells. A major concern is urban expansion onto the major recharge area south of the city.

Inadequate solid waste disposal is both a threat to water quality and an immediate health risk. An estimated 40% of the solid waste generated daily does not reach official disposal sites. Much finds its way to surface waters, resulting in pollution and flooding from channel blockage, and the remainder is dumped informally or left on vacant land to contaminate groundwater (Hadiwinoto & Leitmann, 1994).

Foster et al. (1998) highlight the importance of regulatory and economic instruments to reducing groundwater abstractions in Jakarta. The public water supply provides water to 46% of the population, largely from surface sources. The remaining population rely mainly on groundwater derived from the shallow phreatic aquifer. Historically there have been no controls on abstraction. Industrial abstractions are largely from the underlying confined aquifer, and overpumping has led to de-watering of the phreatic aquifer and saline intrusion, and is illustrated by Figure 2.3. An additional problem is subsidence, estimated at $3–6\,\mathrm{cm/y^{-1}}$. Subsidence results in an increased risk of tidal flooding of the lower (northern) parts of Jakarta. Retrospective well licensing and tiered abstraction-charging has been introduced to confront the situation. However, illegal abstraction persists (Foster et al., 1998).

Nairobi, Kenya: A rapidly developing city

According to Stren (1994), the African continent has been affected for at least the past decade by a severe multi-faceted crisis. Marginalisation of the technological revolution, the bypassing of the continent by international trade and investment, and primary agricultural commodities losing their relative value have all combined to create a virtual collapse of the economic systems of many African countries. Urban centres in Africa do not fit the restructured global economy and their capacity to maintain basic infrastructure and social services has been impaired.

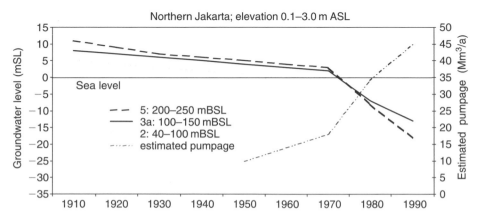

Figure 2.3. Historically falling groundwater levels and rising abstraction rates in the Jakarta aquifer system (after Foster et al., 1998).

Until the end of the 19th century, Nairobi did not exist. This is the case for many sub-Saharan African cities, unlike either mature cities with long histories of strategic and manufacturing centres, or the Asian "megacities", many of which also have long histories as ports and trading centres. In 1899 the colonial railway authorities decided to make Nairobi their headquarters and the first buildings were constructed in what had previously been a Maasai grazing area (Macharia, 1992). Like other cities such as Harare and Lusaka, Nairobi began as a European city, and remained so until independence (1963). It grew primarily as a place for Europeans to live in, as well as becoming a centre that would provide urban services for the growing numbers of settlers in the surrounding country.

Nairobi had a population of 343 500 in 1962, 509 300 in 1969, and 827 800 in 1979; an annual growth rate in excess of 5% (Milukas, 1993). By 1990 the estimated population was 1.5 million with an annual growth rate of 5–7%, (Karue et al., 1992; Werna, 1997). In fact there are a number of secondary towns growing at even faster rates, such as Kisumu with a growth rate of 16%, and Meru with a rate of more than 31% between 1962 and 1979. Nairobi accounts for over 50% of the employment sector.

Kenya has a well-established national electric grid, distributing electricity to all major urban areas, thus eliminating any regional advantages or disadvantages a particular urban centre may have previously suffered. Rapid urbanisation is leading to increased demand for electricity (including the pumping of water), and consequently a demand for sources to feed the grid. These are largely petroleum products, although Nairobi also utilizes hydropower. All petroleum products used in Kenya are imported, either as crude oil for refining at Mombasa, or as finished products. They are piped to Nairobi for further distribution.

The city council has seven main departments. The Department of Health is responsible for solid refuse collection (Werna, 1998). Water and sewerage is the responsibility of the Department of Water and Sewerage (DWS-NCC). According to Werna, there was sufficient water to meet demand in 1989 (180 900 m^3 per day), but demand had outstripped supply by 1992. The third Nairobi Water Supply Project aims to increase supply up to 2008–2010, to allow for further population growth. In the late 1980s, 89% of the population were thought to be obtaining supplies from the WSD-NCC through house connections, communal watering points or water kiosks, with the remaining 11% obtaining water from wells. There is a private sector involvement comprising street vendors who purchase water from kiosks and sell it on to customers. Most areas served by communal taps and street vendors are low-income districts, and less than 12% of the plots in squatter settlements have water connections. In 1990, a squatter settlement in central Nairobi was identified as having no domestic connections despite a main water pipe passing through the settlement to some formal housing estates. Squatter settlements have not been included in the public piped system due to their illegal status.

Approximately 50% of the population of Nairobi are serviced by the public sewerage system, with the remaining population using on-site sanitation such as septic tanks and pit-latrines. In addition to the WSD-NCC, voluntary organisations play a role in sewerage provision through the construction of communal toilet buildings, open sewer canals and pit-latrines. Despite the involvement of voluntary organisations, demand still outstrips supply of sanitary facilities, with high person to toilet ratios (e.g. 500:1 in Langata). A large number of the slum/squatter dwellers do not use toilets and dispose of faeces in public areas (Werna, 1997). As with the water supply network, there are illegal settlements adjacent to sewerage pipes, but without connections.

Solid waste collection in Nairobi is the responsibility of the cleansing section of the Department of Health of the NCC. In 1992, 60% of the city's refuse remained uncollected, with 30% collected by the NCC and 10% by other agents (Werna, 1998). This has resulted in part from a decline in collection vehicles (96 in 1972 to 44 in 1990), poor equipment maintenance and lack of foreign currency to purchase spare parts. This decrease on the part of the NCC has created an opportunity for private disposal companies, but at present privatised collection is exclusive to wealthier areas because of high collection charges.

During recent years there has been an increase in typhoid, dysentery and cholera in Nairobi (Ngecu & Gaciri, 1998). These outbreaks can be linked to the state of the water system. A large section of the city's population in low-income areas have no access to piped or treated water, nor to proper sanitation and sewage systems; discharges are directed to rivers, which in turn are utilised in low-income areas as water sources. The main rock types of the area comprise alkaline lava and pyroclastics, extruded in the Rift Valley during the Pliocene. Ngecu and Gaciri report a survey of river water quality in Nairobi. They surveyed quality upstream, within and downstream of the city. They found the main contaminants to be organic in origin resulting from point sources such as waste disposal sites or urban run-off. The poorest quality waters were found in the industrial area of Nairobi, and near sewage discharge points in slum areas. In the industrial area, raised levels of heavy metals are found (cadmium, chromium and lead).

As a result of urban expansion, the piped water network has not kept pace with the urbanised area. In the suburbs, and peri-urban areas, groundwater is an important alternative to piped water. It is often used directly from the borehole. A study by Gikunju et al. (1992) revealed that 42% of boreholes sampled in Nairobi contained coliforms. Additionally fluoride concentrations pose a problem, although the source is not thought to be anthropogenic, but from dissolution of fluoride bearing minerals within the aquifer system.

Nairobi's informal settlements date back to the early part of the city's history. Kibera, for example, dates back to 1912 when the government allowed Sudanese Nubian soldiers who had served with the King's African Rifles to settle in the area. Informal settlements in Nairobi house over half the city's population, although they only occupy 5.8% of all land area used for residential purposes (Wegelin-Schuringa & Kodo, 1997). They lack basic infrastructure, services and legal recognition and range from small squatter villages to crowded slums in excess of 50 000 people (Lamba, 1994). All are characterised by very high population densities (up to 63 000 persons per sq km; the city average in 1994 was 2 200 per sq km) and all have a history of neglect by public authorities. Informal settlements are considered illegal and therefore do not qualify for city services. To provide services would, it is perceived, give legitimacy to the existence of the illegally settled areas. According to Lamba (1994), the city had never seriously attempted to provide adequate water supply, sanitation, drainage, health care or other services. The result is that water supply is inadequate, drainage consists of channels formed in paths and roads, sanitation facilities are insufficient and waste disposal services do not exist. Informal settlements often exist on unsuitable land such as swamp, flood plains, steep slopes or near hazardous industrial activities. As informal settlements are illegal, it is prohibited to build permanent structures. This generally results in mud and wattle houses (shown in Fig. 2.4) with galvanised iron roofs at best. The most common form of tenancy is room rental from illegal, often absentee, landlords. The landlords wish to maximise their income from the area of land, and therefore construct as many rooms as possible, resulting in little open space or few roads. Such outside space that exists is used for washing, bathing and playing. As there is a

Figure 2.4.   Houses in East Africa.

permanent risk of these settlements being torn down, should the government wish to utilise the land for other activities, there is little incentive for landlords to carry out improvements to structures.

Lack of sanitation is a major problem. This results from a lack of space for latrines, the view that latrines are the responsibility of the landlord and a reluctance of tenants to invest in construction due the likelihood of short term residency, a lack of money, and the possibility that the landlord would increase the rent if a latrine was present. In some areas (e.g. Kitui), public latrine blocks have been built, either connected to a main sewer and flushed with a mains water supply which is not often available, or consisting of large dry pits; and these are managed by the local community.

A report on slum conditions in 1992 revealed abandoned toilets with gaping holes, other latrines being filthy and inaccessible, and evidence that residents were exposed to numerous hazards originating from nearby factories, such as industrial effluent including animal wastes, caustic soda, bleaching agents, lime and other unknown by-products which were emitted into the sewer system that emptied into an open drain running through the slums (Lamba, 1994).

According to Lamba (1994), population growth is not the key issue, nor is urbanisation per se, but rather how it is taking place. Lack of access to basic resources by the urban poor is critical. Urbanisation in Nairobi and many other sub-Saharan African cities is characterised by deepening disparities between the wealthy and the urban poor.

Informal urban agriculture is widespread in Nairobi, but barely tolerated by policy makers (Freeman, 1991, and reviewed by Nyongo, 1994)). According to Freeman, the majority of urban farmers are women (65%). Private residential land is most frequently utilised, with roadside verges a close second, followed by riverbanks, drainage channels, parks, railway lines and empty private and industrial plots. He also concludes that almost half (47%) of urban farmers have no other means of earning a living, and 70% are purely subsistence farmers. Squatters tend to cultivate private and public land without paying for it. Quick-growing crop varieties (particularly maize and bananas) are planted during the short rainy season from October–December, with major cultivation during the long rains (April–June). Urban agriculture is important in Nairobi as in many other African cities, providing a living for people who may otherwise be unemployed, and utilising otherwise unused urban spaces as well as providing fresh food to urban workers.

Urban growth and development in Nairobi is also impacting air quality, and therefore may indirectly affect groundwater quality through rainfall. Values of lead in Nairobi's air were found to be about 5 times higher than levels reported in European countries, and Kenyan petrol was found to contain as much as 0.4 g of lead per litre. Bromine/lead ratios measured in the air were indicative of vehicle emissions. Analysis for organic compounds was not carried out.

2.2   URBAN INFRASTRUCTURE

The previous section described urbanisation in three contrasting settings, all highlighting the common need for water supply, sanitation and drainage. As indicated by Foster et al. (1998), the subsurface plays an important role in these elements of infrastructure, as well as in the disposal of industrial effluent and solid waste, illustrated in Figure 2.2.

Figure 1.1 in Chapter 1 (Yang et al., 1999) demonstrates how urbanisation modifies recharge pathways, creating by-pass flows that reduce the effects of surface impermeabilisation. The impact of urban infrastructure on the quantity and quality of underlying groundwater depends largely upon the geology and hydrogeology of the area, particularly the depth of water table and permeability of the aquifer and overlying materials. It also depends upon the design of systems such as water supply, sanitation measures and drainage systems.

Water supply

In the early stages of urban development, urban groundwater is often utilised for supply from relatively shallow boreholes. It is often supplied at source through hand-pumps, or the use of a bucket and rope, without any treatment. This mode of water supply is still commonly used in the lower-income and peri-urban areas of rapidly developing cities, particularly in Asia and Africa, where piped water supply system expansion does not keep pace with urban expansion. As the city grows and undergoes industrialisation, deeper mechanised boreholes are sunk in the urban area and used for both public and industrial supply, many industrial sites owning their own mechanised borehole. These mechanised boreholes are used to feed piped water supply systems, generally buried beneath roads close to the ground surface in mains water systems. Development has thus replaced the need for people to visit a well with the provision of tap water in the home, or in lower-income communities, the provision of standpipes adjacent to housing development. On a small scale, pumped groundwater is fed to metal storage tanks mounted on stilts, which then drain by gravity into the piped system. These storage tanks are a feature of the skyline in many small, but rapidly expanding, towns in sub-Saharan Africa. This style of water supply is utilised in Iganga, Uganda, in combination with shallow hand-pumped wells in areas not yet reached by the mains water system, and discussed further in Chapter 6. On a larger scale, mechanised deep boreholes may be used to pump water up to large holding reservoirs (covered if small, uncovered if large) which drain into the mains system by gravity. These reservoirs must be situated on hills in or around the city to generate sufficient pressure in the mains system. Generally such systems employ chlorination at the wellhead as a minimum form of treatment, although in lower-income countries, chlorination may be sporadic depending upon the availability of chemicals. Even in developed countries such as the UK and US, failure of chlorination systems on boreholes is not unknown, occasionally leading to disease outbreaks.

A characteristic of more mature cities, as demonstrated in Nottingham, UK, is the abandonment of urban water supply boreholes due to perceived poor water quality. In such cases, development of rural wellfields takes place, or surface water is piped in from sources outside the city and stored in reservoirs. In Nottingham, a combination of rural wellfields and surface water is now used, where the two sources are blended to mitigate problems with high nitrate in some rural areas.

Mains water supply systems contribute to urban recharge (discussed further in section 2.3), as these networks are not watertight, and leakage from the pressurised pipes directly into the subsurface takes place. In many developing countries, mains water systems only supply more central areas and the more wealthy suburbs. The pace of infrastructure development does not keep up with population growth and urban expansion, leaving people in low income and peri-urban areas dependent upon local water sources. In many mature cities, 100% mains water coverage to houses is considered normal. Industry, however, often still relies on its own private urban boreholes for water supply.

Sanitation provision

In the early stages of urban development of now mature cities in the UK, human waste was often disposed of into cesspits or directly onto the street or courtyard, eventually to drain to surface water bodies that effectively were open sewers, or to accumulate and putrefy until it formed a manure that was collected for agricultural use. By the 19th century, pail closets were in use. These were eventually replaced by flush water closets discharging to sewerage systems, although not entirely until the early 20th century in places. Such sewerage systems generally function by gravity drainage, relying on solid wastes to be flushed through the system by greywater (e.g. wash water) and water from flush closets, discharging at a topographically low point. Early systems discharged directly to surface water bodies without treatment, but many now employ some degree of treatment prior to eventual discharge.

A general pattern of sanitation provision emerges as cities develop. Initially there is an absence of sanitation, then the provision of on-site sanitation and eventually the provision of waterborne sewerage, although the latter stage may be limited in arid environments where fresh water is scarce. As these systems progress there is a general improvement in public health and reduction in environmental contamination. However, the provision of waterborne sewerage systems does not eliminate environmental contamination. Sewers will inevitably leak due to poor construction, or damage due to subsidence, tree root ingress or construction works, although volumes of leakage will be limited by the fact that sewerage systems are generally unpressurised, and some sections may be pumped where topographic conditions prevent gravity drainage.

Urban drainage

The stormwater drainage system in a city may have a significant effect on recharge to underlying groundwater. In mature cities, stormwater runoff is often collected in shallow surface drainage systems by covered channels at the side of roads. These systems may either drain directly to surface water bodies (separate systems), or may connect to the foul sewerage system (combined system). Generally such systems reduce direct recharge from rainfall. In less developed cities, the roadside drainage systems simply consist of unlined ditches. These may allow stormwater to accumulate and percolate down into the subsurface. Unfortunately, whilst rainfall is generally of high quality, it may become contaminated when in contact with impermeabilised urban surfaces, such as human or animal wastes, and oil/petrol leaks from vehicles.

In areas where water resources are scarce, attempts may be made at harvesting rainfall through roof collection systems or at increasing direct recharge through the use of permeable materials for pavement or parking lot construction.

Industrial effluents

Foster et al. (1998) highlight the fact that the subsurface is often a major receptor for industrial effluents. These may enter directly from casual disposal or indirectly as seepage from treatment lagoons, or from storage tanks. There are many published studies that demonstrate the quality of urban groundwater to reflect overlying industrial processes (e.g. Rivett et al., 1990; Ford & Tellam, 1994).

Solid waste disposal

Urbanisation and growing populations result in the generation of proportionate quantities of solid waste. In mature cities this waste is collected by either municipal authorities or private companies, and generally disposed of to landfills. These may generate leachates that can infiltrate the subsurface and contaminate underlying groundwater. In lower income cities, waste disposal may be to open dumps, such as in Manila, or may be absent in some areas altogether, leading to scattered waste within the urban area (noted by Barrett et al., 2000, as being a contributing factor to poor shallow groundwater quality).

2.3   HOW CITIES CHANGE RECHARGE

The simplistic view of urbanisation is that recharge is reduced by waterproofing surfaces, or impermeabilisation (Lerner & Barrett, 1996). However, this concept is not always correct (Lerner, 1989; Price & Reed, 1989). Urbanisation may result in a net change in overall groundwater recharge, with anything from a major reduction to a modest increase being possible (Foster et al., 1998). If a significant proportion of the local water supply is obtained from underlying groundwater, large-scale recycling will occur. Networks of water-carrying pipes exist under most cities, and the extent of these generally relate to the stage of infrastructure development of the city. None of these are watertight. Some of these are pressured, such as mains water supply, and some are not, such as the non-pumped sewerage systems. For example, the water supply to the Birmingham conurbation (UK) is equivalent to around 600 mm per year, and leakage rates are estimated at 25% (Lerner & Barrett, 1996). Leakage from water supply pipes and sewerage systems may be supplemented by discharge from septic tanks, pit-latrines, storm water drains. According to Foster et al. (1998), unsewered sanitation greatly increases the rate of urban groundwater recharge. This may have a major impact on the urban groundwater balance, particularly in high-density population areas. If the significant part of the local water supply is imported from outside the urban area, for example from surface water bodies or groundwater well fields, the net groundwater recharge will increase substantially. The balance may also vary with time as abstraction patterns vary. For example, in the city of Nottingham, groundwater was initially abstracted from urban boreholes for public and industrial supply. This led to a decline in water levels. The perception that urban groundwater was of poor quality led to a switch to rural boreholes north of the city. Additionally, as manufacturing industries have declined since the 1950s, industrial abstractions in the urban area have also reduced. This is a pattern seen in many UK urban areas. This, combined with leakage from water imported from the rural area to the urban area, has resulted in a rise of water levels, above "natural" levels, giving rise to the flooding of building basements.

Urbanisation does result in a reduction in direct infiltration of rainfall. Surface impermeabilisation processes include construction of roofs, roads, parking lots, industrial premises and airport aprons. The proportion of land covered is a key factor in the degree of impermeabilisation. Some types of urban pavement such as tiles, bricks and porous asphalt are quite permeable (Foster et al., 1998). The proportion of impermeabilised surfaces may be considerably lower in less-developed cities, where many roads are unsurfaced, although compaction may result in a reduced infiltration capacity. In many developing cities, runoff from impermeable surfaces such as roads, (surfaced or otherwise) may be directly into drainage ditches, allowing seepage to groundwater to occur. In more developed cities, surface runoff is directed to storm sewers. These may either form a discrete network, discharging to surface water bodies such as rivers or canals, or may link into the foul sewerage system (a combined system), with eventual discharge to a treatment works. Direction of surface runoff to surface watercourses and modification of the condition of surface watercourses, such as sealing the bed to reduce erosion, or entirely re-routing to aid other construction projects, may also affect recharge to underlying groundwater.

Irrigation of amenity areas may also contribute to groundwater recharge. Irrigation of parks, gardens and landscaped areas for aesthetic purposes may sometimes be excessive, particularly if water is applied from irrigation channels or hose pipes. The amount of water applied is rarely related to horticultural needs (Foster et al., 1998). Over-irrigation may lead to localised extremely high recharge rates.

Quantification of individual components of recharge is essential if a full understanding of the interaction between an urban environment and its underlying groundwater is to be achieved. This process is, however, extremely complex. An approach to establishing the spatial and temporal amounts of the three major sources of recharge (precipitation, sewers and mains water supply) in the Nottingham aquifer was developed by Yang et al. (1999). A calibrated groundwater flow model was supplemented by calibrated solute balances for 3 conservative species (Cl, SO_4 and total N). Nottingham has annual precipitation of 700 mm/y (with 230 mm/y effective precipitation in non-urbanised areas). Using the models, current urban recharge was estimated at 211 mm/y, of which 138 mm/y ($\pm 40\%$) is mains leakage and 10 mm/y ($\pm 100\%$) is sewer leakage. The wide confidence intervals result from the scarcity of historical field data for calibration. The two dominant urban recharge sources (in terms of quantity) were found to be precipitation and leakage from mains water supply, with mains water being the single largest contributor. Direct recharge from precipitation is clearly reduced by impermeabilisation. Total recharge to the unconfined urban aquifer has been calculated to reduce slightly during the process of urbanisation (by around 8%), although this is within the limits of error. Leakage from mains water supply pipes was calculated by Yang et al. to have increased significantly between 1850 (34 mm/y) and 1995 (138 mm/y). This probably reflects the expansion of the water supply network and volume of supply, driven by increased consumption. It was noted that there was a correlation between the extent of mains water leakage and the age of the urban development in particular urban areas; greatest leakage was associated with older development, and therefore possibly older supply networks. However, the opposite was observed for the sewerage system. Higher rates of sewer leakage were calculated under more modern developments, suggesting the older 19th century sewerage systems to be less prone to leakage. The quantity of sewage recharge was not calculated to be very significant in any part of the urban area (averaging 10 mm/y). However, the impact of this sewage leakage on underlying groundwater quality is more significant. Monitoring of shallow groundwater quality revealed contamination by nitrate (determined through the use of nitrogen isotopes), faecal coliforms (Barrett et al., 1999) and enteric viruses (Powell et al., 2000).

An excellent summary of the overall effects of urbanisation on groundwater recharge is given by Foster et al. (1998) and is shown by Figure 2.5, while Lerner (2002) gives an up-to-date review of the methods of estimating recharge in urban areas. Figure 2.5 demonstrates the overall effect of urbanisation on groundwater recharge, indicating the normal direct precipitation for non-urbanised areas, and potential recharge from mains leakage,

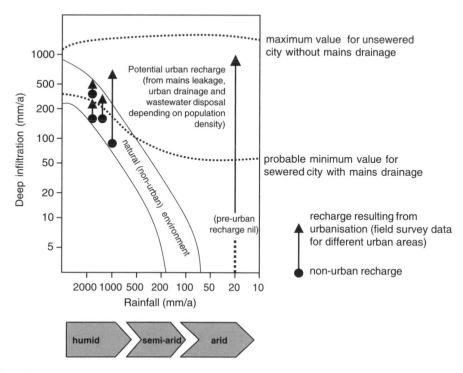

Figure 2.5.   Potential range of subsurface infiltration caused by urbanisation (after Foster et al., 1998).

sewage and urban drainage (storm water runoff). Foster et al. recognised the dependence of the volumetric contribution of mains and sewage upon population density and the stage of urban development. They find that in unsewered cities, 90% of gross water production may eventually enter the subsurface, with consumptive use (not normally exceeding 10%) being the only loss. This is most likely in arid areas, as in more humid regions, some urban drainage and wastewater may be directed to surface watercourses.

2.4   POLLUTION PROBLEMS

Urbanisation not only impacts groundwater quantity through modification of recharge pathways, but it also impacts underlying groundwater quality. Groundwater quality may be impacted by changes in overlying land use such as industrial development, agricultural activity and wastewater generation. Industrial and agricultural impacts are dependent upon the degree of industrial development of the city. However, sewage contamination is a global issue. The sources and pathways will differ, depending upon whether waterborne sewerage or on-site sanitation is employed. As discussed earlier, whilst sewerage systems reduce sewage impacts on underlying groundwater, they certainly do not eliminate them, as demonstrated in the city of Nottingham, UK. Analysis of data produced from a variety of largely unsewered African cities is presented later in this publication, and clearly demonstrates high levels of sewage contamination. Sewage contamination of groundwater results not only from direct infiltration from on-site sanitation and leaking sewerage systems to the subsurface, but also from surface runoff entering poorly designed or maintained springs and boreholes, for example through cracked concrete plinths. This latter pathway is of most significance where sanitary coverage is poor, and faeces are discarded on the land surface (Foster et al., 1998; Barrett et al., 2000). Where urban sewage contains human waste as the dominant contaminant source such as in unsewered cities, faecal bacteria and viruses and nitrogen species are the main pollutants that may reach the underlying groundwater. In sewered cities, sewage may also contain a variety of domestic and industrial effluents, including heavy metals and organic compounds such as solvents, petroleum products and pesticides. Storm water run-off may also enter the subsurface. This water may also contain contaminants such as heavy metals and organic compounds washed from building and road surfaces. As demonstrated in the Nottingham study (Yang et al., 1999) and highlighted by Foster et al. (1998), mains supply water may form a significant part of total urban recharge. It is unlikely that this water will have a negative impact on underlying groundwater quality, particularly if it complies with drinking water quality standards. However, if the water supply is imported from outside the city, from either surface water sources or aquifers adjacent to the city, it is possible that a modification of the underlying groundwater may occur. It is unlikely that the chemistry of the imported water and that of the underlying urban groundwater will be the same.

Industrial contamination of urban groundwater is controlled by a combination of the stage of industrial development of the overlying city, the type of industries present, and the underlying geology, particularly with regard to the natural attenuation capacity. Table 2.1 and Table 2.2 (from Barrett et al., 1999) list potential contaminants that may be found in urban recharge; clearly many of these may be discharged from industrial processes. Heavy metal and organic solvent contamination of urban groundwater is widespread in urban aquifers in industrialised countries. The importance of underlying geology, however, is demonstrated by

Table 2.1. Potential markers for urban recharge.

Category	Group	Species
Inorganic	Major cations	Ca Mg K Na
	Major anions	HCO_3 SO_4 Cl
	Nitrogen species	NO_3 NH_4 (plus organic nitrogen)
	Other minor ions	B PO_4 Sr F Br CN
	Metals	Fe Mn Trace metals
Organic	Atmospheric	chlorofluorocarbons (CFCs)
	Chlorination by-products	trihalomethanes (THMs)
	Faecal	coprostanol, 1-aminopropanone
	Detergent related	optical brighteners, EDTA, limonene
	Industrial	chlorinated solvents, hydrocarbons etc.
Particulate	Faecal microbiological	*E. coli*, faecal streptococci, enterovirus, bacteriophage
	Colloidal	Organic, inorganic
Isotopes	Stable isotopes	^2H ^{15}N ^{18}O ^{35}S

Table 2.2. Sources of possible marker species.

	Atmosphere	Geological materials	Agriculture	Mains water	Sewage	Industrial and commercial sites
Majors	✓	✓	✓	✓	✓	✓
N species	✓		✓	✓	✓	✓
B & PO_4		✓			✓	✓
Other minors	✓	✓		✓	✓	✓
Metals		✓			✓	✓
CFCs	✓			✓	✓	✓
THMs				✓	✓	✓
Faecal					✓	
Detergent					✓	✓
Industrial					✓	✓
Microbiological					✓	
Colloidal					✓	✓

studies of the cities of Birmingham (Ford & Tellam, 1994) and Nottingham (Barrett et al., 1997) in the UK. Both cities overlie Triassic Sandstone aquifers. Both have long histories of industrial activities, including metal working and manufacturing. However, Nottingham's underlying groundwater is less contaminated by heavy metals. The difference appears to be a result of carbonate cementing present in the Nottingham sandstone, but largely absent in the upper part of the Birmingham aquifer. The presence of the carbonate in the Nottingham sandstone results in higher groundwater pH due to carbonate dissolution, and a consequential reduction in both solubility and mobility of metal contaminants.

Whilst data on inorganic and organic groundwater quality in industrialised countries is widely available, there is an absence from many developing urban areas, particularly in sub-Saharan Africa. This may be a result of the perception that levels of industrialisation are low, and that industrial contamination is unlikely to be significant. It is also the case that because groundwater is often used untreated for public consumption, monitoring for sewage contamination (principally faecal bacteria) is of higher priority, particularly where funds for monitoring and analysis are limited. Health impacts from microbiological contamination are immediate and severe; whereas inorganic and organic contamination is more likely to be chronic. However, many developing cities are now rapidly urbanising, often with little enforceable environmental control. A recent limited study in Kampala revealed organic solvent contamination of shallow groundwater (unpublished data from Barrett, 1999).

Land-use in some developing cities can be additionally complicated by the presence of urban agricultural activities. Access to arable land in urban areas, no matter how small, may provide the possibility of better living in low-income areas. In Kampala it is estimated that 30% of all residents take part in urban agriculture, and these households show a significantly improved level of health. The largest urban agriculture sector is livestock. In 1993 it was estimated that 70% of all chickens and eggs consumed in Kampala were produced within the city (IDRC, 1993). This type of urban agriculture may vary from small-scale commercial activities down to individual ownership of goats, cows, pigs or chickens for individual subsistence purposes. The larger scale commercial operations may present a threat to underlying groundwater quality through the production of slurry that may be channelled into drainage ditches or surface water bodies. Small-scale activities may also pose a threat as individual animals are allowed to scavenge within residential areas, resulting in scattered faecal waste. Barrett et al. (2000) demonstrate that the presence of faecal waste on the land surface in urban areas is a common cause of groundwater contamination where the sanitary protection of springs or wells is faulty or absent, allowing contaminated surface runoff to directly enter the groundwater system.

2.5 CONCLUSIONS

The Nottingham case study demonstrates how urban impacts on underlying groundwater quantity and quality vary with time as urban infrastructure and industry develop and populations grow. Groundwater impacts beneath developing urban areas, particularly sewage contamination, are comparable with historical conditions in more mature urban centres, when rapid population growth outstrips sanitation coverage and early stages of industrialisation. Asian "megacities" provide an interesting juxtaposition of the two conditions. Here, highly developed commercial and residential areas with waterborne sewerage systems and piped water supply systems are surrounded both by highly industrialised zones and also rapidly expanding lower income housing, where expansion of informal settlements may outstrip the expansion of infrastructure of both water and sewerage systems.

Foster et al. (1998) highlight the fact that urban groundwater problems evolve over many years or decades, partly reflecting the growth of urban population and increased water demand and waste generation, but more often a consequence of the large storage capacity and slow response rates of most groundwater systems, illustrated by Figure 2.6. The latter point was also made by Yang et al. (1999) as being a constraining factor in trying to accurately quantify individual urban groundwater recharge sources.

During initial urban development, groundwater is usually abstracted from wells that penetrate shallow groundwater within the central urban area. In Nottingham, evidence of such means of abstraction remains in the form of shallow wells, thought to date back to the Middle Ages. Some are still well preserved, and located in the basements of old buildings and in caves beneath the city centre. Similar means of abstraction are seen in developing

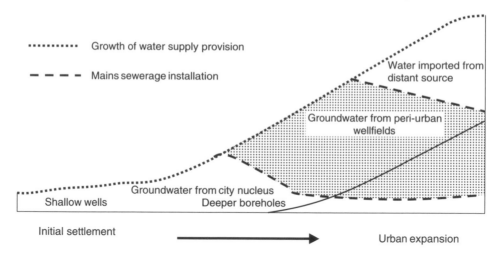

Figure 2.6.   Urban evolution from the perspective of groundwater resources (after Foster et al., 1998).

urban centres today such as the boreholes in Nairobi and protected springs in Kampala (as discussed earlier and in Chapter 6).

The consequence of urban expansion and population growth is that water tables are lowered and contamination of shallow groundwater occurs through disposal of residential and industrial waste to the ground. This results in contamination of the shallow groundwater used for drinking water supply, which again is currently the case in Nairobi and Kampala. The combination of falling water levels and increasing contamination may lead to the municipal water agency installing deeper, mechanised boreholes, producing a greater yield of better quality water. However, this may be a temporary solution, as downward flow of contaminants is induced. The alternative (as demonstrated in the Nottingham case study) is that additional supplies from peri-urban wellfields or surface water sources may be imported to the city. As clearly demonstrated by Yang et al. (1999) in Nottingham, the importation of large volumes of water from outside the city results in increased urban recharge (leaking water supply pipes, sewerage systems or on-site sanitation systems) that may eventually result in water table rebound. Such scenarios may be exacerbated by changes in industrial activity, demonstrated in Nottingham, illustrated by a post-industrial shift away from heavy engineering and large scale manufacturing towards activities such as banking and finance, communications and retail. This results in reduced abstraction in the urban area from industrial boreholes, further shifting the balance in favour of recharge.

CHAPTER 3

Sources, types, characteristics and investigation of urban groundwater pollutants

Sheng Zhang, Ken Howard, Claus Otto, Veronica Ritchie,
Oliver T.N. Sililo and Steve Appleyard

3.1   INTRODUCTION

About a third of the US population, or 86 million people, receive their drinking water from groundwater sources. Industrial solvents and related chemicals are present in the groundwater used to provide drinking water for 35 to 50 million Americans, according to a recent survey of drinking water by the United States Geological Survey (Squillace et al., 1999). One or more volatile organic compounds (chemicals which contain carbon and evaporate easily) were found in 47% of the urban wells and 14% of rural wells. US EPA (US Environmental Protection Agency) standards were violated in 6.4% of urban wells.

Contaminants that may be present in urban groundwater include the following:

- Microbial contaminants, such as viruses and bacteria, which can get into the water from sewage treatment plants, underground septic systems, agricultural livestock operations, and even from creatures living in the wild. The microbial contaminants of greatest concern in drinking water are usually of faecal origin.
- Inorganic contaminants, such as salts and metal ions, which may be present naturally in the soil or rocks, or may enter the water as a result of urban storm water runoff, industrial or domestic wastewater discharges, landfill leakage, etc.
- Organic chemical contaminants, including synthetic and volatile organic chemicals, which are by-products of industrial processes and petroleum production, and which can also arise from gas stations, urban storm water runoff, residential uses such as lawn care or spraying for insects, and septic systems.
- Radioactive contaminants, which can be naturally occurring or resulting from waste disposal.

A full list of contaminants regulated by US EPA regarding a public water supply system, along with a summary of their MCL (Maximum Contaminant Level), MCLG (Maximum Contaminant Level Goal), potential health effects, and typical sources, is available in table form (http://www.epa.gov/safewater/mcl.html).

The ingestion of low levels of chemical contaminants in drinking water can affect human health. The contaminants for which epidemiological studies have suggested a risk associated with their presence in potable water include: aluminium, arsenic, disinfection by-products (DBPs), fluoride, lead, nitrate, radon, pesticides, hydrocarbons and chlorinated hydrocarbons. The contaminants are of both inorganic and organic origin. However, very few have been studied, and there is little documented proof of their health effects on humans via ingestion of contaminated water. Where a body of epidemiological literature does exist, the interpretation of the data is often confusing and controversial (Calderon, 2000). The effect on individuals can vary according to their state of health, age, diet, body weight and other factors. Where direct medical evidence is lacking, maximum recommended concentrations of many elements and compounds have been set by reference to general biochemical considerations. Exposure to multiple contaminants has generally not been considered, nor adequate epidemiological studies conducted for populations drinking contaminated groundwater (Craun, 1984). Potential adverse health effects and risks are usually based on available data from acute and chronic toxicity studies on laboratory animals, poisonings and accidental ingestion, clinical exposures, and occupational exposure.

The disease-producing potential of micro-organisms is well known, and the impact of waterborne illnesses throughout the world immense (Craun, 1984; Keswick, 1984; Powelson & Gerba, 1985). Diseases that can be contracted as a result of drinking contaminated groundwater may be bacterial (*Shigellosis*, *Salmonellosis*, and Typhoid Fever), parasitic (*Giardiasis*) or viral (Hepatitis A). Keswick (1984) reported 264 outbreaks and 62 373 reported cases of waterborne diseases due to contaminated groundwater in the United States between 1946 and 1977.

This chapter principally describes urban groundwater pollutants and their sources. Pollutants such as hydrocarbons, nutrients, pesticides and salts may stem from sewage, solid waste, industrial liquid waste, urban stormwater, or the use of chemicals for industrial and domestic purposes. The excessive salting of icy roads is an example of this, as it may lead to sodium chloride (NaCl) contamination. In some instances, the discussion is related to sources of pollution (such as sewage, Section 3.2). In other cases, a more coherent description is obtained by focussing on pollutants which have some common properties (such as hydrocarbons, Section 3.6).

The chapter closes with some discussion of the issues of investigation of urban groundwater. Section 3.10 covers more classical investigations of dissolved pollutants, although this is only brief as such material is well covered in standard texts. Section 3.11 explains more about the investigation of non-aqueous phase pollutants, a major problem which has only come to light in the last two decades but which is not well covered by traditional hydrogeological textbooks.

3.2   SEWAGE

Introduction

Sewage comprises a complex mixture of natural inorganic and organic matter with a small proportion of man-made substances. It originates from domestic, commercial and industrial sources, and is transported within a system of pipes to outlets such as septic tanks, wastewater treatment plants, or natural watercourses. In certain developing countries, sewage is discharged directly into soak pits (Bodhankar & Chatterjee, 1994).

In a septic tank system, the sewage flow is slowed in the tank to allow solids to settle to the bottom. Liquid from the middle of the tank passes out of the tank to a subsurface soak pit or drain field for infiltration into soil. The percolation of septic wastes through soil is required for the purification of the effluent before it replenishes groundwater. In the United States, 3 billion m^3 of wastewater per year are discharged into the soil via individual on-site disposal units, which include septic tanks and cesspools (Keswick, 1984).

Most urban cities are serviced by water-borne sewerage. For example, in the UK, 97% of all households are sewered (Lerner, 1997). Unfortunately, sewers sometimes leak. According to Eiswirth and Hotzl (1997) approximately 100 million m^3 of wastewater leak annually from partly damaged sewerage systems to soil and groundwater in the Federal Republic of Germany. Mull et al. (1992) estimated that between 5 and 8 million m^3 of sewage was contaminating groundwater in the Hannover area of Germany. In the United States, it is estimated that 950×10^6 m^3 of wastewater is lost each year from leaking sewers (Pedley & Howard, 1997). Misstear and Bishop (1997) reported that most of the major aquifers in England and Wales have been impacted locally by sewer-related groundwater contamination. Barrett et al. (1997) used nitrogen isotopes and microbiological indicators to establish the sewage contamination of groundwater in Nottingham, and found evidence of multi-point source sewage contamination of shallow groundwater.

Health impacts of sewage

Sewage can be a significant factor in the pollution of urban groundwater. Craun (1984) describes several case histories of outbreaks caused by drinking contaminated groundwater in the United States. Among these was a case in Alaska where 89 people who were exposed to sewage contaminated water from a well fell ill with symptoms such as nausea, vomiting, abdominal pain, fever and diarrhoea caused by *S. sonnei*.

Several cases of outbreaks resulting from leaking sewers have been reported. Anon (1985) reported a case where leakage from a broken sewer in Haifa, Israel resulted in epidemics of typhoid and dysentery. These affected 6000 people, with one fatality. Misstear and Bishop (1997) provided a summary of incidences where sewage contamination of groundwater affected communities in England and Wales. One such incidence happened in Bath in 1928, where a leaking sewer led to contamination of a well in Jurassic limestone aquifer, which resulted in a typhoid outbreak and 7 deaths. Another incidence was that at Bramham, Yorkshire, where leakage from a surcharged sewer in 1980 contaminated a borehole and caused in 3000 cases of gastro-enteritis (Short, 1988). In the town of Naas, near Dublin, the public water supply serving approximately 1500 households in the town became grossly contaminated in 1991 by sewage. About 4000 people became ill, suffering varying degrees of gastro-intestinal disorders (Misstear & Bishop, 1997).

Outbreaks have also been reported in developing countries. Bodhankar and Chatterjee (1993) reported the occurrence of various waterborne diseases including cholera, dysentery, gastroenteritis, and typhoid fever, as a result of drinking contaminated groundwater in Pradesh, India.

The risk of infection, disease, and mortality from drinking water contaminated with a virus has been calculated. For example, drinking water with a concentration of one hepatitis A virus particle in 10^4 L over a lifetime can result in a significant risk ($>$1%) of contracting the disease (Powelson & Gerba, 1995). These authors suggest that sometimes even a single virus particle may cause infection. More than 10^6 of infectious virus particles may be excreted per gram of faeces by an infected person, and concentrations as high as 10^5 infectious virus particles per litre have been reported in raw sewage (WHO, 1979).

Table 3.1.   Pollution loads from various plumbing fixtures in the US (mg/d) (from Dillon, 1997).

Wastewater source	Biochemical oxygen demand (BOD)		Chemical oxygen demand		NO-N		NH-N		Phosphates	
	Mean	%	Mean	%	Mean	%	Mean	%	Mean	%
Bathroom sink	1 860	4	3 250	2	2	3	9	0.3	386	3
Bathtub	6 180	13	9 080	8	12	16	43	1	30	0.3
Kitchen sink	9 200	19	18 800	16	8	10	74	2	173	2
Laundry machine	7 900	16	20 300	17	35	49	316	10	4 790	40
Toilet	23 540	48	67 780	57	16	22	2 782	87	6 473	55
Total	48 690	100	119 410	100	73	100	3 224	100	11 862	100

Composition of sewage

Domestic sewage consists mainly of human excreta, with some contribution from food preparation, personal washing and laundry (Table 3.1). Human excreta is characterized by high BOD, suspended solids, faecal bacteria, chloride and ammonia content.

Most of the organic carbon in sewage can be attributed to carbohydrates, fats, proteins, amino acids and volatile acids. Other organic molecules such as hormones, vitamins, chlorinated hydrocarbons and pesticides may also be present. Many inorganic constituents are present in sewage. These include compounds containing sodium, potassium, calcium, magnesium, boron, chloride, sulphate, phosphate, bicarbonate and ammonia. Synthetic detergents are a major source of phosphate, chloride, sulphate and boron.

Micro-organisms found in sewage include bacteria, viruses, and parasites. Bacteria are the most common. Potential pathogenic bacteria present in faeces include *salmonella*, *shigella*, *vibriocholerae* and *Escherichia coli*. More than 100 different types of viruses have been reported in sewage (Keswick, 1984), including polio virus, hepatitis A, *echo*, *coxsackie*, *rota*, *adeno*, and *norwalk*-like viruses.

The composition of sewage can be further complicated by the inclusion of industrial wastes, composed of strong spent liquors from industrial processes, and the comparatively weak wastewaters from rinsing, washing, condensing and contaminated surface runoff. Table 3.2 shows the mean sewage composition from damaged sewers in Rastatt, Germany. Barber (1992) found more than 200 organic compounds in sewage-contaminated groundwater, which included chlorinated hydrocarbons (aliphatic and aromatic), alkyl-substituted hydrocarbons (aliphatic and aromatic), alkylphenols, aldehydes, and phthalate esters. TCE and PCE were found to be the major contaminants, and occurred at greater concentrations than those of other compounds.

Reduction in the concentration of sewage components may be achieved by treatment at sewage treatment works. Here, sewage may undergo several stages of treatment. Including primary treatment, sewage is passed through sedimentation tanks to separate the solids (sludge) from the liquid. This is usually followed by secondary treatment, in which micro-organisms are used to take up organic matter out of solution. Tertiary treatment can be used to reduce the organic matter content even further. Treatment is sometimes conducted in waste stabilization ponds to reduce pathogens.

Table 3.2. Mean sewage composition and mean annual input from damaged sewers into soil and groundwater in Rastatt and Hannover, Germany (from Eiswirth & Hotzl, 1997).

	Mean concentration in sewage (mg/L)	Mean annual input (kg/ha/a)
Potassium	38	55
Sodium	111	153
Ammonium-N	55	80
Organic-N	23	33
Chloride	101	150
Nitrate	7	10
Sulphate	38	55
Boron	1.96	3
Phosphate	11	16
Lead	0.034	0.48
Cadmium	0.005	0.007
Chromium	0.001	0.001
Copper	0.062	0.09
Nickel	0.027	0.05
Zinc	0.85	1.2

Migration and attenuation

Microbiological contaminants
Microbiological contaminants move through the subsurface as suspended particles in water. Although normally associated with a host, pathogenic bacteria are able to survive and multiply in the environment provided that the proper conditions are present (Keswick, 1984). Viruses have also been found to survive for long periods outside their host, although they do not replicate in the environment (Keswick, 1984). Several cases have been reported in literature where bacteria and viruses have travelled in excess of 400 m in sand, gravel and limestone aquifers (Keswick, 1984; Hagedorn, 1984; Yates & Yates, 1988). The longest distance reported is 1 600 m for a coliphage in a limestone aquifer (Yates & Yates, 1988).

In general terms, both the survival and migration of micro-organisms are controlled by the specific micro-organism type, the nature of the soil, and the climate. Specific factors affecting fate include temperature, organic matter, moisture content, pH, and the presence of other organisms (Yates & Yates, 1988). Migration is controlled by variable factors such as moisture content, pH, salt species and concentration, soil texture, organic matter and hydraulic conditions (Table 3.3 and Table 3.4).

Rates of removal of micro-organisms have been studied in the field, laboratory, and slow filters at water treatment plants by a variety of techniques (Table 3.5). Removal rates measured range from 0.7 to 33 days. For a pathogen present in wastewater at a concentration of say 10^4 organisms per litre, it would take four times the log removal rate reported in Table 3.5 to reduce this number to 1 organism per litre, without taking dilution into account (Dillon, 1997).

A number of cases have been documented where microbes have been observed to travel faster than the average groundwater flow (Gerba et al., 1991; Gerba & Bitton, 1984). Two mechanisms that can be used to explain this phenomenon are anion exclusion or pore-size exclusion. In anion exclusion, the negatively charged microbial particles are pushed to the centre of the pore where water velocity is higher than average. Pore-size exclusion refers

Table 3.3.   Factors influencing bacterial fate in the subsurface (from Yates & Yates, 1988).

Factor	Survival	Influence on movement
Temperature	Bacteria survive longer at low temperatures	
Microbial activity	Increased survival time in sterile soil	
Moisture content	Greater survival time in most soils and during times of high rainfall	Migration increases under saturated flow conditions
pH	Increased survival time in alkaline soils (pH > 5) than in acid soils	Low pH enhances bacterial retention
Salt species and concentration		Increases the concentration of ionic salts and cation valencies enhance bacterial adsorption
Soil properties		Greater bacterial migration in coarse-textured soils; bacteria are retained by the clay fraction of soil
Bacterium type	Different bacteria vary in their susceptibility to inactivation by physical, chemical, and biological factors	Filtration and adsorption are affected by the specific physical and chemical characteristics of the bacterium
Organic matter	Increased survival and possible regrowth when sufficient amounts of organic matter are present	Accumulation can improve the filtration process
Hydraulic conditions		Migration increases with increasing hydraulic loads and flow rate

Table 3.4.   Factors influencing virus fate in the subsurface (from Yates & Yates, 1988).

Factor	Survival	Influence on movement
Temperature	Viruses survive longer at lower temperatures	Unknown
Microbial activity	Some viruses are inactivated more readily in the presence of certain micro-organisms; however, adsorption to the surface of bacteria can be protective	Unknown
Moisture content	Some viruses persist longer in moist soils than dry soils	Migration increases under saturated flow conditions
pH	Most enteric viruses are stable over a pH range of 3 to 9; survival may be prolonged at near-neutral pH values	Low pH favours virus adsorption; high pH results in virus desorption from soil particles
Salt species and concentration	Some viruses are protected from inactivation by certain cations; the reverse is also true	Increasing concentration of ionic salts and cation valencies enhance virus adsorption

(*continued*)

Table 3.4.   (*continued*).

Factor	Survival	Influence on movement
Virus association with soil	In many cases, survival is prolonged by adsorption to soil; however, the opposite has also been observed	Virus movement through the soil is slowed or prevented by association with soil
Virus aggregation	Enhances survival	Retards movement
Soil properties	Effects on survival are probably related to the degree of virus adsorption	Greater virus migration in coarse-textured soils: there is a high degree of virus retention by the clay fraction of soil
Virus type	Different virus types vary in their susceptibility to inactivation by physical, chemical and biological factors	Virus adsorption to soils is probably related to physicochemical differences in virus capsid surfaces
Organic matter	Presence of organic matter may protect viruses from inactivation; others have found that it may reversibly retard virus infectivity	Soluble organic matter competes with viruses for adsorption sites on soil particles
Hydraulic conditions	Unknown	Migration increases with increasing hydraulic loads and flow rates

Table 3.5.   Times for one log removal of viruses, bacteriophage and bacteria in groundwater (from Dillon, 1997).

Micro-organism	Decay rate[a] (day^{-1})	Removal time[b] (days)	Reference	Type of experiment
Polio virus 1	0.046	22	Britton et al. (1983)	Lab
	0.21	4.8	Keswick et al. (1982a)	McF
	0.03–0.09	11–33	Jansons et al. (1989a)	Field
Adenovirus 40	0.04–0.05	??	Enriquez et al. (1995)	Lab
Coxsackie virus	0.11	9.1	Keswick et al. (1982a)	McF
	0.05	20	Jansons et al. (1989a)	Field
Echovirus 11	0.10	10	Jansons et al. (1989a)	Field
Rotavirus SA-11	0.36	2.8	Keswick et al. (1982a)	McF
Coliphage f2	1.42	0.7	Britton et al. (1983)	Lab
	0.39	2.6	Keswick et al. (1982a)	McF
Escherichia coli	0.32	3.1	Keswick et al. (1982a)	McF
	0.26	3.9	Mazzeo et al. (1989)	Lab
	0.05–0.11[c]	9.1–20	Dillon et al. (1995)	Field
Faecal	0.23	4.3	Keswick et al. (1982a)	McF
Streptococci	0.03	33	Britton et al. (1983)	Lab
	0.12	8.3	Dillon et al. (1995)	Field
S. faecalis	0.31	3.2	Mazzeo et al. (1989)	Lab
Salmonella typhimurium	0.13	7.7	Keswick et al. (1982a)	McF

[a]Expressed as $\log_{10}(Ct/Co)$, where Ct is the concentration of organisms after 24 hours and Co is the initial concentration of organisms; [b]Time for one log removal; [c]Faecal coliforms; Lab: flasks in laboratory conditions; McF: McFeter's type chambers immersed in flowing groundwater; Field: sampling from well under field conditions.

to a process whereby micro-organisms can only be transported through large pores, where the average pore water velocity is higher than the average for the entire porous medium.

Soluble components of sewage

The fate of nutrients in the subsurface will depend on the oxidation status and the cation exchange capacity of the porous medium. In general, nitrogen in sewage is predominantly in the form of ammonium or organic nitrogen. Nitrification occurs once wastewater escapes from the zone of high oxygen demand into the surrounding unsaturated zone or aerobic groundwater. In the absence of anaerobic conditions denitrification does not occur, and the nitrate behaves as a conservative species in groundwater (Dillon, 1997). An anaerobic core of the plume may develop close to the source of the effluent if the rate of transport of oxygen in the aquifer is exceeded by the oxygen demand imposed by the effluent. Usually this core will be devoid of nitrate with all the nitrogen present in its reduced forms (Dillon, 1997). Denitrification can occur if groundwater is anaerobic, and there is a suitable reducing agent.

Eiswirth and Hotzl (1997) reported results of an investigation in Plittersdorf, Germany, where the following major transformation processes were identified during seepage of wastewater effluents through the subsurface below damaged sewers (Fig. 3.1):

- Precipitation of iron sulphides, anaerobic oxidation, fermentation and ammonification within a thin anaerobic zone immediately below the sewer leakages.
- Biodegradation (oxidation) of organic matter, dissimilatory nitrate reduction and bicarbonate buffering within an aerobic unsaturated zone above the capillary fringe.

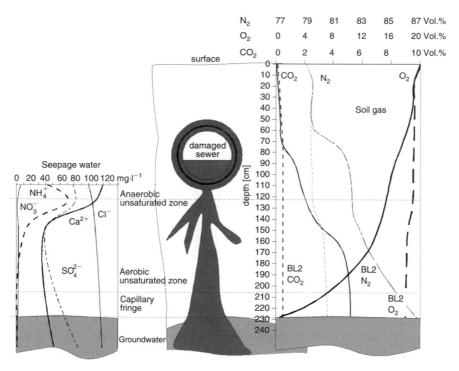

Figure 3.1.   Variation in wastewater chemistry and soil gas composition during the effluent seepage through the subsurface below a sewer (after Eiswirth & Hotzl, 1997).

Phosphate movement through soils is quite limited due to various adsorption and precipitation reactions. Lance (1984) gave numerous citations which indicate that the adsorptive capacity of soils for phosphate is quite high. However, there are instances when phosphate can be quite mobile. Dillon (1997) quoted studies which indicate that phosphorus movement can be very variable with leachate concentration varying between 1% and 100% of source concentrations, depending on the age of the sanitation system and the adsorptive properties of soils. In these studies, the breakthrough of phosphate was always retarded in comparison to nitrate.

The concentration of heavy metals in sewage is usually low and not considered a serious potential pollution problem. However, contamination of groundwater can occur in situations where high concentrations are present from industrial releases, and this will be discussed later in Section 3.4.

The readily degradable organic carbon (proteins, carbohydrates, etc.) in sewage effluent is almost completely decomposed in the soil environment. However, chlorinated organics, detergents, pesticides and other trace organics found in the effluent are not all degraded in soil and can move down to groundwater and persist in the aquifer. Muszkat et al. (1993) compared the distribution of organic contaminants in the unsaturated zone between a site irrigated with sewage effluents and two control sites which received moisture from unpolluted water sources. They found that, at the effluent irrigated site, organic pollutants from the sewage had migrated through the 20 m thick unsaturated zone to the water table. Included in this were several pollutants that are considered to be biodegradable, such as toluene and phthalates. On the other hand, most of the organic contaminants at the control sites did not penetrate into the unsaturated zone. Muszkat (1993) suggested that surface-active components of sewage or stable colloidal fractions may have enhanced vertical migration of organic moieties at the sewage irrigated site.

3.3 SOLID WASTE

Introduction

Solid waste is one of the main sources of potential contamination of groundwater in urban areas and surroundings. Pollutants can be leached from solid waste on manufacturing sites and others, penetrating through the topsoil and through to groundwater.

Urbanized areas throughout the world are characterized by a high production of solid waste. Waste generation rates have been estimated to vary between 0.3–0.6 kg/person/day in some cities of developing countries and 0.7–1.8 kg/person/day for developed cities (Palmer Development Group, 1995). A useful approach is to classify them according to source, and Table 3.6 shows four categories.

The composition of domestic solid waste is influenced by many factors, which include land use and stage of urban development. In general terms, the waste from developed countries is composed of relatively large amounts of paper, metals, glass and plastics, and fewer putrescibles, while waste from developing countries is generally composed of relatively large quantities of putrescibles, dust and ash, and comparatively little paper, metals, glass and plastic (Table 3.7). For example, in Delhi, India 47% of the waste is vegetable matter, whereas in the UK the comparable amount is only 25%.

The composition of solid waste originating from the commercial sector depends entirely on the nature of the source. In general, the composition is similar to domestic wastes,

Table 3.6.   Characteristics of domestic, commercial, industrial and construction/demolition waste.

Type	Source	Examples of components in waste
Domestic	Single family dwellings, multi-family dwellings, low, medium and high-rise apartments	Food waste, paper, cans, cardboard, plastics, textiles, rubber, leather, garden trimmings, wood, glass, non-ferrous metal, ferrous metal, dirt, ash, brick, bone
Commercial	Shops, restaurants, markets, office buildings, hotels, institutions	Food waste, paper, cans, cardboard, plastics, textiles, rubber, leather, garden trimmings, wood, glass, non-ferrous metal, ferrous metal, dirt, ash, brick, bone
Industrial	Light and heavy manufacturing, refineries, chemical plants, mining, power generation	Industrial sludge, dye stuff, paint, empty containers, tannery waste, tars, waxes, cellulose waste
Construction/ demolition	Urban and industrial construction and demolition sites	Soil, stones, concrete, bricks, plaster, timber, paper, plastic, piping, electrical components

Table 3.7.   Comparison of waste composition (%) from different countries (after Palmer Development Group, 1995).

Waste composition	Developed countries		Middle-income countries	Developing countries		
	US	UK	Singapore	Delhi, India	Kathmandu, Nepal	Wuhan, PRC
Vegetable	22	25	5	47	67	16
Paper	34	29	43	6.3	6.5	2.1
Metals	13	8	3	1.2	4.9	0.55
Glass	9	10	1	0.6	1.3	0.61
Textiles	4	3	9	na	6.5	0.62
Plastic/leather/rubber	10	7	6	0.9	0.3	0.5
Wood	4	na	na	na	2.7	1.76
Dust/ash/other material	4	18	32	36	10	77
Moisture content (%)	22	20–30	40	15–40	na	30

but may contain quantities of oils, phenols and hydrocarbon solvents. The composition of industrial waste also varies with the source. This may range from cyanide wastes from metallurgical operations, through sulphite-rich paper and pulp manufacturing wastes, mercury-rich materials from the electrical industry, to solid residues from the petro-chemical industries (such as polychlorinated biphenyls, pesticide or herbicide residues and phenol-rich tar wastes), and pulverized fuel ash from coal burning at power stations. The composition of waste from construction and demolition activities is less variable, and will include soil, stones, concrete, bricks, timber, plastic, piping and electrical components.

Solid waste disposal

Disposal of waste on land has been practised for centuries. Historically, waste was disposed on the land surface or in existing excavations such as quarries, gravel, sand and clay

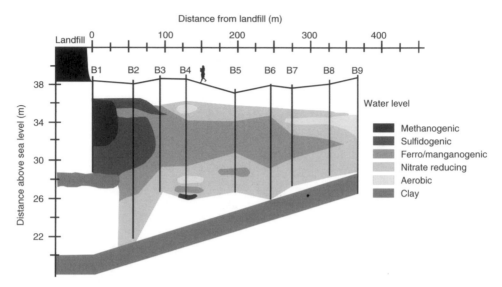

Figure 3.2. Attenuate and disperse landfill (after WHO, 1991).

pits, often without permission or control. In many countries, waste disposal activities are now controlled. Disposal in the form of sanitary landfill has proved to be the most accept-able and economical method for the disposal of solid waste. The term "sanitary landfill" describes an operation in which the wastes are compacted and covered at the end of each day's operation and a final layer of cover material is applied when the disposal site reaches its ultimate capacity. There are two types:

- Attenuate and disperse landfills. The design concept of this type allows the leachate to percolate through the landfill base with the expectation that the attenuation mechanisms operating in the underlying strata will reduce the polluting characteristics of the leachate before it reaches any receptors (Fig. 3.2). Natural geological formations best suited for these types of sites are those with significantly high amounts of clay miner-als present and where leachate movement will be slow. In the past, attenuate and dis-perse landfills were the sole method used for disposal of all types of wastes (Bagchi, 1990). It was believed that the unsaturated zone was capable of completely attenuating the potential contaminants in the leachate. This concept of attenuation has changed significantly. In many countries, only nonhazardous wastes are currently disposed of in attenuate and disperse landfills. In some countries (e.g. Germany) and in some states of the United States (e.g. Wisconsin) these types of landfills are not allowed, regardless of volume or waste type (Bagchi, 1990).
- Containment landfills are intended to restrict leachate seepage into the aquifer so as to minimize groundwater degradation. This can be done by using strata with permeability low enough to prevent significant seepage. Alternatively, the base may be lined with clay or synthetic membrane or both and a leachate collection system installed (Fig. 3.3).

Leachate characteristics

Leachate is generated as a result of the percolation of liquids through solid waste and the squeezing of the waste due to compaction. Thus, leachate can be defined as a liquid that is

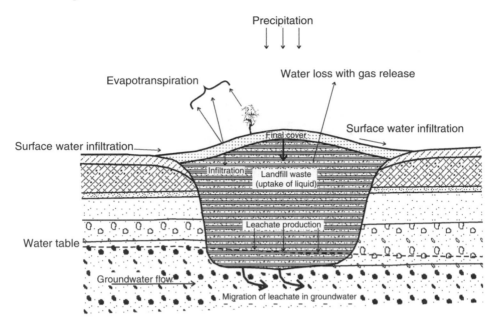

Figure 3.3.   Containment landfill.

produced when water or any other liquid comes in contact with solid waste. Liquid enters the waste mainly from external sources such as surface drainage, precipitation and groundwater. As water percolates through the waste, dissolution of waste substances occurs. The main mechanisms by which waste mass is transferred to percolating water are (Lu et al., 1985):
- Dissolution of soluble salts in refuse.
- Conversion of biodegradable organic material to gaseous and soluble forms (stabilization).

The mass of refuse stored in a landfill represents a finite source of pollutants (Lu et al., 1985; Reitzel et al., 1991; Howard et al., 1996). The amount available for leaching is largely a function of the physicochemical character of the solid waste, rates and extent of solid waste stabilization, and the volume of liquid infiltration into the landfill (Lu et al., 1985). Howard et al. (1996) estimate that only about 2.9% of the landfill waste in urban Toronto, Canada will enter the landfill leachate.

Factors that will affect the composition of leachate include (Lu et al., 1985; Bagchi, 1990; Reitzel et al., 1991):
- Solid waste composition.
- Processes that occur in a landfill (i.e. physical, chemical, and biological activities).
- Conditions within the landfill (e.g. redox conditions, pH, moisture content).
- Age of refuse.

Since these factors vary considerably, significant variability exists in leachate composition. The composition of solid waste will affect the quality of leachate that forms (Table 3.8). Leachate from household wastes contains high concentrations of total organic

Table 3.8. Composition of leachates from solid wastes in landfills (mg/L except pH-value) (after WHO, 1991).

Determinant	Household waste (UK)	Pisea (UK) (43% industrial)	Rainham (UK)* (Industrial/ household)	Granmo (Norway) (66% industrial)	Ceder Hills (US) (Industrial/ household)
pH-value	5.8–7.5	8.0–8.5	6.9–8.0	6.8	5.4
COD	100–62 400	850–1 350		470	38 000
BOD	2–38 000	80–250		320	24 500
TOC	20–19 000	200–650	77–10 000	100	
Volatile acids (C1–C6)	ND-3 700	20	600–10 000	10	7 100
Ammoniacal-N	5–1 000	200–600	90–1 700	120	
Organic-N	ND-770	5.20			62
Nitrate-N	0.5–5			0.04	
Nitrate-N	0.2–2	0.10–10	8.0		
o-Phosphate	0.02–3	0.20		(Total) 0.6	(Total) 11.3
Chloride	100–3 000	3 400	400–1 300	680	
Sulphate	60–460	340	150–1 100	30	
Sodium (Na)	40–2 800	2 185	200	462	
Potassium (K)	20–2 050	888	50–125	200	
Magnesium (Mg)	10–480	214			66
Calcium (Cu)	0.05–1.0	0.05	0.5	0.02	1.05
Manganese (Mn)	0.3–250	0.5			
Iron (Fe)	0.1–2 050	10	0.6–1 000	70	810
Nickel (Ni)	0.05–1.70	0.04	0.5	0.1	1.20
Copper (Cu)	0.01–0.15	0.09	0.5	0.09	1.30
Zinc (Zn)	0.05–130	0.16	1.0–10	0.06	155
Cadmium (Cd)	0.005–0.01	0.02		0.000	0.03
Lead (Pb)	0.05–0.60	0.10	0.5	0.004	1.40
Monohydric phenols		0.01	ND-2.0		
Total cyanide		0.01	0.09–0.52		
Organochlorine pesticides			0.01		
Organophosphorus pesticides		0.05			
PCBs		0.05			

*Samples obtained from boreholes within the fill.

carbon (TOC), of which more than 80% is in the form of volatile fatty acids (Jackson, 1980). The organic carbon component of the leachate changes with time to higher molecular weight substances such as carbohydrates. Sulphate, chloride and ammonia concentrations from household wastes are normally high, reaching a peak value within a year or two of disposal of the wastes, but then decrease irregularly over a longer period of time (Jackson, 1980). Leachates from commercial and household solid wastes are similar in composition.

As indicated above, the composition of industrial wastes varies with the source. In many cases, toxic substances are present. The pulverised fuel ash resulting from coal burning at power stations is relatively inert, containing only about 2% of soluble material, principally as sulphate, but small traces of metals such as germanium and selenium can be found (Jackson, 1980).

A variety of physical, chemical, and biological activities occurring within the refuse will influence the quality of the leachate that will form. Biological processes have the greatest impact. The decomposition of organic waste material takes place as a result of biological activity and is shown diagrammatically in Figure 3.4. It is essentially a three-phase process. In the first phase, aerobic processes occur whereby aerobic bacteria break down some of the complex organic compounds to more simple substances. The aerobic phase of decomposition is short because of the high biochemical oxygen demand (BOD) of the refuse and limited amount of oxygen present in the waste mass (Lu et al., 1985).

Once the oxygen has been consumed, the degradation process is continued by two major groups of anaerobic organisms. During the initial stage of anaerobic biodegradation process, acid fermentation will prevail. Organic compounds are converted to simple fatty acids, together with ammonia, hydrogen, carbon dioxide and water. The second anaerobic phase is characterized by methane fermentation, whereby methanogenic bacteria break down the fatty acids (such as acetic, propionic and butyric acids) into methane and carbon dioxide. In addition, hydrogen and carbon dioxide are converted to methane and water.

Important chemical processes that will occur in a refuse mass include redox, adsorption and complexation reactions. The pH will influence chemical processes and will affect speciation of most of the constituents in the system. In general, acid pH conditions will increase solubilisation of oxides, hydroxides while decreasing the sorptive capacity of the refuse (Lu et al., 1985). It will also increase ion exchange between leachate and organic matter. The redox potential will affect the oxidation state and chemical form of many constituents in the system. Reducing conditions will influence the solubility of nutrients and metallic solids, resulting in precipitation or dissolution of these constituents.

Adsorption and complexation reactions are probably the most important processes influencing the attenuation or mobility of trace metal constituents in the refuse mass (Lu et al., 1985). Under oxidising conditions, adsorption can regulate the concentration of constituents well below the level controlled by precipitation effects. Lignin-type aromatic compounds

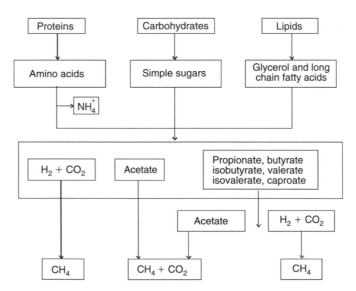

Figure 3.4.   Decomposition of organic materials occurring in household waste (after WHO, 1991).

can adsorb trace metal constituents from the leachate, as can iron and manganese hydrous oxide solids, hydrated aluminium oxides and clay minerals from daily soil cover (Lu et al., 1985). In complexation, metal ions combine with ligands such as chloride, ammonia, phosphate and sulphate as well as an array of organic compounds. In general, complexation increases concentration of metals in solution to levels far in excess of their normal solubilities. Of great importance, however, is the impact of sulphide solubility equilibria on the levels of trace metal complexes that can exist in the presence of sulphide. In essence, sulphide effectively competes with most complexing agents, and consequently many heavy metals will precipitate as sulphides rather than remain in solution as complexes (Lu et al., 1985).

Conditions such as temperature, moisture content and available oxygen in a landfill will have an effect on leachate quality. Temperature affects both bacterial growth and chemical reactions. Optimum mesophyllic temperature for methanogenesis is between 30 and 40 °C (Reitzel et al., 1991). Elevated temperatures also favour the solubility of most salts and the kinetics of most chemically driven reactions (Lu et al., 1985; Reitzel et al., 1991).

Moisture plays a significant role in solute dissolution, microbial decay and the leaching of chemicals out of a waste. Optimum moisture conditions for microbial activity are reported to be 50% and ideally 100% of the refuse dry weight (Reitzel et al., 1991).

The effect of available oxygen is notable especially for putrescible waste. Chemicals released as a result of aerobic decomposition are significantly different from those released by anaerobic decomposition. The anaerobic condition in a waste develops due to the frequent covering of waste with soil or with fresh waste. A predominantly anaerobic condition develops in thicker refuse beds (Bagchi, 1990).

Time will play an important role in determining the variation in leachate quality (Table 3.9). Generally, leachate quality reaches a peak value after a few years and then gradually declines (Bagchi, 1990).

Table 3.9.   Typical composition of leachates from recent and aged domestic wastes (mg/L except pH-value) (after WHO, 1991).

Determinant	Leachate from recent wastes	Leachate from aged wastes
pH-value	6.2	7.5
COD (chemical oxygen demand)	23 800	1 160
BOD (biochemical oxygen demand)	11 900	260
TOC (total organic carbon)	8 000	465
Fatty acids (as C)	5 688	5
Ammoniacal-N	790	370
Oxidized-N	3	1
o-Phosphate	0.73	1.4
Chloride	1 315	2 080
Sodium (Na)	960	1 300
Magnesium (Mg)	252	185
Potassium (K)	780	590
Calcium (Ca)	1 820	250
Manganese (Mn)	27	2.1
Iron (Fe)	540	23
Nickel (Ni)	0.6	0.1
Copper (Cu)	0.12	0.3
Zinc (Zn)	21.5	0.4
Lead (Pb)	8.4	0.14

Migration and attenuation of leachate in the subsurface

Leachate migration processes include filtration, dispersion, gas movement, adsorption, acid–base reactions, oxidation–reduction, precipitation/dissolution and microbiological degradation, which alter the concentration of individual contaminants. The major attenuation mechanisms are shown in Table 3.10.

Although natural attenuation can lessen the severity, a number of occurrences of pollution from solid waste disposal sites have been reported in literature (Borden & Yanoschak, 1990; Kerndoff et al., 1992; Barber et al., 1992; Kjeldesen, 1993; Howard et al., 1996; Grischek et al., 1996). Borden and Yanoschak (1990) found that 71% of the landfills examined in North Carolina had caused groundwater pollution in one or more monitoring wells. Statistically significant increases were detected in the average concentrations of inorganic contaminants. The largest percentage increases were observed for zinc, turbidity, conductivity, total dissolved solids and lead. The organic compounds of greatest threat to groundwater were chlorinated solvents, petroleum-derived hydrocarbons and pesticides. In an investigation on groundwater pollutants associated with leakage at waste sites

Table 3.10.   Major attenuation mechanisms of landfill leachate constituents (after Bagchi, 1990).

Leachate constituent	Major attenuation mechanism	Mobility in clayey environment
Aluminium	Precipitation	Low
Ammonium	Exchange, biological uptake	Moderate
Arsenic	Precipitation, adsorption	Moderate
Barium	Adsorption, exchange, precipitation	Low
Beryllium	Precipitation, exchange	Low
Boron	Adsorption, precipitation	High
Cadmium	Precipitation, adsorption	Moderate
Calcium	Precipitation, exchange	High
Chemical oxygen demand	Biological uptake, filtration	Moderate
Chloride	Dilution	High
Chromium	Precipitation, exchange, adsorption	Low (Cr^{3+}); high (Cr^{6+})
Copper	Adsorption, exchange, precipitation	Low
Cyanide	Adsorption	High
Fluoride	Exchange	High
Iron	Precipitation, exchange adsorption	Moderate
Lead	Adsorption, exchange precipitation	Low
Magnesium	Exchange, precipitation	Moderate
Manganese	Precipitation, exchange	High
Mercury	Adsorption, precipitation	High
Nickel	Adsorption, precipitation	Moderate
Nitrate	Biological uptake, dilution	High
Potassium	Adsorption, exchange	Moderate
Selenium	Adsorption, exchange	Moderate
Silica	Precipitation	Moderate
Sodium	Exchange	Low to high
Sulphate	Exchange, dilution	High
Zinc	Exchange, adsorption, precipitation	Low
Virus	Unknown	Low
Volatile organic compound	Biological uptake dilution	Moderate

in Germany, Kerndoff et al. (1992) found that the most frequently reported inorganic contaminants were arsenic, cadmium, sodium, magnesium, zinc, chloride, ammonium, boron, nickel and chromium. In contrast to the relatively small number of inorganic constituents that have been associated with waste sites, the number of organic substances that can cause groundwater contamination at waste disposal sites is quite large. In Germany and the United States, approximately 1 200 organic contaminants have been reported in disposal site groundwater (Kerndoff et al., 1992). However, only a small number of these compounds are present on a frequent basis and at a concentration clearly above established detection limits (approximately 1 μg/L). The majority of the identified contaminants (>1 000) have a detection frequency of less than 0.1% (Kerndoff et al., 1992). Figure 3.5 shows an example of organic compounds that have been detected in groundwater at waste sites.

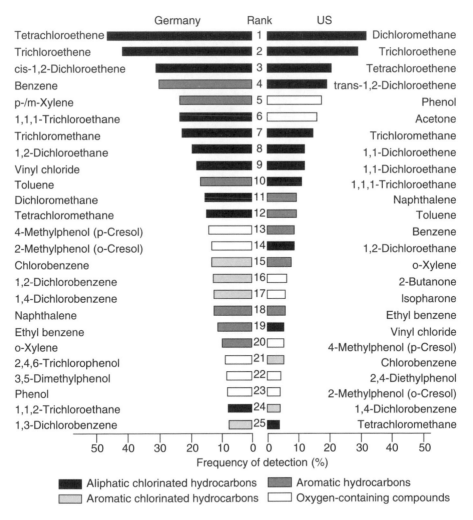

Figure 3.5.   Ranking of frequency (>1 μg/L) of principal organic contaminants in groundwaters downgradient from waste sites in Germany and the United States (after Kerndorff et al., 1992).

Migration trends of selected parameters

A simplified view of migration trends of general parameters (pH and redox potential) and organic compounds is given in this Section.

pH levels in leachates are usually acidic due to biodegradation reactions. The pH levels can increase when leachate percolates in underlying soils through dissolution of solids in the soil that contain a buffering capacity.

The redox condition of leachates from active biodegradation landfills is usually reducing (Lu et al., 1985). Soluble sulphides produced in the refuse mass are the main species which control the redox level. Numerous redox reactions can be expected during leachate migration through underlying soils and the aquifer. Lyngkilde and Christensen (1992a,b) identified a sequence of redox zones in groundwater downgradient from an old municipal landfill. Over a 370 m downgradient distance, the redox zones changed from methanogenic, sulphidogenic, over ferrogenic, to nitrate reducing and aerobic zones (Fig. 3.6). Manganogenic zones were small and scattered. The ferrogenic zone was by far the largest zone in the leachate plume, stretching for nearly 350 m. Lyngkilde and Christensen (1992b) found that the most significant degradation of organic compounds took place in this zone.

When water comes into contact with refuse mass, water-soluble compounds are rapidly dissolved in water and leached. Because polar organic compounds such as acids and phenols are more water-soluble than non-polar compounds such as hydrocarbons, they are leached out more rapidly and efficiently, and comprise the bulk of the organic load of the initial plume (Swallow, 1992). The initial surge of water-soluble components in the plume moves away from the site fairly quickly, whereas the hydrophobic substances move much more slowly. The latter effect is due to sorption. In the second stage of leachate formation, the

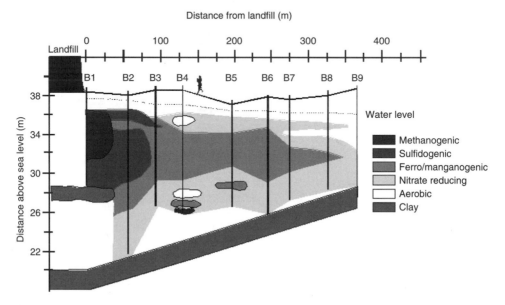

Figure 3.6.   Distribution of redox zones in the groundwater downgradient of the Vejen landfill, Denmark (after Lyngkilde & Christensen, 1992a).

water-soluble compounds in the plume are continually replenished by chemical hydrolysis and biodegradation. In chemical hydrolysis, molecules react with water and acquire polar functional groups that increase their solubility. In the process of biodegradation, soil organisms convert non-polar, hydrophobic organic compounds into water-soluble compounds as a first step in the mineralisation process. The production of water-soluble compounds from hydrophobic compounds continually renews the leachate plume and allows the partially degraded, polar, water-soluble contaminants to move away from the site as they dissolve in groundwater (Swallow, 1992).

3.4   INDUSTRIAL LIQUID WASTES

Introduction

Of all pollution sources, industrial liquid wastes are potentially the most dangerous because their volumes are often large and in many cases they are toxic. The composition of industrial effluents varies with the source, as shown in Table 3.11.

Table 3.11.   Characteristics of industrial liquid wastes (from Jackson, 1980).

Source	Potential characteristics of effluent
Food and drink manufacturing	High BOD. Suspended solids often high, colloidal and dissolved organic substances. Odours.
Textile and clothing	High suspended solids. High BOD. Alkaline effluent.
Tanneries	High BOD, total solids, hardness, chlorides, sulphides, chromium.
Adhesive/sealant	High in organic solvents.
Pulp and paper	High in inorganic salts.
Chemicals	
Detergents	High BOD. Saponified soap residues.
Explosives	Low pH, high organic acids, alcohols, oils.
Insecticides/herbicides	High TOC, toxic benzene derivatives, low pH.
Synthetic resins and fibres	High BOD.
Ink and printing paste	High in solvents.
Acids	Low pH.
Paint and coating	High in organic solvents, some chlorinated; heavy metals including Pb, Zn, Cr.
Petroleum, petrochemical	
Refining	High BOD, chloride, phenols, sulphur compounds.
Process	High BOD, suspended solids, chloride, variable pH.
Foundries	Low pH. High suspended solids, phenols, oil.
Electronic manufacturing	High in copper and other heavy metals, methanol, isopropanol, fluoro- and chlorofluoro-carbon.
Plating and metal finishing	Low pH. High content of toxic heavy metals, sometimes as sludges.
Engineering works	High suspended solids, soluble cutting oils, trace heavy metals. Variable BOD, pH.
Wood treatment	Creosote, pentachlorophenol, some copper and chromium compounds.
Thermal power	Increased water temperature. Slight increase in dissolved solids by evaporation of cooling wastes.

Table 3.12. Industrial sources of selected chemicals (compiled from Eckenfelder, 1989; Craun, 1984).

Chemical	Source of wastewater/effluent
Arsenic	Metallurgical industry, glassware and ceramic production, tannery operation, dye stuff, pesticide manufacture, some organic and inorganic chemicals manufacture, petroleum refining, rare-earth industry
Barium	Paint and pigment industry, metallurgical industry, glass, ceramics, dye manufacturing, vulcanizing of rubber, explosives manufacturing
Cadmium	Metallurgical alloying, ceramics, electroplating, photography, pigment works, textile printing, chemical industries
Chromium	Metal-plating industries
Copper	Metal-process pickling baths and plating baths; chemical manufacturing process
Fluorides	Glass manufacturing, electroplating, steel and aluminium, and pesticide and fertiliser manufacture
Iron	Chemical industrial wastewater, dye manufacture, metal processing, textile mills, petroleum refining
Lead	Storage-battery manufacture
Manganese	Steel alloy, dry-cell battery manufacture, glass and ceramics, paint, vanish, ink and dye works
Mercury	Chlor-alkali industry, electrical and electronics industry, explosives manufacturing, photographic industry, pesticide and some preservative industry, chemical and petrochemical industry
Nickel	Metal-processing industries, steel foundries, motor vehicle and aircraft industries, printing and chemical industries
Silver	Porcelain, photographic, electroplating, and ink manufacturing industries
Zinc	Steelworks, fibre manufacture, plating and metal processing industry
TCE	Industrial solvent used in dry cleaning and metal degreasing, organic synthesis
Carbon tetrachloride	Manufacture of chloro-fluoromethanes, grain fumigants, cleaning agents and solvents
Tetrachloroethylene	Industrial solvent used in dry cleaning and degreasing operations
1,1,1-Trichloroethane	Industrial cleaning, metal degreasing
1,2-Dichloroethane	Manufacture of vinyl chloride and tetraethyl lead. It is also used as a constituent of paint, varnish and finish removers
Methylene chloride	Manufacture of paint and varnish removers, insecticides, solvents

An alternative view of the types of pollution resulting from industrial liquid wastes in groundwater is given in Table 3.12. Here industrial pollutants are linked to the sources of wastewaters that most commonly contain them.

Groundwater contamination by industrial liquid wastes

The disposal of liquid wastes onto land can cause groundwater contamination. Various methods that are often used for disposal include:

- Disposal by flooding on industrial sites.
- Disposal in pits, ponds or lagoons.
- Co-disposal in sanitary landfills.

- Disposal in sealed containers in containment landfills.
- Disposal at land treatment sites.
- Irrigation of effluent on agricultural lands.
- Deep well injection.

Contaminants may also enter the aquifer through a variety of other pathways, including accidental spillages at industrial sites, accidental spillages during transport by road, rail, or in pipelines, and leakages from storage tanks, impoundments and treatment ponds. All these sources can have a serious impact on groundwater quality.

Liquid wastes are usually highly mobile because of their liquid nature. The volumes are often large and their application in the form of shock loads often prevents sufficient adsorption by soils, biodegradation, or dilution by groundwater (Hirschberg, 1986). The potential of groundwater contamination from liquid wastes is therefore quite high. Liquid wastes disposed by deep well injection should not pose a threat to overlying aquifers as long as a thorough hydrogeological investigation has been carried out (Jackson, 1980). Groundwater contamination from industrial liquid wastes has been well documented (Keswick, 1984; Craun, 1984; Hirschberg, 1986; Goerlitz, 1992; Baedecker & Cozzarelli, 1992; Appleyard, 1993; Grischek, 1996; Stuart & Milne, 1997; Dragisic et al., 1997).

Keswick (1984) reported a case where, in 1971, a large chemical company contracted a private hauler to dispose of 6 000 barrels of liquid chemical wastes at a landfill. Instead, the barrels were dumped at an abandoned chicken farm, and evidence of contamination in local wells appeared by 1974. This resulted in the loss of 148 private wells. Goerlitz (1992) reviewed studies of contaminated groundwater conducted by the US Geological Survey Organics Project, Menlo Park, California. In one of the studies, comprehensive sampling down-gradient from an ammunition processing facility identified significant groundwater contamination. The source of the pollution was attributed to waste fluids from the processing of explosives and demilitarisation of munitions, which were disposed of in unlined beds or pits within the ammunition depot.

Appleyard (1993) reported a case where groundwater contamination had occurred at a herbicide manufacturing plant near Perth, Western Australia. Effluent from the plant was disposed in an unlined pond on the site, and groundwater nearby became contaminated with the herbicides 2,4-D and 2,4,5-T, and with chlorinated phenols. Another case study reported by Appleyard (1996) involved groundwater contamination at a disused liquid waste disposal site at Gnangara, near Perth. Liquid wastes were disposed in three unlined ponds at the site over a 13-year period. The wastes included trade and industrial wastes, mainly chicken offal and brewery wastes. The groundwater contamination plume caused by waste disposal at the site extended more than 1 000 m from the site. Chemical and microbiological analyses of contaminated groundwater indicated elevated concentrations of ammonia, iron and bacteria.

Stuart and Milne (1997) reported contamination of groundwater in Leon, Mexico, from irrigation with wastewater. The wastewater contained industrial effluents from the tanning industry, which included high concentrations of salt and hexavalent chromium compounds, and is used untreated. In Birmingham, UK, sampling studies of groundwater revealed that most boreholes were contaminated with chlorinated hydrocarbon solvents such as trichloroethene (TCE), trichloroethane and trichloromethane (Lerner & Barrett, 1996). A close correlation was observed between land use and the pollutants present, implying that the sites were fouling their own groundwater. In Dresden, Germany, Grischek et al. (1996) found that all observation boreholes investigated were contaminated with tetrachloroethene (PCE).

3.5   URBAN STORMWATER

Introduction

Before urbanisation, groundwater was recharged by rainwater infiltrating through pervious surfaces, including native grasslands and woods. The recharging water was relatively uncontaminated. Urbanisation reduced the permeable soil surface area through which recharge by infiltration could occur, resulting in less net recharge to groundwater and increased surface runoff. Stormwater could no longer infiltrate into the ground but had to be drained through sewer systems. With the increase in human activity through urbanisation, the waters available for recharge generally carry increased quantities of contaminants. Contaminated waters recharging groundwater include effluent from septic tanks, waste waters from percolation basins, infiltrating stormwater, and infiltrating water from agricultural irrigation. Stormwater drains in cities are generally positioned to take storm runoff from roadways and surrounding land, and from the roofs of buildings. In the unsewered areas that exist in our cities, stormwater drains also take the sullage (e.g. household wastes) and, in several municipalities, sewage effluent after it has passed through a septic tank and sand filter. Stormwater outlets and overflows from combined systems may pollute receiving waters.

Important pollutants in urban stormwater

Urban stormwater research has quantified some inorganic and organic hazardous and toxic substances frequently found in urban runoff (Table 3.13).
    Pollutants in urban stormwater result largely from two sources:
- Washing of substances deposited on impervious surfaces, such as motor vehicle emissions (including lead oil and fuel leakage), atmospheric fallout, vegetation, and litter. Spills during transportation including dust, sand and petroleum products would also be included in this category, as would animal faeces, particularly those from dogs and scavenger birds such as seagulls. Street salts during winter conditions may also be a factor.
- Erosion of drainage channels caused by high volume of runoff generated by increased areas of impervious surface.

Table 3.13.   Hazardous and toxic substances found in selected urban stormwater runoff (sources: US EPA, 1983, 1995; Pitt & McLean, 1986; Ellis, 1997).

	Residential areas	Industrial areas
Halogenated aliphatics		×
Phthalates	×	×
PAH		×
BTEX compounds	×	×
Metals		
Aluminium	×	×
Chromium		×
Copper	×	×
Lead	×	×
Zinc	×	×
Nickel	×	×
Pesticides and phenols	×	×

Other important sources of urban diffuse pollution can include leaking septic tanks, illegal discharges into storm drains, and cross-contamination from sewer systems.

Processes that generate such loads include:

* The accumulation of contaminated solids on impervious surfaces during dry periods.
* The washing-off of pollutants by storm runoff.
* The dissolution of pollutants due to acidity of rainwater.

The term "first flush" is used to denote the period of the start of a storm. It describes the initial washing action of stormwater on accumulated pollutants, where incremental load exceeds incremental flow (Novotny, 1995; Duncan, 1995). The first flush phenomenon is related to both catchment and storm characteristics. It is restricted to small catchments and frequently, although not exclusively, occurs in urban areas (Duncan, 1995). Urban areas provide high runoff coefficients due to the area of impervious materials. Pollutants suspended by a first flush create shock loadings to receiving waters and can have characteristics similar to raw sewage (Pugh & McIntosh, 1991).

Data from the Nationwide Urban Runoff Project carried out in 1983 by the US EPA reveals significant differences in toxic urban runoff conditions. Hydrocarbon products in the form of refined oils often produce urban stormwater pollution. Pesticides such as dieldrin, chlordane, endrin, endosulfan and isophorone were mostly found in runoff from residential areas, while heavy metals and other hazardous materials were more prevalent from industrial areas. Urban runoff dry-weather base flows may also be contributors of these pollutants.

Ellis and Hvitved-Jacobsen (1996) found the type and range of pollutants associated with urban stormwater to be extremely variable. The pollution loads due to urban stormwater are intermittent, since the major flows occur in times of heavy rainfall. Table 3.14 and Table 3.15 show examples of major constituents in urban stormwater runoff.

The suspended solid concentration in stormwater runoff can vary over two orders of magnitude, as shown by Table 3.14 and Table 3.15. A major proportion of the total

Table 3.14. Range of typical pollutant concentrations in Australian stormwater (NSW Environment Protection Authority, 1996).

Pollutant	Dry weather concentrations	Wet weather event mean concentrations
Suspended solids (mg/L)	1–350	20–1 000
Nutrients (mg/L)		
Total phosphorus	0.001–2.2	0.12–1.6
Filterable phosphorus		0.01–0.63
Total nitrogen	0.1–11.6	0.6–8.6
Oxidized nitrogen		0.07–2.8
Ammonia		0.01–9.8
Micro-organisms		
Faecal coliforms (cfu/100 mL)	40–40 000	4 000–200 000
Heavy metals (mg/L)		
Cadmium		0.01–0.09
Chromium		0.006–0.025
Copper		0.027–0.094
Lead		0.19–0.53
Nickel		0.014–0.025
Zinc		0.27–1.10

Table 3.15.   Water quality in urban runoff from different sources in Denmark (from Mikkelsen et al., 1994).

Pollutant	Urban runoff	Highway runoff	Roof runoff	Atmospheric deposition (%)
SS (mg/L)	30–100	30–60	5–50	
COD	40–60	25–60	Less	19
Total N	2	1–2	Less	70
Total P	0.5	0.2–0.5	Less	23
Pb (μg/L)	50–150	50–125	10–100	40
Zn	300–500	125–400	100–1 000	30
Cd	0.5–3			
Cu	5–40	5–25	10–100	7

polluting load arising from urban surfaces is associated with the fine particulate fraction of the wash-off. This fraction may contain over 90% of inorganic lead as well as 70% of the copper, chromium and hydrocarbons (Ellis, 1997).

Impact of stormwater disposal on groundwater

We have seen that pollutants in stormwater can contaminate groundwater. They are more difficult to treat with the use of conventional stormwater control practices such as detention basins, infiltration basins, screens, drainage systems and percolation ponds, as these generally rely on sedimentation and filtration principles. Most of the toxic organics and metals are associated with the non-filterable suspended solid fractions of runoff during the wet weather. Pollutants in dry-weather storm drainage flows, however, tend to be much more associated with filtered sample fractions.

Detention ponds and stormwater compensation basins are probably the most common management practice for the control of stormwater runoff. The monitored performance of wet detention ponds indicates 80% removal of suspended solids, 70% of BOD and COD, 60–70% for nutrients, and 60–95% for heavy metals (Pitt et al., 1994). Mechanisms such as volatilisation and biodegradation, which are viable in wet detention ponds, are not active in closed sump devices. Soakwells are generally associated with on-site wastewater disposal systems and are connected to leach drains. Upland infiltration trenches, percolation ponds and grass roadside drainage swales are sited in situ in urban areas. Infiltration basins are usually located at stormwater outfalls or adjacent to large paved areas, and are prescribed areas that provide stormwater detention and sub-surface storage of incoming water in the soil. Infiltration is a function of soil and subsoil permeability, soil moisture content and vegetation. The basins can filter flow and pollutants from up-gradient urban sources and deliver a large fraction of the surface flow to groundwater, if carefully designed and located. Areas with a shallow water table, steep slopes, sandy soils and nearby groundwater use, are inappropriate locations for infiltration basins.

In Perth, Western Australia, Appleyard (1993) showed that the impact of infiltration basins reduced natural groundwater salinity and increased dissolved oxygen concentrations in the upper part of the aquifer down-gradient of the site. Concentrations of toxic heavy metals, nutrient, pesticides, and phenolic compounds in groundwater near the basins were low and within Australian drinking water standards. Sediment in the base of an infiltration basin draining a major road contained in excess of 3 500 ppm of lead.

Phthalates (from a possible source of plastic litter) were detected in almost all monitoring bores near the basins.

Another study of change in groundwater quality caused by stormwater infiltration was conducted by Malmquist and Hard (1981) in Sweden. The three investigation sites were located at a roof water runoff infiltration trench at an industrial building, an infiltration trench in a residential area, and a ditch alongside a highway. The concentration of nitrogen in groundwater down-gradient from the infiltration points decreased due to dilution. No increase in total nitrogen was observed in the residential area, possibly due to reduction processes in the trenches. Phosphorus was much higher in the stormwater and snow than in the groundwater. The concentrations of heavy metals in groundwater did not increase except for copper in the residential area. Bacterial contamination of groundwater was only small. PAHs concentrations increased only in groundwater at the highway sampling site. The study concluded that stormwater infiltration only affects groundwater quality to a small extent. However, these findings were indicative only, and the study did not monitor the long-term performance and impact of the stormwater infiltration devices.

Where stormwater sewers are used, it is likely that there will be groundwater infiltration into the sewers in areas of high groundwater levels together with some recharge through exfiltration where the water table is below that of the sewer. The construction of storm-water conveyancing systems often incorporates permeable support material such as gravel and sand. This can provide a permeable pathway for lateral flow of water trying to enter the sewer, and in addition serves as a route for exfiltrated stormwater, enabling penetration deeper into the aquifer at such sites.

A summary of pollutants commonly detected in urban stormwater that may cause groundwater contamination problems is presented in Table 3.16. A high mobility indicates a low sorption potential in the vadose zone. A high abundance means high concentrations and detection frequencies in stormwater. A high filterable fraction indicates that a high fraction of the pollutant associated with particulates would be removed using conventional stormwater sedimentation controls, meaning a high soluble fraction has a low removal potential. This table assumes worst-case mobility conditions, such as sands. Clayey soils and high organic contents decrease the mobility of organic compounds.

3.6 HYDROCARBONS

Introduction

Hydrocarbon usage is a major source of groundwater contamination in urban areas. Many hydrocarbons are toxic at low concentrations, and some have either proven to be, or are suspected of being, carcinogens. Urban areas throughout the world use hydrocarbons for fuel for vehicles. Other types are used as industrial solvents and degreasing agents, as inert carriers for the application of pesticides and other chemicals, in industrial manufacturing processes, and are produced as waste products of manufacturing processes.

Hydrocarbons are chemical compounds of carbon and hydrogen. They can be long straight or branched chain molecules, or a variety of ring-like structures that may be connected to other hydrocarbon chains. This unique ability of carbon to form a large combination of molecular structures with hydrogen allows these molecules to have a wide range of physical and chemical behaviours. The substitution of some of the hydrogen atoms by other elements such as chlorine can change the physical and chemical properties of compounds still further,

Table 3.16.   Potential for groundwater contamination by urban stormwater pollutants (after Pitt et al., 1994).

Class of compound	Compound	Groundwater vulnerability			
		Abundance in stormwater runoff	Fraction filterable	Mobility in sandy low organic soils	Surface infiltration with/without pre-treatment
Nutrients	Nitrates	Low/mod	High	Mobile	Low/mod
Pesticides	2,4-D	Low	Low	Mobile	Low
	Lindane	Mod	Low	Intermed.	Low
	Malathion	Low	Low	Mobile	Low
	Atrazine	Low	Low	Mobile	Low
	Chlordane	Mod	V. low	Intermed.	Mod
	Diazinon	Low	Low	Mobile	Low
Hydrocarbons	VOCs	Low	V. high	Mobile	Low
	BTEX	Low	Mod	Mobile	Low
	PAHs	Low	V. low	Mobile	Low
	MTBE	Low/mod	Low	Mobile	Low
	Phthalates	Low/mod	Low	Low	Low
	Phenols	Mod	Low	Intermed.	Low
Pathogens	Enteroviruse	Present	High	Mobile	High
	Shigella	Present	Mod	Low/inter.	Low/mod
	Protozoa	Present	Mod	Low/inter.	Low/mod
	Pseudomonas aeruginosa	V. high	Mod	Low/inter.	Low/mod
Heavy metals	Nickel	High	Low	Low	Low
	Cadmium	Low	Mod	Low	Low
	Chromium	Mod	V. low	Intermed.	Low/mod
	Lead	Mod	V. low	V. low	Low
	Zinc	High	High	Low	Low
Salts	Chloride	High (seasonal)	High	Mobile	High

and can create compounds which are extremely useful in industrial processes, but which may be toxic or carcinogenic to humans or animals. Although there is an almost infinite number of chemical compounds containing carbon and hydrogen, compounds with similar chemical structures tend to have similar physical and chemical properties, and are grouped accordingly. There are a number of types of hydrocarbon compounds that have had a major impact on groundwater quality in cities throughout the world, and these are detailed below.

Petroleum hydrocarbons
Hydrocarbons produced by the refining of natural oil deposits consist mainly of long chains of carbon atoms chemically bonded to hydrogen atoms. The hydrocarbon chains are mostly unbranched, but some branched compounds are produced in the refining process to improve the burning characteristics of resultant fuels. These sorts of organic compounds are known as *aliphatic compounds*.

The physical properties of petroleum hydrocarbons vary greatly with the average length of hydrocarbon chains. Hydrocarbons with a chain length of less than four carbon atoms are gases at room temperature, and compounds with a chain length of greater than about 35 carbon atoms are mostly solids. Hydrocarbons commonly used as fuels for vehicles have carbon chains that fall between these extremes.

Table 3.17. Commonly used chlorinated solvents and their properties (after Fetter, 1988).

Compound	Specific gravity	Solubility in water (mg/L)
Carbon tetrachloride	1.59	800 (20 °C)
Chloroform	1.48	8 000 (20 °C)
Methylene chloride	1.33	20 000 (20 °C)
Ethylene chloride	1.24	9 200 (0 °C)
1,1,1-trichloroethane (TCA)	1.34	4 400 (20 °C)
1,1,2-trichloroethane	1.44	4 500 (20 °C)
Trichloroethene (TCE) (Trichloroethylene)	1.46	1 100 (25 °C)
Tetrachloroethene (PCE) (Perchloroethylene)	1.62	150 (25 °C)

Petroleum hydrocarbons are sparingly soluble in water, are less dense than water, and are volatile. The volatility decreases with increasing carbon chain length. Most soil and groundwater contamination problems are caused by petrol, kerosene and diesel, and are generally due to leaks from large storage tanks.

Petrol also contains a number of ring-like hydrocarbons known as *aromatic compounds*, including the compounds benzene, toluene, ethyl benzene and the xylenes (collectively known as the BTEX group of compounds). These compounds are sufficiently soluble in water to cause groundwater contamination problems, and are toxic at low concentrations. Benzene in particular is a carcinogen, and is the compound that will usually restrict the use of a groundwater resource for water supply after a petrol spill.

Chlorinated hydrocarbon solvents
Chlorinated solvents are a group of hydrocarbon compounds with a low molecular weight (generally less than five carbon atoms), and where one or more of the hydrogen atoms are replaced with chlorine atoms. These compounds are highly volatile, are denser than water, and are sparingly soluble in water. They are excellent solvents for many other organic compounds, and are widely used in industrial processes as solvents and degreasing agents. These solvents are also used widely in small commercial enterprises such as dry-cleaning and car service centres, and have caused severe contamination problems despite the size of such operations.

The most commonly used chlorinated solvents are shown in Table 3.17, together with some of their physical properties.

Groundwater in many industrialized areas of the United States and Europe was found to be contaminated in the 1970s by these chemicals, particularly trichloroethene (TCE). This chemical is a suspected carcinogen, and even low concentrations can make groundwater unsuitable for water supply. Contamination of groundwater by this chemical is so extensive beneath some old industrial cities that boreholes drilled almost anywhere in the city area are likely to encounter TCE contamination (Burston et al., 1993; Lerner, 1994).

Contamination from gasworks and wood preserving sites
Many established urban areas have a legacy of soil and groundwater contamination left from old gaswork plants. These were mostly operational before the use of electricity and natural gas for heating and lighting became widespread in urban areas. There are estimated to be between 3 000 and 4 000 gaswork sites in the United Kingdom, and up to 1 500 sites in the US (Naidu et al., 1996). These plants heated coal in large retort vessels at high temperatures, which produced a gas mixture containing carbon monoxide, methane and hydrogen.

Table 3.18.   Common contaminants from gaswork sites and their properties (after US EPA, 1990).

Compound	Specific gravity	Solubility in water (mg/L)
Benzene	0.88	1 780 (20 °C)
Phenol	1.06	84 000
Naphthalene	1.10	Very low (20 °C)
Pyrene	1.27	0.16 (26 °C)
Chrysene	1.27	0.06 (25 °C)
Benzo(α)anthracene	1.35	0.03 (25 °C)

The process also created large amounts of tar and creosote liquors, which were often disposed of onsite.

Most of the problems at gaswork sites have been caused by the contamination of soil and groundwater by heavy metals and cyanide, and organic compounds including phenols, benzene and creosote liquors produced at these sites. Table 3.18 lists common contaminants. Creosote is a complex mixture of over 250 different chemical compounds, but consists mostly of aliphatic hydrocarbons and polycyclic aromatic hydrocarbons (PAHs), including naphthalene, acenapthene, fluoranthene, pyrene, chrysene, and carbazole (US EPA, 1990). Soil may also be contaminated by the carcinogenic PAH benzo(α)pyrene.

One of the major problems confronting redevelopers of gaswork sites is the extremely variable distribution of contaminants over the area, depending on the layout and the location of the waste dumps. This is often a problem at older sites where gas plants were dismantled many years ago and site records are now virtually non-existent.

Contamination by creosote liquors is also a problem at wood treatment sites, where wood is treated with creosote to protect it from rotting. Pentachlorophenol and salts of copper, chromium and arsenic are also commonly used, and there are many instances of soil and groundwater becoming contaminated by these compounds at wood preservation sites (Naidu et al., 1996).

Factors influencing the mobility of hydrocarbons in soils

The fact that hydrocarbons generally have a low solubility in water and are volatile strongly influences the way that these contaminants become distributed in soils at a spill site, and the extent to which they will leach through the unsaturated zone and contaminate groundwater. These physical features can allow hydrocarbons entrained in soils to be present in three phases: as a non-aqueous phase liquid (NAPL, for example liquid hydrocarbon), an aqueous phase (hydrocarbons dissolved in water), and a vapour phase (Fig. 3.7).

When an NAPL is introduced into the sub-surface, it will move downwards through the unsaturated zone as an immiscible liquid due to gravitational and hydraulic potential gradients. Some lateral spreading also takes place due to capillary forces (Schwille, 1984). As the hydrocarbon liquid progresses downward, part of it is trapped in pore spaces and is immobilized due to capillary forces, and the portion that remains in soil is referred to as the residual saturation of NAPL. Whether the NAPL reaches the watertable as a free-moving liquid depends on the amount introduced, the physical properties of the soil, the area of infiltration, and the residual saturation in the unsaturated zone. Some of the physical factors that control the movement of NAPLs are as follows.

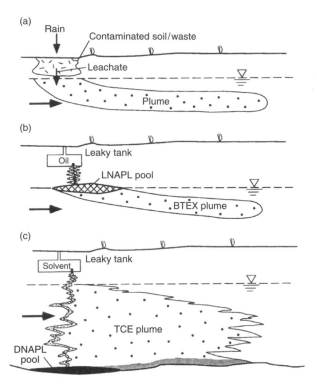

Figure 3.7. Typical distribution of LNAPL, dissolved hydrocarbons and hydrocarbon vapour in the unsaturated and saturated zones at a hydrocarbon spill site.

Saturation and wettability

Saturation describes the relative abundance of fluid in a porous medium as the volume of the fluid per unit void volume:

$$S_i = \text{(part of the porosity occupied by fluid } i)/\text{(total porosity)} \qquad (3.1)$$

Wettability describes the tendency for one fluid to be attracted to a surface in preference to another (Domenico & Schwartz, 1990), and is generally measured by determining the contact angle between the fluid and the solid surface. A contact angle between the fluid and solid surface of more than 90° indicates that the fluid is non-wetting, and a contact angle of less than 90° indicates that the fluid is wetting the solid surface.

Wettability varies depending on the type of solids and fluids, but the following generalisations hold (Domenico & Schwartz, 1990):

- Water is generally the wetting fluid with respect to NAPLs and air on rock-forming minerals.
- NAPLs are wetting fluids when air is the other fluid, but a non-wetting fluid when water is present.
- NAPLs are the wetting fluids on organic matter, such as humus in soil, in preference to either water or air.

These characteristics mean that NAPLs move at different rates depending on whether soils are wet or dry, and depending on the amount of organic material in the soil profile.

Imbibition and drainage

Imbibition and drainage are the dynamic processes by which fluids displace each other in a soil profile. Imbibition is the displacement of the non-wetting fluid by the wetting fluid, and the reverse case is known as drainage. For example, water added to a dry soil and displacing air is an example of imbibition, and the entry of an NAPL into water-saturated sediments is an example of drainage. The rate of movement of a multifluid system (e.g. NAPL and water) varies depending on whether movement is by imbibition or drainage, largely due to variations in the relative permeability. This is defined as:

$$k_{ri} = \text{(effective permeability of fluid } i)/\text{(intrinsic permeability)} \tag{3.2}$$

There are two important features to note about the rate of movement of a mixture of two or more immiscible fluids:

- The rate of movement of the combination is less than the rate of movement of each individual fluid through a porous medium.
- The relative permeabilities of both the wetting and non-wetting fluids approach zero at particular mixing ratios, that is at a particular proportion of either wetting or non-wetting liquids, the mixture cannot move through pores.

The point at which the fluid mixture cannot move through the porous medium is known as residual saturation, which has already been defined above. However, the distribution of fluids in pore spaces varies considerably depending on whether the wetting fluid is at residual saturation, or the non-wetting fluid is at residual saturation. In cases where the wetting fluid is at residual saturation, the fluid is held by capillary forces in the narrowest parts of the pore spaces (pendular saturation), whereas in cases where the non-wetting fluid is at residual saturation, the fluid is held in larger blobs between pores (Fig. 3.8).

A common situation is for water, air and an oily contaminant to be found together in the unsaturated zone. Here, water is the wetting fluid, and air is the non-wetting fluid. Oil has intermediate wetting properties, being non-wetting with respect to water, but wetting with respect to air. The oil at residual saturation is caught as blobs and pendular rings trapped

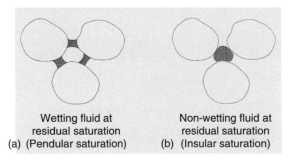

Wetting fluid at residual saturation	Non-wetting fluid at residual saturation
(a) (Pendular saturation)	(b) (Insular saturation)

Figure 3.8.   Fluid saturation when (a) the wetting fluid is at residual saturation, and (b) the non-wetting fluid is at residual saturation.

between the water in the smallest part of the pores (Wilson & Conrad, 1984). The residual saturation of a hydrocarbon is typically between 2 and 5 times larger below the water table than in the unsaturated zone (Wilson & Conrad, 1984).

The depth of penetration of hydrocarbons from a spill site into a soil profile can be estimated by the following relationship:

$$D = (1000\,V)/(A \cdot R \cdot k) \tag{3.3}$$

where

D is maximum depth of penetration (m)
V is volume of infiltrating hydrocarbon (m^3)
R is retention capacity of soil (ranges from $50\,m^{-3}$ for gravels to $400\,m^{-3}$ for fine sand or silt)
"k" is an approximate correction factor for different oil viscosities
  k = 0.5 for low viscosity hydrocarbons (e.g. petrol)
  k = 1.5 for kerosene and diesel
  k = 2 for light fuel oil

Movement of volatile hydrocarbons
The high volatility of many hydrocarbons and the movement of vapours within soil profiles can lead to the redistribution of contamination in the unsaturated zone, and can lead to quite small hydrocarbon spills producing extensive contamination problems. Hydrocarbon vapours can move laterally over distances of tens of metres in a few weeks (Mendoza & Frind, 1990a,b), preferring easy transport routes like trenches excavated for underground services, such as telephone or sewer lines, which may have a higher permeability than the surrounding soil. Vapours can then accumulate in poorly ventilated buildings near hydrocarbon spills, and can provide an additional means of exposure to contamination. In extreme cases, the accumulated hydrocarbon vapours may cause explosions if ignited (Jacobsen, 1983).

The most volatile hydrocarbons commonly found in contamination incidents are chlorinated solvents and benzene, and volatility generally decreases with increasing molecular weight. The maximum vapour concentrations of hydrocarbons in a closed system at equilibrium is given by:

$$c_{vapour} = p/RT \tag{3.4}$$

where

c_{vapour} is vapour concentration (moles/litre)
p is vapour pressure of chemical (atm)
R is gas constant (litre $\cdot$ atm/Kelvin $\cdot$ mole)
T is temperature (Kelvin = degree Celsius + 273)

Many soil and groundwater contamination problems are caused by mixtures rather than by pure compounds. Vapour pressure of a particular compound in the mixture is given by Raoult's Law:

$$p_i = x_i p_i^o \tag{3.5}$$

where

p_i^o = vapour pressure of pure component at a given temperature
x_i = mole fraction of component in the mixture

This factor can greatly reduce the volatility of hydrocarbon mixtures at spill sites, and if not taken into account, can lead to the extent of contamination being under-estimated in soil gas surveys. Low soil vapour concentrations do not necessarily indicate low NAPL concentrations in soils. Vapour pressure reduction due to hydrocarbon mixtures can also reduce the effectiveness of soil venting as a remediation option for some contamination problems.

Retention of hydrocarbons by soil organic matter
Soils containing large amounts of organic matter can often retain hydrocarbons, and minimize the amount of contamination that is leached to groundwater. The extent to which hydrocarbons and other organic compounds adsorb onto humus and other organic matter in soils depends on the chemical properties of the organic compound, and on whether the compound contains structures like hydroxyl groups which may allow it to dissolve more easily in water.

The extent to which an organic compound will be retained in soil profiles is reflected by its octanol–water partitioning coefficient, K_{ow}, and the partition coefficient is usually reported as $\log K_{ow}$. Values of $\log K_{ow}$ less than 1 usually indicate that the compound is hydrophilic, meaning it will leach readily with water. Values greater than about 4 indicate that the compound will be hydrophobic and will not leach easily from soil but will rather be strongly adsorbed by soil organic matter. Many hydrocarbons have intermediate octanol–water partition coefficients.

Groundwater contamination by hydrocarbons

The leaching of soluble constituents from a soil contaminated with an NAPL can cause the development of a groundwater contamination plume, even if the NAPL is at residual saturation and there is no direct movement of liquid hydrocarbons to the water table (Fig. 3.9). If the NAPL is highly volatile, a very broad contamination plume may develop due to vapour transport of hydrocarbons in the unsaturated zone and subsequent leaching to the water table.

If the amount of hydrocarbons held in the soil exceeds residual saturation levels, free NAPL can migrate downwards through the soil profile until it reaches the capillary fringe near the water table. Once the NAPL reaches the saturated zone in an aquifer, further migration is determined by whether its density is lower than, or greater than water.

Non-aqueous phase liquids that are less dense than water are known as Light Non-Aqueous Phase Liquids (LNAPLs) or "floaters". LNAPLs will accumulate on top of the capillary fringe and spread laterally, shown by Figure 3.7. If the mass of liquid hydrocarbons becomes sufficiently large, a depression in the water table may be created. As the water table moves up and down, hydrocarbons are smeared over sediments above and below the water table, and a substantial amount of LNAPL may be trapped in pore spaces below the average depth of the water table (Hunt et al., 1988). The NAPL may only periodically be remobilized and appear in monitoring bores (Steffy et al., 1995). Leaks of petroleum fuels from storage tanks are the most commonly encountered cases of LNAPL contamination.

Non-aqueous phase liquids which are denser than water are known as Dense Non-Aqueous Phase Liquids (DNAPLs) or "sinkers". DNAPLs moving downwards through a

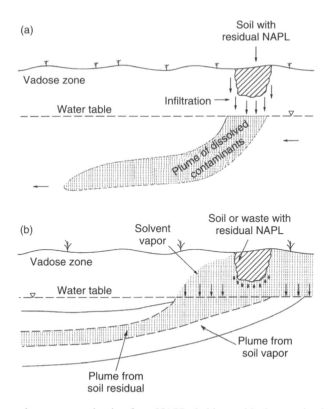

Figure 3.9. Groundwater contamination from NAPLs held at residual saturation in soils.

soil profile will accumulate on top of the capillary fringe until sufficient pressure is developed to overcome surface tension effects, and they will penetrate downward through the water table. As groundwater is displaced by the pressure of the moving DNAPL, the hydrocarbon–water interface becomes unstable and "fingers" of DNAPL form in the aquifer (Schwille, 1988). If sufficient DNAPL is present, liquid hydrocarbons will continue to move downwards through the aquifer until they reach a low permeability layer, and leaving a trail of DNAPL at residual saturation up to the water table (Fig. 3.7). DNAPL may accumulate in pools in hollows on aquitards or at the base of the aquifer, and may move under gravity along dips on geological strata (which may be in the opposite direction to groundwater flow). DNAPL may even cascade over the edge of clay lenses in aquifers and form new pools deeper in the aquifer. DNAPL may also penetrate micro-fractures in clay lenses that are otherwise impermeable, and contaminate groundwater at depth in an aquifer. Spills of chlorinated solvents and some industrial wastes such as creosotes and coal tars are common examples of DNAPL contamination.

Hydrocarbon NAPL present at residual saturation or as free pools in the aquifer provides an ongoing source of soluble contaminants that may affect groundwater many hundreds of metres down-gradient in the direction of groundwater flow. It is this feature that makes the management and clean-up of hydrocarbon spills so difficult. This is particularly the case for DNAPLs where pools of free hydrocarbons may be located at great depth in an aquifer and be difficult to detect in investigation programmes.

Factors influencing the development of dissolved contamination plumes
Contact of NAPL and groundwater causes soluble constituents within the NAPL mixture to be dissolved, giving rise to groundwater contamination plumes. The concentration of dissolved contaminants produced varies depending on the aqueous solubility of hydrocarbons and the composition of the hydrocarbon mixture. Under equilibrium conditions, the concentration is given by (Bannerjee, 1984):

$$C_i = x_i S_i \tag{3.6}$$

where

C_i is equilibrium concentration of hydrocarbon in water
x_i is mole fraction of hydrocarbon in NAPL
S_i is aqueous solubility of hydrocarbon

The most soluble constituents are leached out of an NAPL first, leaving an NAPL mixture that becomes progressively enriched in chemical compounds with a low solubility. In practice, equilibrium conditions are rarely attained, as the solubility is controlled by the rate of diffusion of individual hydrocarbons across the NAPL–water interface, particularly when the NAPL immediately in contact with water becomes depleted in very soluble hydrocarbons. It is this rate controlled process which is probably responsible for producing much lower aqueous concentrations of volatile hydrocarbons than predicted by theoretical solubilities (Mercer & Cohen, 1990).

With fuel spills, contact of the free hydrocarbon phase and groundwater takes place near the water table. In general, the solubility of aliphatic hydrocarbons is low, and decreases with increasing molecular weight of the compounds. Aromatic hydrocarbons including benzene, toluene, ethyl benzene and the xylenes (BTEX compounds) are more soluble, and contamination by these compounds can make groundwater unsuitable for use up to several hundred metres down-gradient in the direction of groundwater flow from a fuel spill (Fig. 3.7). The distribution of BTEX compounds in the plume depends on the degree to which sorption on aquifer sediments takes place, and their resistance to biodegradation processes in the aquifer. Commonly, toluene, ethyl benzene and the xylenes have a more restricted distribution than benzene, which is less biodegradable than the other BTEX compounds (Rao et al., 1996). Concentrations of hydrocarbons in contamination plumes may also vary greatly with time due to seasonal inundation of residual contaminated soil by changes in water table elevation, and due to seasonal variations in groundwater flow direction (Rao et al., 1996).

Spills of DNAPLs can also produce extensive dissolved contamination plumes, particularly spills of chlorinated solvents, which have caused groundwater contamination plumes in excess of a kilometre in length. Groundwater contamination from these chemical compounds can be very persistent.

Chemical transformation of organic compounds in groundwater
Hydrocarbons and other organic compounds in groundwater may be involved in a number of chemical processes which reduce their concentrations, but which may produce products that may be more toxic, persistent or mobile than the parent compounds. Some of these processes can take place under sterile conditions, but most chemical processes in groundwater are accelerated by the presence of micro-organisms that utilise specific chemical reactions as a source of energy.

There are two main classes of chemical reaction that take place without the presence of micro-organisms.

- Substitution reactions are chemical reactions of the form:

$$RX + N^- = RN + X^- \tag{3.7}$$

where

R is an alkyli group or other organic molecule
X is an inorganic group including Cl^-, or Br^-
N is a "nucleophile", typically a chemical species like HS^-

The most common substitution reactions that take place in groundwater are hydrolysis reactions:

$$RX + H_2O = ROH + HX \tag{3.8}$$

That is, a hydrocarbon molecule with an inorganic group reacts with water to form an alcohol, which may in turn be progressively broken down.

- Elimination reactions involve the removal of a component of an organic compound to form a new compound with a double bond between carbon atoms. The most common elimination reactions are those involved with the removal of chloride. For example, chlorinated solvents such as TCA (trichloroethane) can undergo elimination reactions like:

$$Cl_3C\text{—}CH_3 \text{ (TCA)} \rightarrow Cl_2C\text{=}CH_2 \text{ (1,1 DCE – a toxic compound)} + HCl \tag{3.9}$$

Substitution and elimination reactions may take place in parallel in groundwater, and simple hydrocarbon molecules may be transformed through a number of steps into quite complex molecules. This is particularly the case where groundwater contains high concentrations of hydrogen sulphide. In this case, simple compounds such as ethylene dibromide (a soil fumigant) may react with sulphide as polymerise to form long chain and ring compounds containing sulphur (Mackay, 1991).

Although many of the above chemical reactions can take place abiotically, the rate of reaction is commonly accelerated by the presence of micro-organisms which are present in aquifer materials. Under suitable conditions, these organisms are able to metabolise contaminants and gain energy from the breakdown of a variety of chemical compounds. The rate of biological breakdown or biodegradation of hydrocarbons in groundwater is governed to a large extent by the concentration of oxygen in groundwater.

Most shallow aquifers contain dissolved oxygen near the water table. These aquifers can support aerobic micro-organisms which can degrade a wide variety of organic compounds including petroleum hydrocarbons. The extent of biodegradation depends on oxygen concentration. Generally, about two parts by weight of oxygen are required to completely metabolise one part by weight of the organic compound. For example, oxygenated groundwater containing about 4 mg/L of oxygen can degrade only 2 mg/L of benzene. However, the solubility of benzene (1 780 mg/L) is such that water saturated with benzene will greatly exceed the capacity of groundwater for natural aerobic degradation (Wilson, 1991).

Biological activity in the vicinity of a hydrocarbon spill will also deplete groundwater of oxygen, and greatly reduce the rate of biodegradation. Other oxidation/reduction reactions yield less energy for micro-organisms than aerobic respiration, but many micro-organisms are able to use nitrate, iron and sulphate reduction in particular to breakdown organic contaminants in groundwater in the absence of oxygen. Commonly, dissolved groundwater contamination plumes containing petroleum hydrocarbons and some other contaminants are zoned on the basis of the depletion of oxygen, nitrate, iron and sulphate. Large amounts of hydrogen sulphide may be produced as a product of sulphate reduction near the margins of a plume, and this can often be an early warning of the arrival of a hydrocarbon plume at a well.

Micro-organisms can produce methane by the fermentation of some organic compounds in groundwater which is extremely deficient in oxygen. Although the micro-organisms that actually produce methane can only metabolise a limited number of organic compounds, they can act in consort with other micro-organisms which can break down more complex organic compounds down to substances that can be utilised by the methane producers (Wilson, 1991). These partnerships or consortia can totally degrade a large number of natural and synthetic organic compounds, and are particularly important for the degradation of chlorinated solvents.

The degradation of chlorinated solvents in anaerobic groundwater takes place by progressive elimination of chloride. For example, tetrachloroethene is progressively transformed to tri- and dichloroethene, and finally to vinyl chloride (monochloroethene). In some situations, large amounts of vinyl chloride can accumulate in groundwater, causing concern as this compound is much more toxic and carcinogenic than its parent compounds. In other situations, vinyl chloride is further metabolised. The factors that control the fate of vinyl chloride are not clearly understood (Wilson, 1991).

Although oxygen is not involved in the breakdown of chlorinated solvents, the process of elimination reactions has some similarity to aerobic respiration in the way that electrons are transferred in the process. In aerobic respiration, molecular oxygen accepts an electron and is reduced to the hydrogenated compound, water. The chlorinated solvents undergoing chloride removal also accept electrons and are reduced to the corresponding hydrogenated compound. Whether chloride removal benefits the micro-organisms that carry out these reactions is not known. However, the active micro-organisms must have a source of hydrogen or some other organic compounds to provide electrons for chloride removal. This could be a co-occurring contaminant (e.g. fatty acids in landfill leachate), or naturally occurring organic matter in aquifer sediments (e.g. peat layers, coal or black shale).

The rate at which hydrocarbons are degraded in groundwater depends both on chemical conditions within the aquifer, and on the chemical structure of the hydrocarbons. In general, aliphatic compounds are more readily degraded than aromatic compounds. The rate of degradation is usually measured by the half-life, or the time taken for the concentration of chemical compound to degrade to half of its original concentration. The half-lives of hydrocarbon concentrations usually range from tens of days for aliphatic compounds, to hundreds of days for aromatic compounds (Rao et al., 1996; Acton & Barker, 1992; Beller et al., 1992; Edwards et al., 1992). Benzene may be especially persistent in groundwater, and may have a half-life in excess of 800 days in some aquifers (Rao et al., 1996).

3.7 NUTRIENTS

Introduction

Nutrients are chemical substances that promote growth in living organisms. In the context of water pollution studies, the term is usually applied to simple dissolved inorganic ions containing nitrogen (N), phosphorus (P), and potassium (K); however, this definition may be extended to include other essential and trace elements.

When considering the impact of nutrients on ground and surface water quality, only compounds of nitrogen and phosphorus merit serious consideration. Potassium is harmless to humans and is of negligible environmental concern (Lerner et al., 1999). It is one of the most commonly found groundwater constituents.

Conditions in aquifers are generally oxidising, so nitrogen is generally found as nitrate, but it can occur in recharge as organic or ammonium before oxidation takes place in the aquifers. Similarly, sulphur is normally found as sulphate. Nitrate is often seen as an agricultural pollution of groundwater, and might be expected to occur at higher concentrations in the groundwater surrounding a city than in that beneath it. However, several studies have shown small differences between rural and urban nitrate concentrations because of the non-agricultural sources of nitrogen that are concentrated in cities (Lerner et al., 1999).

Sources and contaminant pathways

Today, the major sources of nutrients are the result of human need, both in agricultural and urban areas. In agricultural areas, commercial fertilisers and applied animal manures represent the primary sources of both nitrogen and phosphorus. In fact, between 1945 and 1985, the use of nitrogen fertilisers in the United States has increased twenty fold, while the use of phosphorus fertilisers has increased about four fold. Comparable increases have occurred in the more affluent countries worldwide. In urban areas, fertilisers, soaps, detergents, sewage wastes, industrial wastes and landfills all represent important sources of both nitrogen and phosphorus. Large amounts of nitrogen are also created by industrial furnaces and automobile engines burning fossil fuels, predominantly in the form of nitric acid (acid rain). On a global scale, anthropogenic emissions account for just 8% of nitrogen oxides produced in the atmosphere. In urban areas, anthropogenic emissions may outweigh natural production by a factor of several hundreds.

Natural sources of phosphorus are confined almost entirely to phosphate minerals, decaying organic matter and animal wastes. Humans excrete approximately 500 g of phosphorus per year (Train, 1979) and represent an important source. Mass fluxes from natural sources are difficult to estimate due to the low solubility of the phosphorus compounds and their predilection for transport via suspended sediment. Large amounts of phosphorus may be mobilised in flowing water, but natural background levels in solution rarely exceed 0.1 mg/L total phosphorus.

Natural sources of nitrogen include mineral deposits, organic matter and animal wastes; however, the atmosphere provides an important additional non-point source with reactions involving dinitrogen and oxygen naturally producing significant amounts of nitrous and nitric oxides (NO_x). These oxides combine with water to form nitric acid (HNO_3), which is either washed out by rainwater or combines with ammonia to form ammonium nitrate (NH_4NO_3). Background concentrations of nitrate-N are normally <0.6 mg/L in streams and <2 mg/L in groundwater. Nitrite-N and ammonia-N concentrations are normally significantly lower and rarely exceed 0.1 mg/L.

Some root zone leaching will occur, as happens in agricultural areas. An urban feature of soil processes is the disturbance from housing development. This is equivalent to ploughing up grassland which is known to be a major input of nitrogen in agricultural areas (Lerner et al., 1999).

Contaminant impacts

A consequence of the escalating use of fertilisers, both in the United States and worldwide, is widespread contamination of both ground and surface water quality, notably but not exclusively in rural areas. Phosphates are virtually immobile in soils, but a significant proportion of applied phosphate fertilisers are released into streams by soil erosion. Nitrates, on the other hand, are a very mobile form of nitrogen as they are not retained by the soil and are highly soluble. Overland runoff from agricultural fields readily contributes excess nitrate to surface water bodies; nitrates are also easily leached from the soil profile to contaminate groundwater in underlying aquifers. In many cases, less than 50% of the applied nitrate is utilised by the plant (Cannel & Burford, 1976), whilst the remainder largely contributes to water pollution.

Nitrate contamination of groundwater is now a serious problem in agricultural areas throughout the world. Recent surveys in the United States have shown that 1% of public water supplies, 9% of all domestic wells and 21% of all farm wells exceed the maximum acceptable concentration of 10 mg/L NO_3-N (Mueller et al., 1995; Macilwain, 1995). Serious problems have also been encountered in the UK (Young et al., 1976; Young & Gray, 1978; Howard, 1985) and in Canada (Hill, 1982; Priddle et al., 1989).

In the urban setting, most problems of contamination from phosphorus and nitrogen nutrients relate to sewage, fertilisers, contaminated lands and landfills. Phosphorus is much less of a concern. The release of phosphates to urban springs, rivers and lakes can be a local problem but contamination of groundwater is rare. Nitrate contamination represents the primary worry in urban areas, particularly where underlying groundwater is used for potable supply. Where point sources of nitrate contamination are involved, contaminant levels in groundwater can far exceed those experienced in the agricultural setting.

The most serious potential source of nitrate contamination in unsewered urban areas is sewage waste (Lerner, 1994; Howard, 1997). In North America, Wilhelm et al. (1994) have suggested that septic tanks designed to treat domestic sewage may constitute as many as 20 million potential point sources of groundwater contamination. Certainly the failure rate of such systems is uncomfortably high, and in Canada, nitrate plumes over 100 m in length have been observed (Robertson et al., 1991). In a study of septic systems at two sites in Australia, Hoxley and Dudding (1994) additionally report contamination of groundwater by faecal bacteria. In one case, *E. coli* was detected at distances greater than 500 m from the suspected source.

While contamination by leaking sewers has been reported as a serious potential problem in heavily urbanised regions of Europe (Eiswirth & Hotzl, 1994), the most serious impact from sewage wastes has been reported in the densely populated, less affluent parts of the world, where drainage ditches, unlined open canals and rivers are frequently used to convey a large variety of urban waste. In parts of Mexico, for example (Howard, 1993; British Geological Survey, 1995), widespread contamination of groundwater by nitrate and faecal coliforms has been reported in urban groundwater. Similar problems have also been reported in Brazil, where a population of over 250 000 in the industrial towns of Barra Mansa and Volta Redonda contributes 14 200 kg of BOD (biochemical oxygen demand) and 1 790 kg of nitrogen each day to the Paraiba do Sul (Foster, 1990; Hydroscience Inc., 1977). In India many large towns face severe nitrate contamination of groundwater as a result of inadequate sewer systems. In Madras, Somasundaram et al. (1993) describe nitrate concentrations in underlying groundwater in excess of 200 mg/L NO_3-N, with over 70% of the wells showing nitrate above the 10 mg/L NO_3-N drinking water standard. Other sites in India with similar problems include Lucknow (maximum nitrate 147 mg/L NO_3-N) (Sahgal et al., 1989), Bhopal (maximum nitrate 434 mg/L NO_3-N) (Sharma, 1988) and Tirupathi (maximum nitrate 79 mg/L NO_3-N) (Sukhija et al., 1989).

In the more affluent countries, landfills (Howard et al., 1996) and fertilisers (Morton et al., 1988) are the more likely causes of groundwater contamination in heavily urbanised areas. In Long Island, New York, Flipse et al. (1985) showed that fertiliser use accounted for over 70% of the nitrate detected in groundwater beneath a sewered housing development in Suffolk County. In Perth, Australia, Sharma et al. (1994) installed suction lysimeters beneath fertilised urban lawns and showed that between 16 and 47% of incident water passed below the root zone carrying nutrients with flow-weighted nitrate-N concentrations ranging up to 5.37 mg/L. In Toronto, Canada, shallow urban springs regularly reveal elevated nitrate concentrations, locally in excess of 29 mg/L NO_3-N (Howard & Taylor, 1998). While the use of fertilisers is suspected to be the primary source of the nitrate contamination in Toronto, a chemical audit of potential contaminant sources in the area (Howard & Livingstone, 1997) shows that landfills and septic tanks may be equally to blame as they release comparable amounts of nitrogen to the subsurface.

A study result in Nottingham, UK, shows that nitrogen loads are similar in both urban and rural areas, but they have very different sources. In the urban area, at about 20 kg/ha/y, the origins of nitrogen in groundwater were:

Leaking mains	37%
Leaking sewers	13%
Soil leaching	9%
Others (contaminated land, industry, etc.)	41%

Sewers contributed less than 15% of the nitrogen load, while contaminated land and industry were major contributors (Lerner et al., 1999).

Health and environmental concerns

Surface water and groundwater respond to an excessive influx of nutrients in quite different ways. In surface water the nutrients are utilised by living organisms and the environmental problems that result are a consequence of enhanced biological activity. In groundwater, where life form is normally limited to microbes and the occasional micro-invertebrate, nutrients will simply accumulate in solution until the groundwater becomes too contaminated for its intended use.

In surface water, the primary concern is that excessive nutrient concentrations will cause serious and sometimes irreversible damage to the delicate ecosystems found in ponds, lakes, rivers and shallow seas. The problem can also affect the hyporheic zone at the groundwater/surface water interface beneath the stream bed (Fraser et al., 1996). Under natural conditions, most surface water bodies are oligotrophic: the supply of nutrients is low, photosynthetic production is minimal, and the water remains oxygenated at all depths (Drever, 1997). When nutrients are introduced in excessive amounts, the water body becomes eutrophic and the enhanced biological activity often produces massive algal blooms (Harris, 1994). The rapid growth in biomass is followed by an equally rapid death, and the process of organic decomposition consumes large amounts of oxygen. This commonly creates anaerobic conditions, particularly in the bottom sediments. In many parts of the world, deeper lakes are turned over twice a year when the air temperature moves close to 4 °C and water at the surface of the lake reaches its maximum density. In eutrophic lakes, lake turnover can bring hydrogen sulphide and other toxic compounds to shallower levels and cause widespread mortality to fish. Turnover can also return column phosphate to the water that had previously been adsorbed on to undecomposed organic matter, and this additional influx of phosphate can seriously exacerbate the problem of eutrophication.

Groundwater can contribute to eutrophication by transporting nutrient-rich water and sup-plying it to surface water bodies; however, groundwater is not affected directly as the bio-logical activity is very low. Unfortunately, this lack of biological activity also means that groundwater is unable to assimilate nutrients and excessive amounts eventually lead to a deterioration of water quality. When contamination occurs, it is the presence of nitrogen compounds that normally represents the greatest cause for alarm. Unlike potassium and most inorganic phosphorus compounds, which are relatively immobile in the subsurface and harmless to humans, nitrogen compounds are readily transported in the subsurface and can prove lethal.

A particular concern is the association between elevated nitrate concentrations in ingested water and a disease known as methaemoglobinaemia, nitrate cyanosis or simply "blue baby syndrome" (Comley, 1945; Spalding & Exner, 1993). The disease is caused by the bacterial reduction of nitrate to nitrite in the intestinal tract. The nitrite enters the blood stream and combines with the haemoglobin to form methaemoglobin, which reduces the ability of the blood to transport oxygen. The reduction of nitrate to nitrite occurs primarily in very young children because the lower acidity of their gastric juices provides a better environment for nitrate-reducing bacteria. Cattle also have digestive systems capable of supporting these sorts of bacteria and can also be affected by methaemoglobinaemia if their intake of nitrate in drinking water is excessive. While older children and adults can appar-ently tolerate relatively high nitrate levels in drinking water, concerns have been raised that nitrate may play a role in the production of nitrosamines in the stomach which are known carcinogens (Hill et al., 1973).

In its uncomplicated form, methaemoglobinaemia is usually easily diagnosed and its treatment understood. However, to minimise health risk, most jurisdictions throughout the world recommend that levels of NO_3-N in drinking water should not exceed 10 mg/L. In North America, for example, the maximum contaminant level (MCL) for NO_3^--N and NO_3^--N + NO_2^--N is maintained at 10 mg/L (US EPA, 1996) despite opposition from those who believe that the standard is both too rigorous and too costly to enforce. The MCL for NO_2^--N alone is set at 1 mg/L to recognise the fact that nitrite is the direct causative factor in methaemoglobinaemia. Ammonium nitrogen is also undesirable in water for public supplies primarily because of taste and odour that are detectable above concentra-tions of about 0.5 mg/L. However, no MCL has been established for this ion in North America.

3.8 PESTICIDES

Introduction

The term "pesticide" is used to describe those chemicals that control insects, weeds, and a wide variety of life forms that negatively influence the production of crops. The term may also include chemicals used by consumers for such purposes as the extermination of termites and roaches, cleansing of mould from shower curtains, destroying crab grass, killing fleas on pets and disinfecting swimming pools. By design, pesticides tend to be toxic to living organisms, either selectively (narrow-spectrum pesticides) or non-selectively (broad-spectrum pesticides). As a consequence, their chemical behaviour and environmental fate have become popular foci for research. Until recent years, remarkably little of this research has concerned groundwater.

The vast majority of pesticides are designed to prevent, destroy or simply control undesirable insects (insecticides), plant life including algae and weeds (herbicides), rodents (rodenticides), and moulds, mildew and fungi (fungicides). Other pesticides of interest include nematicides, for the control of nematodes, defoliants designed to strip the leaves from trees and woody plants, and various poisons or repellents for the control of undesirable amphibians, reptiles, birds, fish, mammals and invertebrates.

The use of chemicals to eliminate perceived pests dates back several thousands of years. The world-wide production of pesticides grew from a scant 100 000 tonnes in 1945 to over 1 million tonnes in 1965 and 1.8 million tonnes by 1975. In the past twenty years this explosive increase has stabilised, and today approximately 2.5 million tonnes of chemical pesticides are used annually throughout the world (Chiras, 1998).

Over the years, many thousands of chemical pesticides have been synthesised and tested. Whilst these have altered little in recent years, the list of approved chemicals has changed considerably in response to a growing volume of environmental and toxicological research data. Chlorinated hydrocarbons such as DDT, aldrin, kepone, dieldrin, chlordane, heptachlor, endrin, lindane, toxaphene and mirex (Moore & Ramamoorthy, 1984) have dominated the market in the past. However, their chemical persistence in the environment, their tendency to accumulate in the food chain and their ability to cause cancer, birth defects and neurological disorders, has led to their strict control or outright banishment. As an illustration of their persistence, concentrations of DDT, lindane dieldrin, and endrin in Lake Ontario, Canada, remained marginally below established drinking water quality standards in 1983 despite being banned or severely restricted over ten years earlier (Biberhofer & Stevens, 1987).

In more recent times, there has been a shift towards to the use of organophospahates (e.g. malathion and parathion) and carbamates (e.g. carbaryl distributed as "Sevin"). Both these groups tend to degrade very rapidly (in a matter of days or weeks) and thus do not persist in the environment. They are, however, water soluble and highly toxic at very low levels. Questions have been raised as to their ability to break down under conditions of low temperature, low light, low dissolved oxygen and low biological activity that are commonly encountered in groundwater.

Drinking water quality standards established by the US EPA for specific pesticides can be found through its website. Typically, the US EPA maximum contaminant levels (MCLs) are slightly more cautious than maximum acceptable levels in effect for Australia (NHMRC/AWRC, 1987), but are 20–30 times less stringent than standards established by the European Union. Irrespective of the regulatory authority, it should be noted that drinking water quality standards have not been established for the great majority of pesticides presently in use.

Sources and pathways

Pesticide sources and pathways into the environment are numerous and diverse. Crop protection represents by far the most important use of pesticides, and thus agricultural sources significantly outnumber domestic and industrial sources. In most cases, pesticides can be categorized as "non-point" or "distributed" source contaminants, capable of causing low-level degradation of groundwater quality over large areas (Priddle et al., 1989; MacRitchie et al., 1994; Kolpin et al., 1997). However, point source contamination of groundwater can be a serious problem where pesticides are manufactured, stored or discarded, and at sites where pesticides are mixed and application equipment is loaded or rinsed (Fetter, 1993). Point source pollution can also occur when pesticides are applied locally in excessive amounts, such as may occur with aerial spraying.

In urban areas, pesticides are used domestically, in and around the house and garden. They are also used by golf course personnel to maintain greens and fairways and by municipalities to control or eliminate weeds along roads and pathways and to help maintain parkland in a manicured condition. While the amount of pesticide used in urban areas pales in comparison to agricultural applications, its use does represent a serious concern when the chemicals are inadequately stored or disposed of, or used indiscriminately and in excess of recommended application rates. In practice, the greatest threat of pesticides in urban areas comes from the direct ingestion of the chemical, perhaps by inhalation of chemical dust or vapour, by dermal contact with contaminated soils or by consumption of garden produce containing pesticide residues. Until recently, the contamination of groundwater in urban areas by pesticides has not been regarded as a potential problem; in fact the vast majority of pesticide studies have ignored wells in urban areas and have focussed exclusively on agricultural sources of pesticide and impacts on rural well water.

It is now recognised that a significant potential for the pollution of urban groundwater exists in the vicinity of urban parkland, golf courses, leaking sewer pipes and landfills, as well as in areas where large amounts of insecticide or weed control products are used domestically. Serious problems can also occur where urban areas have encroached on to land where earlier generation, chemically persistent pesticides were once used. In Australia, for example, sites used for dipping sheep can be heavily contaminated with potentially leachable arsenic and organochlorine compounds (Knight, 1993). In some areas, as in Canberra, these sites now lie beneath urban development.

In general, four factors determine whether a pesticide is likely to contaminate the underlying groundwater:
- The volatility, solubility, degradation potential of the pesticide (Smith, 1988), and its affinity for chemical adsorption.
- The organic content of the soil and rock materials.
- Hydrogeological conditions of the site.
- Pesticide management practices.

Ironically, many of the earliest pesticides developed (organochlorines) which are now banned or heavily restricted, display the least tendency to contaminate groundwater. The majority are poorly soluble in water and are readily adsorbed onto organic matter in the soils and sediments (Hassall, 1982; Langmuir, 1997; Tindall et al., 1999). They are effectively immobile under conditions of intergranular groundwater flow but have been reported in groundwater where fracture flow is the dominant transport mechanism. In Perth, Western Australia, Gerritse et al. (1990) report negligible contamination of the Bassendean sands aquifer despite the extensive treatment of urban soils for the control of termites with organochlorine pesticides ($50\,kg\,ha^{-1}$ for a housing density of $10\,ha^{-1}$ over a period of 10 years). Vertical transport velocities for lindane dieldrin, heptachlor and chlordane were estimated to be 2.1, 0.2, 0.1 and 0.1 cm/y respectively. The organophosphate chloropyrifos was also monitored during the study. Levels were found to be low and no association with urbanisation was found. However, Appleyard (1995a) found pesticide contamination of the Perth aquifer in association with its storage and preparation facilities.

Some of the more comprehensive pesticide studies have been carried out in the United States where pesticide usage is extremely high and groundwater provides drinking water for over 50% of the population (Moody et al., 1988; Barbash et al., 1996). Between April, 1988 and February, 1990, samples from 566 community water well systems and 783 rural domestic wells were analysed for 101 pesticides and 25 pesticide degradates (US EPA, 1992). Approximately 10% of CWS wells and 4% of rural domestic wells were found to contain pesticides or pesticide residues above the minimum reporting limits used in the survey. The most commonly detected pesticides were DCPA acid metabolites (DCPA is used extensively on home lawns, golf courses and farms to control annual grasses and broadleaf weeds) and atrazine, a herbicide used almost exclusively on agricultural land. Six pesticides including atrazine, lindane, alachlor and EDB were found in rural domestic wells at concentrations in excess of US EPA maximum contaminant levels.

In 1993, the United States Geological Survey (USGS) initiated the United States National Water Quality Assessment, a survey of groundwater quality in 20 of the country's most important hydrologic basins. The study focussed specifically on the quality of shallow, recently recharged groundwater (generally younger than ten years) and tested for the presence of 46 pesticide compounds (25 herbicides, 17 insecticides, 2 herbicide transformation products and 2 insecticide degradates) in water from 1 012 wells and 22 springs (Kolpin et al., 1998). A particular feature of this study was the inclusion of 221 urban well site samples amongst the 1 034 samples collected. Significantly, 46.6% of the urban samples showed evidence of pesticide contamination, a value only slightly less than the 56.4% recorded in rural samples. Of the 46 pesticide compounds examined, 39 were detected, with atrazine, simazine and prometon being the most commonly encountered. Only atrazine exceeded its maximum contaminant level (MCL) of 0.003 mg/L and this occurred in just one well. In the urban samples, prometon, which is a herbicide popularly used as an asphalt additive and a treatment for driveways, fence lines, lawns and gardens, was especially evident. In addition, the urban samples tended to show a relatively high incidence of contamination by insecticides.

Impact on groundwater

There is no clear world-wide picture of the true nature and extent of pesticide contamination in groundwater (Stauffer, 1998). The vast majority of countries do not include the common pesticides in routine analyses of groundwater. In addition, there is no consensus on the health hazards posed by pesticides and what constitutes a safe level in drinking water. It was once believed that the vast majority of pesticides released into the subsurface would become immobilised in the upper levels of the soil or degrade before reaching the water table. This is clearly not the case. Where detailed studies have been undertaken, notably in the United States, they often reveal widespread degradation of groundwater, albeit at very low concentrations. The studies also show that the threat of pesticide contamination is not confined to rural areas and that urban groundwater is at comparable risk.

3.9   ROAD DE-ICING SALTS

Introduction

Throughout snow-belt regions of the world, winter maintenance of roads and highways involves the annual application of millions of tonnes of road de-icing chemicals. These chemicals take several different forms (Nystén & Suokko, 1998) and are applied using a variety of methods. With rare exceptions, however, sodium chloride (NaCl) and calcium chloride ($CaCl_2$) are the chemicals of choice. Sodium chloride is the cheapest and is therefore most commonly used. It is particularly cost-effective at temperatures above $-12\,°C$. Calcium chloride is more effective at temperatures in the range $-12\,°C$ to $-34\,°C$ but is less frequently used because it is two to four times more expensive and has been known to make the road surface slippery when wet (Hanley, 1979). Several studies suggest that the salts used for de-icing purposes are unusually pure and normally contain very few impurities of special environmental concern (Howard & Beck, 1993). Sodium ferrocyanide is commonly added to road salt as a de-caking agent; however, while this can release free cyanide upon exposure to light, the cyanide has very limited mobility in the sub-surface being readily adsorbed onto soil particles during overland flow (Olson & Ohno, 1989).

De-icing chemicals are usually applied directly to the road surface in a pure chemical form, but they can also be used in conjunction with abrasives such as sand. Salt application rates typically range up to $30\,g/m^2$, which over the course of the winter can readily translate to between 10 and 20 tonnes of salt per 2-lane kilometre. Annual totals of salt application are given for several European countries in Table 3.19 The most common practice is to apply pure sodium chloride to highways and primary urban roads, while mixtures of sand and between 5 and 95% sodium chloride are preferred on side streets. Mixtures of sand and 5% sodium chloride are also used on gravel roads since pure sodium chloride tends to

Table 3.19.   Salt usage (NaCl and $CaCl_2$) for representative European countries during the winter season 1986–87 (after OECD, 1989).

Country	Length of road and highway network (km)	Salt used (10^3 t)
Belgium (1985)	14 058	200
Denmark	7 160	122
Germany (1986)	39 722	627
Italy (1986)	300 922	240
Switzerland	70 000	129
Sweden	30 000	141
United Kingdom	347 000	2 085

promote potholing of the road surface during a thaw. Some jurisdictions use a mixture of sand and calcium chloride with success, finding that while 50% less salt is required, melting temperatures below $-20\,^{\circ}$C can still be maintained (Fabricius & Whyte, 1980).

Transport behaviour

Road salts differ from many contaminants released to the sub-surface in that the mass and chemical nature of the source are often relatively well known. Similarly, the timing and distribution of salt release to the subsurface are usually well documented. Thus from the outset, simple mass balance calculations using a knowledge of salt application rates, estimates of release to the subsurface, aquifer recharge and aquifer storage allow potential detrimental effects on groundwater to be established. This approach is best used in heavily urbanised areas where road salt will tend to behave as a distributed or non-point source.

For a more detailed understanding of salt transport behaviour, chemical movement must be studied by focussing on individual roads and highways (Jones & Sroka, 1997). At this scale, salt behaves as a line source of contamination, which generates a plume extending away from the highway in the general direction of groundwater flow. Due to the high frequency of salt application over a typical winter, the relatively slow release of salt to the water table and the extended time frame over which mixing of saline water occurs with water in the aquifer, this plume tends to be continuous even though the source concentration may fluctuate over time. While saline water tends to be more dense than fresh water, density differences are rarely sufficient to cause appreciable sinking of the plume.

Figure 3.10 and Figure 3.11 show the chloride plume predicted for a four-lane highway overlying a sand aquifer in southern Ontario, Canada (transmissivity $20\,\mathrm{m^2/d}$, effective porosity 25%, hydraulic gradient 0.01). The salt is released at a rate of 33 t/km/a and mixes with groundwater in the top 2 m of the aquifer prior to transport. The plume data were generated using an analytical solution for three-dimensional contaminant transport (Howard et al., 1993; Howard, 1998). Figure 3.10 shows how chloride concentration at the top of the aquifer is likely to vary as a function of time in years for six monitoring wells located 100, 200, 300, 500, 1 000 and 1 500 m down the flow-line away from the roadway. Figure 3.11 shows vertical sections through the chloride plume after 50 years and 300 years. As anticipated, sites close to the highway will be most seriously impacted, with chloride concentrations exceeding 300 mg/L at distances up to 300 m within about 50 years. More distant sites also show significant impacts, but in most cases these impacts will not be realised for several hundred years. From a monitoring well perspective it should be noted that wells must be located close to the highway if the plume is to be detected early. However, it should also be noted that wells located close to the highway will rapidly approach chemical steady state, even though the leading edge of the plume continues to migrate in the aquifer. Thus wells located close to the highway may give the false impression that the problem has stabilised even though water quality continues to deteriorate.

In most studies, it is convenient to assume that both the chloride and sodium ions behave conservatively. While this is always true for chloride, sodium ions can get involved in ion exchange reactions of the form:

$$2\mathrm{Na}^+ + \mathrm{Ca_{(adsorbed)}} \rightleftharpoons \mathrm{Ca}^{2+} + 2\mathrm{Na_{(adsorbed)}} \tag{3.10}$$

and as a result can be retarded, usually by a factor of between 1 and 5. Depending on the alkalinity of the water and the prevailing pH, elevated calcium concentrations may also bring

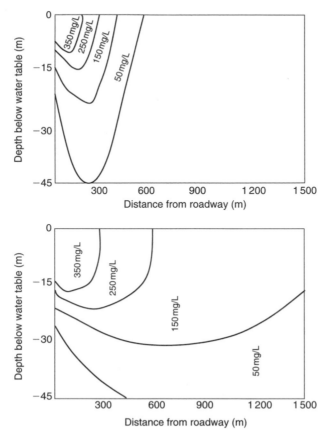

Figure 3.10.   Sections through predicted chloride plume after 50 years (upper diagram) and 300 years (lower diagram).

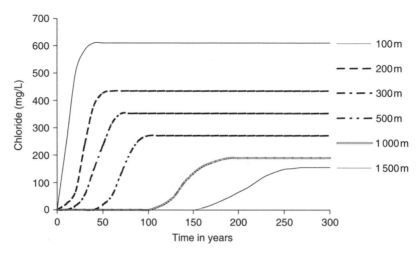

Figure 3.11.   Predicted chloride concentration at the top of the aquifer for six monitoring well locations.

about precipitation of carbonates, notably calcite and dolomite. A consequence of the reaction processes is that the ion ratios of Ca:Na:Cl observed in groundwater monitoring wells will rarely match the respective elemental ratios of the source. This is especially problematical given that elevated concentrations of Ca, Na and Cl in groundwater are not uniquely associated with road de-icing chemicals, and can be derived from numerous sources, both natural (evaporite deposits and fossil seawater) and anthropogenic (landfills, septic systems and snow dumps, to name just a few). While some success has been reported using various isotopes and minor ions, the use of hydrochemistry to provide a definitive indication of source origin remains an elusive goal.

Health and environmental problems

Environmental damage due to NaCl road de-icing chemicals is well documented (Jones et al., 1986; Scott & Wylie, 1980) but most accounts relate to visible impacts such as vehicle corrosion, damaged and dying vegetation and the disintegration of concrete. From a human perspective, the most common concern is an increase in salinity to levels which would make the water unsuitable for consumption. Water quality guidelines for sodium and chloride in drinking water are defined by various interested agencies. In most cases the guideline for chloride is set at 250 mg/L in recognition of a taste threshold that for most people exists between 200 and 300 mg/L. The guideline for sodium is more contentious since the ion has been strongly linked with the development of hypertension, a condition affecting perhaps 20% of the US population (Moses, 1980; Craun, 1984; Tuthill & Calabrese, 1979). Raised sodium intake has also been associated indirectly with hypernatraemia (WHO, 1984). Most agencies retain a guideline of 200 mg/L for purely aesthetic purposes. However, other agencies have been more cautious, the European Union setting a Guide Level of just 20 mg/L.

3.10   INVESTIGATION OF URBAN GROUNDWATER POLLUTION

The investigation of urban groundwater quality and contamination is the procedure by which we develop our understanding of regional groundwater quality in the urban area and surroundings, and/or address local contaminant plume migration within the groundwater flow regime. Evaluation of contaminant level, distribution, and mobility will all be part of any human health or ecological risk assessment, and will facilitate the design of a timely and cost-effective corrective action programme. Only by understanding how groundwater pollution has arisen and is evolving can we manage it. This understanding requires us to have a clear conceptual model, which enables us to design investigation and monitoring schemes, and to predict how pollution patterns will evolve in the future. In all cases, this model of the subsurface environment is constructed of three principal components of information (Bedient et al., 1999):

(a) Geology: the physical framework within which subsurface fluids collect and flow.
(b) Hydrology: the movement of fluids through this physical framework.
(c) Chemistry: the nature of the chemical constituents that are entrained in this flow system and the chemical and physical interactions between the contaminants and the subsurface formation and groundwater that may be occurring.

This section is only a brief introduction to the development of conceptual models through the investigation of urban geology, hydrology, and groundwater quality and contamination. The classical techniques of investigation are well covered in many standard texts such as Fetter (1999), Testa et al. (2000), Bedient et al. (1999), Lanen (1998) and Driscoll (1989), etc.

Hydrogeological investigation

An urban area usually has many boreholes, and geological, hydrological, and groundwater chemistry data are generally available. As well as the hydrogeological maps and reports compiled previously, data can be obtained by drilling new boreholes, conducting geophysical surveys at the ground surface or in boreholes, and hydraulical testing and chemical sampling of boreholes. The expertise required to utilise these different methods varies widely. Some may be so complex that only highly trained hydrogeologists are able to utilise them. The essential geological aim is a description of the principal stratigraphic units underlying the area related to groundwater matters, including their thickness, lateral continuity and water-bearing properties. Geological data must be compiled to define the lateral and vertical configuration of permeable and impermeable strata comprising the framework with which subsurface fluids collect and flow (Bedient et al., 1999). The aim of the whole investigation is to understand the groundwater flow system within the physical framework described by the geology. Flow is controlled by the physical properties of the rocks and by the patterns of recharge and discharge, as discussed in Chapter 1. The key physical properties and issues are:

● Relative hydraulic conductivity of geological units. Which are aquifers or aquitards?
● The character of the rocks. In particular are there preferential flow paths such as fractures or Karstic channels? Are there barriers caused by intrusive dykes or fault throws?
● Connections between hydrogeological units. Are aquitards continuous and of low hydraulic conductivity everywhere, or are there windows where flow can occur more easily between aquifers? What is ease of connection between aquifers and surface water for either recharge or discharge?

Groundwater recharge and discharge has to be studied as part of the flow system. Lerner et al. (1990) provide a comprehensive review of the sources of recharge, and catalogue and review methods of estimation. They identify the following sources of recharge from natural sources of water:

● direct and localised recharge from precipitation
● indirect recharge from streams, rivers and other surface water bodies
● inter-aquifer flow

together with these sources deriving from human activity:

● excess irrigation water
● induce recharge from rivers, wetlands and other surface bodies
● leakage from water mains
● septic tank discharge, baking sewers and related sewerage water
● disposal of liquid effluents outline surface or in wells

The principal routes of discharge for groundwater are, of course, to surface water and through wells. Evapotranspiration by plants and discharge to sabkas are occasionally important. The influence of urban areas on recharge is described in Chapter 2, and a recent review of estimation methods is given by Lerner (2002). However, it is rare for cities to completely cover a whole groundwater system. Recharge and discharge in the full required groundwater system, including the rural parts, will be needed to place the urban part in the correct context.

The groundwater flow system is created by recharge, discharge and rock properties, as discussed above. It is revealed by groundwater heads, and the collection of substantial numbers of water level observations in wells and piezometers is essential. Every effort should be made to visit all existing wells and measure groundwater levels during an investigation, to get a data set which is consistent and recorded over a short period of time. Vertical flows are often neglected, hence it is worth an extra effort to obtain vertical profiles of heads within and between aquifers.

Unfortunately, it is the intermeshing of the geological, hydrogeological and flow data with the groundwater heads information which reveals the conceptual model of the groundwater system. With this model, we can proceed to investigate and manage groundwater quality. Irrespective of whether we are investigating quality at a regional or local scale, we will require a regional understanding of the flow system.

Regional investigation of groundwater quality

The most common method of mapping groundwater quality is to extract and analyse a number of groundwater samples from wells in the area. We need to know the construction and utilisation details of the wells, and to know which aquifers they sample. Where existing wells are insufficient, additional drilling will be needed to obtain groundwater samples. A number of techniques are available which utilise either manual labour or a drilling rig, if available. More information about drilling sampling wells can be found in Driscoll (1986), and Bedient et al. (1999).

Sampling groundwater
Drilling wells is expensive, and the process of collecting and analysing water samples is time-consuming and costly. It is worth making sure the samples are taken properly so that the data is meaningful. There are many textbooks which explain how to collect samples properly (Fetter, 1999; Bedient et al., 1999), and only a few key points are covered here:
- Wherever possible, at least three standing volumes of water should be removed by pumping or bailing before the well is sampled.
- Samples should be collected in appropriate vessels produced with the appropriate preservative. Sample vessels are glass or plastic, depending on the analysis required. They should be chilled on ice and sent to a laboratory for analysis as soon as possible.
- Chlorinated solvents are volatile so, where they are suspected, particular care should be taken to prevent the loss while sampling.
- Several parameters are subject to rapid alteration when groundwater is removed from a well and exposed to oxygen and atmospheric pressure, caused by aeration, oxidation and the loss or gain of dissolved gasses. Water quality measurements that may be subject to rapid change include temperature, pH, dissolved oxygen, Eh and alkalinity. All of these should be measured in the field immediately.

The major constituents of groundwater are those that constitute the bulk of the dissolved solids. Typical major cations are calcium (Ca^{2+}), magnesium (Mg^{2+}), sodium (Na^+), and potassium (K^+); the major anions are chloride (Cl^-), sulphate (SO_4^{2-}), carbonate (CO_3^{2-}), and bicarbonate (HCO_3^-), although nitrate (NO_3^-) is often important in relation to sewage and agriculture. The concentrations of all major dissolved ions in groundwater must always be measured to understand the complete geochemical system, even though they are not pollutants.

The minor ions are indicators of inorganic groundwater quality at a site or facility. Important constituents include iron (Fe^{2+}/Fe^{3+}), manganese (Mn^{2+}/Mn^{4+}), fluoride (F^-), nitrate (NO_3^{2-}). Trace elements such as arsenic (As^{3+}/As^{5+}), lead (Pb^{2+}), chromium (Cr^{3+}/Cr^{6+}) and other heavy metals may be present in amounts of only a few micrograms per litre, but they are very important from a water quality standpoint.

The organic chemicals are often the most serious health risks in groundwater. Unfortunately they are also difficult to sample and analyse, and for this reason are often neglected. They often present risk at concentrations of only a few µg/L, and many are volatile and so easily lost during sampling. Some biodegrade and can be lost between sampling and analysis. However, it is not possible to assess urban groundwater without good data on organic pollutants, and any investigation or monitoring must include them.

Investigation of contaminated sites

The above section focuses on regional investigations. However, much of the pollution of urban groundwater, especially organic chemicals, comes from point sources, such as petrol stations and factories. These cases require intensive investigation, which need much higher densities of wells than those required for regional investigations (Table 3.20).

Contaminant spills, such as hydrocarbons, pesticides, and industrial liquids, may take place in areas with few existing wells. Although there may be little information about the depth and construction details of these wells, they should be sampled to assess possible health risks to the local community, particularly if local groundwater is used for drinking water supply. Additional information about the history of bore or well construction may be obtained by interviewing the community, who may also be able to report on whether there have been any sudden changes in the odour or taste of groundwater, often due to hydrogen sulphide or other sulphur compounds which are early indicators of the arrival of a hydrocarbon contamination plume (Benker et al., 1996).

A useful conceptual model of localised pollution is the source and plume model. The source zone is where the bulk of the pollutants located, as residual NAPL is adsorbed in

Table 3.20.   Number of bores required for monitoring specific landuses that may cause contamination by hydrocarbons (after Davidson, 1995).

Land use	Average number of monitoring bores
Service station	
Contamination on–site	5–10
Contamination off–site	10–25
Small bulk terminal (few storage tanks)	7–10
Large bulk terminal (many storage tanks)	15–50
Small refinery or chemical plant	15–50
Large refinery or chemical plant	30–100

the soil and unsaturated zone. Recharge or groundwater flowing through the source zone dissolves or desorbs the pollutants, forming a plume. In general, the greatest risks are associated with the mobile pollutants in the plume, but the greatest mass is in the source zone. The source must therefore be dealt with in order that the plume is not constantly replenished.

The greatest risk in investigating localized pollution is of mobilising the source. This happens frequently when boreholes are drilled through the zone of DNAPL. Because of its density, the DNAPL can flow down the borehole to greater depths and pollute large volumes of the aquifer. Hence, it is unwise to drill near a suspected DNAPL unless great care is taken. Section 3.11 discusses further means of investigating NAPLs.

Groundwater plume delineation should be conducted in a step-wise procedure in order to minimize the number of groundwater sampling points required. Bedient et al. (1999) suggested a five-step typical work programme, shown in Figure 3.12. Such a programme starts with source delineation by locating potential sources and characterising contaminant properties and proceeds, and moves through to detecting plume dimensions and composition. The first borehole should be drilled outside the source zone and used to establish the geology. New boreholes are added gradually, mainly along the down-gradient direction but also transversely, with positions based on a continuous updating of the conceptual model. At least one up-gradient will be needed to establish background groundwater quality. Investigation of the source zone will usually be needed but, as mentioned above, must proceed cautiously if there is any risk of DNAPL being present.

3.11   INVESTIGATION OF NAPL POLLUTION

The role of monitoring bores is to provide information about the extent and character of the pollution source; to determine the extent and severity of a groundwater contamination plume; to enable changes in the position of the plume and concentration changes to be tracked over a period of time; and to provide information about the efficacy of any remediation programme that is carried out. Standard monitoring bores for water levels and dissolved water quality are well described in many textbooks and in hydrogeology courses, and do not need to be included here. However, when NAPL may be present, extra care must be taken over the design and construction of monitoring bores and interpretation of their data. This section briefly discusses the issues and some of the techniques available. Relevant manuals and textbooks should be consulted before understanding any fieldwork, as it is very easy to spread NAPL pollution more widely just by poor investigation methods.

At sites where NAPLs are present as well as dissolved contamination, the location and depth of monitoring bores is especially important. Well screens are positioned to intersect the top of the aquifer in confined flow systems, or to straddle the expected zone of water table fluctuation in unconfined aquifers. Placement of the screen across the top of the water-bearing zone permits detection of floating accumulations of LNAPLs, while intersection of the screen with the base of the aquifer is more appropriate for investigation of dense DNAPLs and dissolved contaminants. Free phase hydrocarbons are most commonly detected at sites where there is contamination by petroleum hydrocarbons, and an NAPL is present at the water table. If the slotted interval is not constructed at the water table, the LNAPL may be missed, and if there are no sufficient slots both above and below the water table, seasonal fluctuations in the LNAPL layer may not be detected. Where an LNAPL is

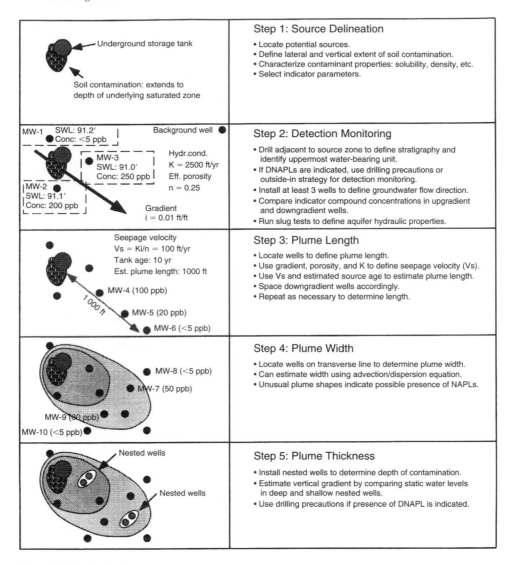

Figure 3.12.   Typical work programme for groundwater plume delineation (after Bedient et al., 1999).

present, additional bores are usually constructed below the water table to monitor dissolved groundwater contamination.

The same principles apply to monitoring DNAPL layers, although they are often difficult to detect because they are usually present in small pools at the base of an aquifer, or are perched on discrete zones of low permeability within the aquifer. These isolated DNAPL pools may not be detected unless there is an extremely intense preliminary groundwater investigation programme before bores are constructed.

A number of monitoring bores are required to fully characterize the extent and severity of a groundwater contamination problem. Additional bores are usually constructed

Table 3.21. Techniques for measuring NAPL layers in monitoring bores.

Detector type	Accuracy	Ease of use	Cost	DNAPL use
Hydrocarbon detecting paste	Moderate	Moderate	Low	No
Clear bailers	Low	Moderate	Low	Yes
Float/conductivity probes	Moderate	Easy	Moderate	Yes
Density sensors	High	Hard	High	Yes
Light absorption	High	Easy	Moderate	Yes
Opto-electric	High	Easy	Moderate	Yes

up-gradient and down-gradient of the contaminated area to provide background information, and to allow the movement of contamination plumes to be detected. Although the bore numbers will vary greatly depending on the resources available for monitoring and on local environmental legislation, Table 3.20 gives a guide to the number commonly used for monitoring specific activities that may cause hydrocarbon contamination.

The detection and eventual removal of NAPL layers in aquifers is extremely important because they generally contain a large proportion of contamination in the sub-surface, and they will act as a continuous source of dissolved groundwater contamination if they are not removed. The assessment of the thickness of free product layers is important in determining the total mass of hydrocarbons present in a hydrocarbon spill, for selecting remediation strategies, and for determining the time that might be required for the recovery of free product.

NAPL layers are often detected after monitoring bores are constructed. There are a number of techniques that can be used to measure the thickness of the NAPL layer and these are tabulated in Table 3.21.

The techniques vary greatly in cost, accuracy and ease of use, with the cheapest method being the use of a paste which changes colour on contact with hydrocarbons, but not with water. The paste can be smeared onto rods and dipped into a monitoring bore. It is then removed, and the width of the zone that has changed colour is measured. This method has a disadvantage in that repeat measurements are difficult to make because the paste must be removed and renewed each time. The pastes also cannot be used to measure DNAPL layers. The use of clear plastic bailers is another simple and cheap method of determining the thickness of NAPL layers, but may suffer from inaccuracies.

By far the most widely used and tested method of detecting NAPL–water interfaces are provided by opto-electric indicators, which combine refraction and conductivity measurements. The liquid level in a monitoring bore is detected when an infra-red beam is refracted away from an infra-red detector. Conductivity measurements are used to distinguish water from free products. Opto-electric indicators can detect the interfaces between air/liquid, LNAPL/water, and DNAPL/water.

A major problem of measuring of LNAPL layers in monitoring bores is relating the thickness of the free product layer in the bore to its thickness in the aquifer. The thickness of an LNAPL in a monitoring bore is often greater than the true aquifer thickness, because the bore acts as a local collection point or sump for free product to drain into. This takes place because the LNAPL is often concentrated in the capillary fringe above the water table, and there is sufficient head difference between this layer and the water table to allow drainage to occur. The increase in LNAPL thickness in the bore is due to the depression of the water table due to the weight of the accumulated free product. This factor can lead to gross overestimates of the volume of free product present in an aquifer if it is not taken into account, and seasonal movement of the water table may cause changes in measured product thickness, which may be misinterpreted as real LNAPL volume changes in the aquifer.

There are a number of ways in which the thickness of LNAPL layers measured in bores can be related to true aquifer thicknesses. As a first approximation, the aquifer thickness of LNAPL can be estimated by (Kinsella, 1995):

$$\Delta h_{aquifer} = \Delta h_{bore}(\rho_{water} - \rho_{LNAPL})/\rho_{LNAPL} \tag{3.11}$$

where

$\Delta h_{aquifer}$ = thickness of LNAPL layer in aquifer
Δh_{bore} = thickness of LNAPL layer in bore
ρ_{water} = density of water
ρ_{LNAPL} = density of LNAPL

The extent to which the water table is depressed in a bore by the accumulated thickness of LNAPL can be estimated by:

$$WT_{CORRECT} = WT_{MEASURE} - (\Delta h_{bore}SG_{LNAPL}) \tag{3.12}$$

where

$WT_{CORRECT}$ = true depth to water table
$WT_{MEASURE}$ = measured depth to water table
SG_{LNAPL} = specific gravity of free product

For example, if the depth to water table in a bore is 10 m with a 1 m thick product layer that has a specific gravity of 0.75, the true water table depth is 0.75 m. That is, the water table has been depressed by 0.75 m due to the accumulated weight of free product in the bore.

The most accurate way of measuring the true LNAPL thickness is by taking sediment cores and directly measuring the distribution of hydrocarbons with depth, but this method is not suitable for continuous monitoring of the changes in distribution of product with time. However, bail-down tests carried out on monitoring bores can also give a good estimate of true product thickness in the absence of sediment core data.

Bail-down tests are carried out by completely bailing out the LNAPL layer in a bore, and then monitoring the depth of the water table and product layer as the bore is allowed to recover. During the recovery of the fluid levels in the bore, the top of the LNAPL layer returns to its original level, but the LNAPL–water interface initially rises and then falls as the accumulation of product depresses the water table, giving a clear inflection point on a depth–time plot. When depth-to-product and depth-to-water measurements are plotted against time on a logarithmic scale (Fig. 3.13), the difference between the two measurements when the water table depth shows an inflection point which indicates the true product thickness (Wilson & Etheridge, 1991).

Dissolved contaminants penetrating to the groundwater table will become entrained in the natural groundwater flow system, and spread both laterally and vertically in accordance with local groundwater flow gradients. Free-phase liquid contaminants may be subject to an additional "density gradient" with light non-aqueous phase liquids (LNAPLs, such as gasoline) floating on top of the zone of saturation and collecting in the structural highs of confined water-bearing units. Alternatively, dense non-aqueous phase liquids

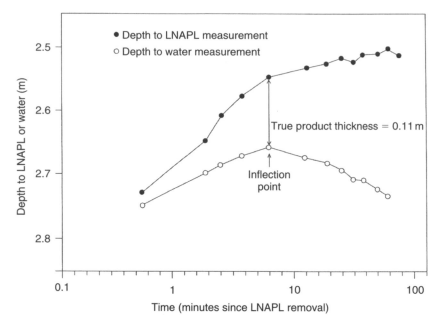

Figure 3.13. Example of a bail-down test to determine the true thickness of an LNAPL layer for measurements made in a monitoring bore.

(DNAPLs) can percolate downward through the water-bearing stratum to spread on top of underlying confined units (Bedient, 1999).

Timing and frequency of data collection are also important considerations. At a site with residual petroleum contamination in soil directly above the water table, concentrations of petroleum in groundwater may be highest soon after rainfall. This results from two effects, as water infiltrating through the soil will dissolve petroleum in the unsaturated zone and, as the water table rises, residual petroleum in the same zone associated with the water tables will then dissolve into the groundwater.

3.12   CONCLUSIONS

It might seem that living in urban areas is particularly hazardous, given the number of pollutants which may contribute to the contamination of groundwater under such areas. Because many people obtain their drinking water from groundwater sources, it is important to set regulatory standards and to monitor supplies regularly to ensure that potential health risks to humans are avoided.

In the developed world, with its well-established cities and a pattern of service infrastructure, these standards mean that groundwater is generally protected and little damage is caused to humans. However, this is not the case where urban expansion is very rapid and where economic conditions preclude proper development of infrastructure. As we shall see in Chapter 6, groundwater under such cities can be easily contaminated by sewage, solid wastes and other pollutants in these conditions, and then infections and diseases become rife.

CHAPTER 4

Mature industrial cities

Aidan A. Cronin and David N. Lerner

4.1   INTRODUCTION

The Industrial Revolution began in Great Britain in the 18th century and led to the rapid expansion and industrialisation of many of the major British urban areas of today. Groundwater resources played an important role in the development of many of these cities. They helped to meet the increasing water demands from the emerging industries as well as providing an abundant and good-quality potable water supply for the growing population. Examples of cities in the UK where groundwater was used for such purposes include London, Birmingham, Nottingham, Liverpool, Coventry and Belfast (Fig. 4.1). These cities

Figure 4.1.   Location map case studies of mature urban UK cities.

will be the principal focus of this chapter because the length of time they have supported major industrial activity means they offer insights into the role of groundwater in mature urban areas. Examples of aquifers affected by urbanisation over a significant period of time, from other countries besides the UK, are also examined.

Urban development has a profound effect on groundwater in several ways. The main effects are changes in recharge and groundwater levels, (Lerner et al., 1990; Ku et al., 1992; Foster & Morris, 1994; Carmon et al., 1997) and in groundwater quality (Nazari et al., 1993; Eiswirth & Hotzl, 1994; Lawrence et al., 1996; Lerner, 1996). The role of groundwater in mature industrial cities often contrasts with other urban groupings such as the "Megacities" of Asia and the rapidly developing cities of sub-Saharan Africa, outlined in other chapters, and so the main issues affecting the quantity and quality of water under mature cities are examined below. This leads to a discussion on future strategies for the protection and sustainable abstraction of groundwater under mature industrial cities.

4.2   BACKGROUND TO GROUNDWATER AND MATURE INDUSTRIAL CITIES

Prior to 1800 towns in West Central Europe developed as marketplaces for agricultural produce, merchant goods or artisan handicraft, but manufacturing powered by coal became the major activity in most cities after the Industrial Revolution (Brunn & Williams, 1983). Retail and service industries grew rapidly to supply a burgeoning urban population. First people moved from rural areas to cities to find employment and then to suburbs as the cities became crowded (Brunn & Williams, 1983). This greatly increased the urbanised area. In Birmingham, for example, inner city overcrowding led to large-scale migration to new industrial districts 3–5 km (from 1860–1910) and 6–8 km (from 1910–1960) from the original city centre industrial nucleus (Ford & Tellam, 1994). Industrial expansion continued throughout the 19th century and most of the 20th century. These cities however experienced massive industrial decline in the 1970s and 1980s with the population suffering from associated social problems. Modern urban regeneration has led to a change of focus from the mass production of goods to service-led systems, which are characterised by more flexible production (Crewe & Beaverstock, 1998).

Many human settlements throughout the world have been located near reliable groundwater sources (Simpson, 1994). Indeed, the provision of water supply, sanitation and drainage are key elements of the urbanisation process (Foster & Morris, 1994). In the early industrial development of mature UK cities such as Birmingham and Nottingham, there was free access to groundwater resources. This led to heavy groundwater pumping for industrial use resulting in large drawdowns in groundwater levels. Declining industrial abstraction came about as industry changed from very water-intensive manufacturing to less water-demanding activities. The relocation of the majority of the population from the city centre to the surrounding areas meant that many of these new suburbs may not have been underlain by a productive aquifer. In addition many new wells needed to be sunk in order to provide water for the expanding city area. Gradually a shift to public water supply took place for potable and industrial use (Lerner, 1997). Liverpool, Manchester and Birmingham all used their great wealth to switch from groundwater to surface water sources in the late 19th and early 20th centuries (Tellam, 1995). Nottingham switched to rural boreholes. In general, groundwater did not figure as prominently a source of potable public supply as surface water due to actual contamination problems with the groundwater or perceived pollution risk problems (Lerner, 1997).

With many of the industrial and public supply wells being discarded, groundwater abstraction went into steady decline. This decline was accelerated with the emergence of various industrial pollution problems. Abstraction licensing began in England in 1965 to avoid over-exploitation of the resource. However, Northern Ireland has only started to examine licensing in the very recent past. Over-abstraction of groundwater in certain areas has led to the drying-up of springs and low river flows and this has become a problem in areas relying on fishing, tourism and recreation (Lerner, 1997). The interaction between surface water and groundwater is difficult to observe and measure but its importance is being recognised more and more (Winter et al., 1998).

On-site sanitation, including pit latrines and septic tank systems, are the main method of disposal for most of the world's urban population (Lerner, 1996). However, cesspools and septic tanks have disappeared from most of the UK, apart from isolated single dwellings, with even small villages being sewered by the late 1970s (Lerner & Barrett, 1996). Faulty sewers have now become an issue in groundwater quality as sewer leakage impacts have been identified under urban areas.

The main aquifers under British urban areas are deep and consolidated. London is underlain by the Chalk, formed in the late Cretaceous period. At the time of deposition the Chalk had a porosity of approximately 70% but cementation has reduced this to about 40% in Southern England (Price, 1994). This is a fissure flow aquifer with transmissivity values often greater than $1\,000\,m^2/d$. All the other cities mentioned above are underlain by Permo-Triassic sandstones. Lithology can be subject to rapid variations in both vertical and horizontal directions (Bennett, 1976). Porosity values are normally in the region of 20 to 35% (Lovelock, 1977). Hydraulic conductivity variations and fissures can affect flow significantly. Usual values are in the region of 0.1 to 10 m/d. Quaternary drift cover complicates recharge estimation in these aquifers. It can vary from in excess of 30 m of low-conductivity glacial clays in Belfast to a complete absence of cover in parts of Nottingham.

Rainfall in the British Isles is essentially cyclonic, arriving from the Atlantic in discrete fronts. Average values vary from 500 to 1 000 mm/y. Aquifers are usually recharged in winter months (September–April) when precipitation exceeds evapo-transpiration. Water levels rise during this period and reach a peak so a strong seasonal pattern is visible. Urbanisation can influence this pattern significantly.

In contrast to the UK, some of the most important aquifers in other EU countries and the USA are unconsolidated sand and gravel aquifers. In the USA these include the Tertiary deposits of the western plains and the sand and gravel of the eastern coastal strip (Domenico & Schwartz, 1990). The latter aquifer is a strip 160 to 320 km in width that stretches from New Jersey into Texas and is an important aquifer on the densely populated east coast. Aquifers along this Atlantic coastal zone supply drinking water to an estimated 30 million people (USGS, 2000). Unlike the UK, substantial areas of the population in North America are not sewered (Lerner, 1996). In such urban environments almost 100% of water supplies will become recharge.

4.3   URBAN RECHARGE

The increase in impermeable area in an urban region changes the surface and groundwater hydrology (Lerner et al., 1990). Infiltration and direct recharge decrease. Surface runoff is increased, although this water may become localised recharge from soakaways and storm drains. Although urbanisation increases storm runoff, there is no direct evidence that the increase is at the expense of recharge (Lerner, 1997; Yang et al., 1998). Recharge is increased by leakage from water mains and sewers, septic tanks and irrigation practices (Lerner et al., 1990; Morris et al., 1994). Urban recharge can be estimated (Lerner et al., 1990) either holistically:

$$\text{Net Recharge} = \text{imports of water} + \text{groundwater abstraction} \\ - \text{consumptive use} - \text{effluent leaving} \tag{4.1}$$

Or as the sum of the components:

$$\text{Net Recharge} = \text{rainfall recharge} + \text{leakage from mains} + \text{leakage from} \\ \text{sewers} + \text{infiltration from septic tanks etc.} \tag{4.2}$$

When aquifer-based water supply systems are substituted by surface or imported water to keep pace with urban growth, a period of groundwater drawdown is often followed by a period of recovery, which may produce waterlogging of underground spaces and excavations (Custodio, 1997). Rising water levels, due to increased infiltration rates, can cause structural damage due to hydrostatic uplift (Morris et al., 1997). In addition, foundations and other underground structures are vulnerable to corrosion when the groundwater has high sulphate, chloride or magnesium values. Other problems include malfunction of septic tanks and the possibility of contamination of water mains in areas of low pressure in the system. Rising groundwater levels also can mobilise pollutants and toxic gases that sorbed in the aquifer unsaturated zone (Foster & Morris, 1994). This may be especially problematic in industrial areas where the contaminants first entered the subsurface.

Falling water tables, due to heavy abstraction, can cause damage to buildings and services due to subsidence. This is due to ground settlement and can happen on a local scale affecting individual buildings or on a regional scale affecting whole conurbations (Simpson, 1994). Groundwater and surface water systems are interdependent and so a fall in groundwater levels can lead to dry lakes and reduced river flows. A reduction in the water table can also lead to the wilting of crops and trees and hence changes in land use (Simpson, 1994).

Case study: London (CIRIA, 1989)

London was an important city in Roman times but had been settled much earlier. It was the first bridging point on the River Thames and high ground to the north of the river made it quite defensible. Alluvial gravel and sand deposits provided building materials and a good water supply. Its population grew from 117 000 in 1800 to 2 365 000 by 1900 due to the stimulation provided by the Industrial Revolution (Brunn & Williams, 1983).

London is underlain by a deep aquifer comprised of Chalk and Basal Sands. It is confined over much of the area by layers of thick low permeability clay, known as the London clay. This separates the deep aquifer from the perched groundwater in the overlying gravels and superficial deposits. Heavy abstraction led to large drops in groundwater levels. In Trafalgar Square water levels dropped by over 50 m in the period 1844 to 1965. Associated subsidence in Central London is in the order of several hundred millimetres. Indeed, parts of London have sunk over 4.5 m since Roman times (Freeman, 1966). This led to an increase in strength in clays and engineering works were designed and constructed taking this into account.

Groundwater abstraction under Central London has dramatically decreased since the late 1960s and this has seen water levels rise steadily. By 1988 levels had risen by 20 m and even more in north-west London. Regional groundwater flow modelling predicted groundwater levels to return to their original levels within the next 30 years. This has serious implications as rising water levels can begin to affect building foundations and services. Damage is possible due to instability caused by differential ground movements. This would be most prevalent in areas where the clay is thin and the foundations are close to the aquifer, and these conditions are met in the lower-lying areas of Central London. In addition, water rising though sands between the chalk and clay is high in sulphate due to the oxidation of pyrite. This water can cause corrosion in concrete. Seepage of groundwater into basements would also render these areas unusable. There is the potential for fires to be caused by oil products floating on groundwater entering these areas. London has over 130 km of tunnels near the base of the clay or in the sandy deposits. Sections of this could suffer increased seepage, chemical attack and increased loading on linings.

Provision is already being made for continued rising groundwater levels in the design of many new buildings. The cost of remedial works to protect buildings and tunnels will be high if groundwater levels rise without control. Control can be carried out either on a regional basis or by individual structure owners. The former would be the most efficient in terms of abstraction and discharge management. Water can be disposed of into a sewerage system or feeding it into the water supply network.

Case study: Nottingham

Nottingham is thought to have its origin in Roman or pre-Roman times and it became a major trading centre in the Middle Ages (Charseley et al., 1990). The Danes saw the significance of the town as a head of navigation where road and river traffic met, while the Normans instigated the main growth on the sandstone (Freeman, 1966). Expansion occurred in the 19th century with local coal mining and the development of major industries such as textiles, lace making, and engineering works. Rapid urban growth commenced in the 1870s. Public water supply is from boreholes in rural areas to the north and west of the city, one borehole in the city and river water from the Derwent in Derbyshire. Nottingham has a long history of public water supply with the first records dating from 1696. By 1850 all the rivers in Nottingham were polluted and public water supply was only from the Sherwood Sandstone and the River Derwent (Edwards, 1966). The geology and hydrogeology are outlined in Figure 4.2.

Rising groundwater levels in Nottingham have led to problems with basement flooding (CIRIA, 1989). Recharge to the Nottingham Sherwood Sandstone was an important yet difficult parameter to quantify. Not only must the change in infiltration due to urbanisation be considered but also the recharge from leaky water mains and sewers must be accounted for. Previous estimates were in the range of 114 to 267 m/y (Lamplugh et al., 1914; Land, 1966; Rushton & Bishop, 1993). Such is the complexity of the system however that a transient groundwater flow and solute transport model was developed to estimate the contribution of the various recharge sources to the overall recharge total (Barrett et al., 1997; Yang et al., 1999). Average recharge to the city was computed as 211 mm/y. Recharge from mains leakage contributed about 138 mm/y. Recharge from sewerage leakage was roughly 9 mm/y. Sensitivity analysis was carried out to estimate the confidence intervals on the recharge estimates from each recharge source and found that mains recharge may be ±40%

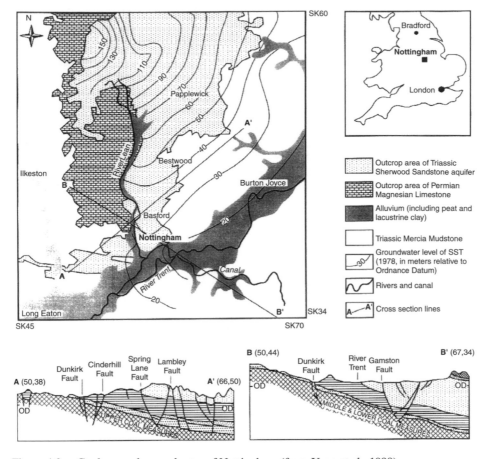

Figure 4.2.   Geology and groundwater of Nottingham (from Yang et al., 1999).

and the sewer recharge value may be $\pm 100\%$. Uncertainty in these estimates arises from scarcity in hydrological data, limited historical geochemical data and also the relative insensitivity of this high volume groundwater system to changes in solute inputs.

Discussion

Falling water levels are not such a problem under mature urban areas. It is usually more of an issue in developing cities. For example, subsidence was a problem in developing London when abstraction volumes from Central London were significant. A recent example of falling water levels is in Mexico City. Here groundwater abstraction is twice recharge rates, and this has led to drawdowns in excess of 3 m/y with associated land subsidence in the region of 0.1 to 0.4 m/y (Sanchez-Diaz & Gutierrez-Ojeda, 1997). Mature urban areas have evolved from these issues to problems associated with rising groundwater levels.

London and Nottingham are not the only mature urban areas affected by rising groundwater levels. In the UK, Birmingham and Liverpool have suffered flooded basements and tunnels while Paris has experienced flooding of underground carparks and New York's subway has suffered operational difficulties with rising water (CIRIA, 1989). In London decreasing abstractions have led to the problem of rising groundwater levels in urban areas. However, it can be seen from Nottingham that leaking sewers and mains water have compensated for the decrease of infiltration in urban areas and so have exacerbated the rising groundwater level problem. However, groundwater levels can be managed by pumping schemes at either local or regional level. Alternatively, vulnerable structures may need to be protected by securing and strengthening. This may be expensive but is the only viable option to prevent damage caused by rising groundwater levels. If pumping is decided to control groundwater levels, it may be the quality of the water that dictates its use.

4.4   MANAGEMENT PROBLEMS WITH URBAN GROUNDWATER QUALITY

Groundwater quality in mature urban areas is threatened by such point sources as industrial and municipal waste discharges, and leaky underground storage facilities, as well as miscellaneous accidental spills of organic or inorganic contaminants. It is particularly important to evaluate groundwater vulnerability in urbanised areas because of the concentration of potential contamination sources that accompany economic development (Eaton & Zaporozec, 1997). Groundwater pollution is controlled by pollutant source properties, groundwater flow and the attenuation capacity of the subsurface. However, in the urban environment, contaminant sources are extremely numerous and very difficult to quantify, whilst closely spaced interfering production boreholes give rise to very complex flow patterns. Hence urban pollution patterns can be extremely difficult to identify (Tellam, 1995). The threats to urban groundwater quality are introduced briefly below followed by case studies illustrating these problems in mature urban settings. Chapter 3 describes the threats to urban groundwater quality in more detail.

Sewage

Pathogens are organisms that are capable of causing disease. There are known to be over 100 microbial pathogens capable of transmission through contaminated water (Powell et al., 2001). Consumption of such pathogens can lead to severe problems of the digestive system, which can be life-threatening to the very young, very old, or those with damaged immune systems. Snow (1854) first demonstrated the association of disease outbreaks with the consumption of untreated water affected by sewage. To this day sewage remains a major issue in the supply of potable water (Lerner & Barrett, 1996). In Germany, several hundred million cubic metres of wastewater leak every year from partly damaged sewerage systems to soil and groundwater, and this represents a serious source of groundwater contamination with elevated pathogen, sulphate, chloride, and nitrogen values (Eiswirth & Hotzl, 1994). Leaky sewers have also caused public water supply contamination and associated gastric illnesses in Britain and Ireland (Misstear et al., 1996).

Case study: Nottingham

Barrett et al. (1997) undertook a study of the Sherwood Sandstone aquifer in order to ascertain the influence of leaky sewers on the groundwater quality. Deep groundwater samples were taken from all of the abstracting industrial boreholes in Nottingham and additionally from the public water supply boreholes situated to the north and east of the city. Shallow groundwater samples were taken from purpose-installed monitoring boreholes in the Meadows housing estate near the city centre and also from flooded caves beneath the city centre. Sampling of raw sewage, rainfall, river water (from the rivers Leen and Trent) was also carried out, and data on mains water quality within all of the Nottingham supply zones were made available by Severn Trent Water plc. In general, the urban inorganic groundwater quality was worse beneath the city than beneath the surrounding rural land, although the degradation was not as serious as expected beneath a city with an industrial history as long as that of Nottingham. As expected, the shallow groundwater was more contaminated than the deep groundwater with the poorest groundwater quality in the industrial Basford area.

Nitrogen isotopes and microbiological indicators proved to be the most useful marker species in establishing the presence of sewage contamination of groundwater. The conclusion regarding the deep boreholes was that the majority have nitrate from a soil organic nitrogen source, presumably originating in the adjacent rural area (upstream), but that at least 3 have a component of sewage-derived nitrogen (Rivers et al., 1996). Many of the shallow groundwater samples were impacted by sewage effluent.

A microbiological survey was carried out using groundwater samples taken from the Meadows boreholes. Coliform bacteria (including *Escherichia coli*), *faecal streptococci* and sulphate reducing clostridia were the bacteria targeted as indicators of faecal (sewage) contamination. The conclusion of the microbiological survey was that 9 of the 11 shallow boreholes were faecally contaminated. The correlation between the microbiological survey results and the nitrogen isotope analyses proved strong. Therefore, there was strong evidence of sewage contamination at all but two sites (Barrett et al., 1999).

Microbiological work on faecally derived micro-organisms in Nottingham groundwater has been on-going. Powell et al. (2000a) have successfully applied a glass wool trap for the concentration of enteroviruses from large volume groundwater samples whilst in the field. The traps have also been used in monitoring groundwaters under Birmingham and have detected enteroviruses in both shallow piezometers and deep industrial abstraction boreholes (Powell et al., 2000a,b). Enteric viruses were detected in over 60% of groundwater samples analysed, most likely due to groundwater recharge from leaky sewers (Powell et al., 2000a).

The profile of contamination with depth remains unresolved because the study outlined above relied upon samples collected from boreholes that were open through much of the vertical extent of the aquifer. Depth-specific sampling of groundwater within the sandstone has recently been possible through installation of 5 dedicated, multilevel piezometers in Birmingham and Nottingham (Taylor et al., 2000). Coliphage (viruses that infect coliforms) and enteroviruses were less frequently detected than *Escherichia Coli* and *faecal streptococci*, but were observed to depths of 50 m (Powell et al., 2001). Such results confirm the presence of faecally derived micro-organisms at significant depths in the Nottingham urban area.

Case study: Liverpool (Whitehead et al., 1999)

Urbanisation of Liverpool began in 1207, and by the 1700s the area was very prosperous with thriving commerce. A mixture of industries developed in the vicinity including chemicals, glass manufacture, tanneries and textiles (Tellam, 1994). The major aquifer is the Triassic Sherwood Sandstone Group, which dip southwards at about 5°. The aquifer's low transmissivity and high storativity mean contamination effects will linger long in the groundwater (Tellam, 1994). Groundwater met water demands until the middle of the 19th century after which surface reservoirs were constructed. Industrial groundwater usage has continued but at declining rates. This, together with leakage from sewers and mains water lines, mean groundwater levels are rising at a rate of 0.3 m/y (Ion, 1996), similar to the other mature urban centres described earlier.

The sewerage system has been extensive since the late 19th century, expanding along with the city. Nitrate and potassium had been found in elevated levels under the city (Tellam, 1996) and had been attributed to leaky sewers. Because of the multiple potential nitrate source, nitrogen isotope and microbiological indicators were used to "fingerprint" sewerage contamination. Indicator micro-organisms were found in four of nine groundwater samples taken and all but one of the samples had nitrogen isotope values indicative of sources originating from human or animal wastes. It was concluded that these effects demonstrated the occurrence of sewer leakage to the groundwater.

Discussion

Mature urban areas are usually served by sewers, in contrast to developing cities which are more likely to be served by on-site systems such as septic tanks or cesspools. In the latter population growth precedes the development to handle wastewater, leading to widespread contamination of shallow groundwater by effluent (Morris et al., 1997). Problems arise in mature urban cities when, as happened in Liverpool, these sewers are often over 100 years old and are liable to leak. This impacts on groundwater quality. Exceptions to mature urban areas being sewered occur regularly in North America. 70 million people in the USA depend on septic tanks for sewage disposal (Hershaft, 1976). Septic tank discharges have been shown to give rise to rapid virus transport in sand and gravel aquifers (DeBorde et al., 1998). This has had public health implications. Craun (1981) states that sewage seepage from cesspools and septic tanks was responsible for 41% of waterborne communicable disease outbreaks and 66% of illnesses causes by groundwater in the USA between 1971 and 1978. Septic tanks have frequently being blamed for reports of localised groundwater pollution and, in cases of high septic tank density, regional groundwater problems have been identified (Chanda et al., 1997). Though localised problems have been identified with sewer leakage in mature cities (as in the case of Nottingham above), regional groundwater microbiological problems are usually much fewer with this method of waste disposal.

Major ions

The most common contaminant identified in groundwater is dissolved nitrogen in the form of nitrate (NO_3^-) (Freeze & Cherry, 1979). Elevated nitrogen levels can cause excessive growth of aquatic plants and algae leading to decreases in dissolved oxygen with its associated implications for aquatic invertebrates in surface waters. Another problem with high nitrogen levels is the risk of methemoglobinemia in infants. The European Union drinking water limit of 50 mg/L NO_3^- has often been exceeded in groundwater and has led to the shutdown of public supply wells (Custodio, 1997). Studies in various countries show nitrate is contributed mainly by the usage of fertilisers (Ehlers & Grieger, 1983; Bernard et al., 1992; Robins, 1996). The large number of gardens, athletic grounds and parks in urban areas are important point sources in urban areas (Mull et al., 1992). Salt used for de-icing roads is a major contributor of sodium and chloride to groundwater (Eisen & Anderson, 1979). Howard (1997) states that chloride derived from road de-icing salts represented the largest chemical loading to the subsurface over the last 50 years in the Toronto metropolitan area. Ammonia is also a serious groundwater pollutant (Lerner, 1996). Phosphorus, associated with eutrophication problems in surface waters, is of less importance than nitrogen in groundwater because of the low solubility of phosphorus compounds in groundwater and the limited mobility of phosphorous due to its tendency to sorb on solids (USGS, 1999; Domenico & Schwartz, 1990).

Residence times of these pollutants depend very much on the hydrogeological features of the vadose zone materials (Vrba & Zaporozec, 1994). The principal attributes used in the assessment of intrinsic groundwater vulnerability are recharge, soil properties and characteristics of the unsaturated and saturated zones. Attributes of secondary importance include topography, groundwater/surface water relations and the nature of the underlying aquifer. The importance of the secondary attributes varies with the area.

Case study: Long Island, New York, from Kimmel (1984), Flipse et al. (1984)

Long Island, New York, includes the counties Kings and Queens in New York City. Though populated for centuries, population grew rapidly in the period 1940 to 1970. Unconsolidated sand aquifers supply water to the entire island, except for Kings and parts of Queens counties. These are underlain by low permeability crystalline bedrock. Streams are an important groundwater discharge with about 40% of recharge coming out through streams. From the 1930s stormwater recharge basins were used to dispose of storm water and street runoff, infiltrating up to 2 million gallons a day. Over-pumping has caused seawater intrusion into the aquifer.

Nitrate levels in the aquifer increased markedly in the period 1950 to 1980, and were of particular concern in the areas totally dependent on groundwater for potable water. The use of cesspools and septic tanks has contributed high levels of nitrate, TDS, sulphate and chloride to groundwater. Waterborne sewerage, which was first installed in Brooklyn about 1850, moved eastward over the island lagging behind population growth. Landfills contribute to nitrate contamination with ammonia oxidised to nitrate.

400 samples were taken from 15 wells in the Medford area of Suffolk County. The vast majority of wells showed a significant regional increasing trend in nitrate levels. Average annual regional increases in concentrations were approximately 1 mg/L. Nitrogen loading in the groundwater was estimated at 2 300 kg/y assuming a 60% leaching rate on a fertiliser application of 0.0107 kg/m^2/y. Precipitation also contributes nitrogen. Average precipitation is 106 mm/y and contains 0.87 mg/L (as N). Another source of nitrate input to the groundwater system came from many homeowners irrigating their lawns with municipal water (average total nitrate concentration of 1.6 mg/L) during summer months. Animal wastes are a source of nitrogen also, entering via storm runoff and recharge basins or by direct infiltration, though as much as 50% may be lost by volatilisation. Nitrogen loading from sewer leakage was not concluded to be significant in Medford after population wastewater production figures were compared with sewage treatment plant inflows. Nitrogen isotope analyses of groundwater in Long Island showed that values increased from the east of the island to the west. This corresponds with a decreasing degree of urbanisation and hence relative contribution of water with a waste content to groundwater. The $\delta^{15}N$ values in the Medford area had a mean value of 4.3‰ and a median of 4.7‰, indicative of non-animal sources of nitrate. This supports the view that the high concentration of nitrate in Medford was due mainly to lawn fertilisers.

Therefore on Long Island elevated nitrate levels in older urbanised areas are due to the historical use of septic tanks. More recently urbanised areas, employing sewers instead of septic tanks, also have elevated nitrate levels but these are primarily due to fertiliser application.

Case study: Nottingham (from Lerner & Wakida, 1999)

Nitrate values in the Nottingham area have increased substantially over the course of the 20th century. A maximum NO_3^- value of 36 mg/L was measured in Farnsfield in 1907 (Lamplugh et al., 1914). Recent data from Barrett et al. (1997) shows many boreholes now with NO_3^- values well in excess of 50 mg/L. Nitrate is often regarded as an agricultural pollutant but differences between urban and rural nitrate concentrations were found to be small in Nottingham city and its surrounding countryside. This was due to the many and varied non-point sources of nitrogen in urban areas such as landfills, gas works, soakaways, fertiliser use in gardens and recreational areas, leaky mains water and sewers and industrial discharges and spillages. Nitrate leaching from housing development is another major potential source, possibly even equivalent to ploughing of pasture land which releases an estimated pulse of 450 kg N/ha. Solute transport modelling estimated the total-N loading in the Nottingham urban area to be 21 kg/ha/y. The estimated proportions coming from different sources were: leaky mains 37%, leaking sewers 13%, soil leaching 9%, contaminated land, industry etc. 41%.

Discussion

Urbanisation has led to increases in various solutes, nitrate being of particular concern. On Long Island elevated nitrate levels in older urbanised areas are due to the historical use of septic tanks. However, sewered urban areas also have elevated nitrate levels. This is mainly from industrial sources and landfills in the Nottingham case but urban fertiliser application is a major source in other areas, such as Medford. Proper management of such industrial sources to minimise emissions and spillages is important. The education of those fertilising green areas is important if nitrate levels are to be reduced, as are the repair of leaky sewers and proper construction of new sewers. Restrictions on the construction of new houses are also needed in vulnerable areas of aquifers, including the timing of building to avoid the recharge season (Lerner & Wakida, 1999). All mature urban areas can suffer from inputs of nitrates from large-scale sources such as atmospheric deposition. This can only be tackled by long term strategies aimed at reducing industrial and car emissions.

Organic contaminants

The large quantities of organic compounds manufactured for agricultural, industrial and municipal purposes have created the greatest potential for groundwater contamination (Bedient et al., 1994). Many of these organics can cause adverse health effects and are carcinogenic and/or toxic. Volatile Organic Compounds (VOCs) are commonly detected in shallow groundwater beneath many urban areas across the United States (Kolpin et al., 1997). VOCs commonly detected in urban waters across the USA include petrol-related compounds (e.g., toluene, xylene) and chlorinated compounds (e.g., chloroform, perchloroethylene [PCE], trichloroethylene [TCE], 111-trichloroethane [TCA]) (USEPA, 1992; Lopes & Bender, 1998).

Chlorinated hydrocarbon solvents (CHS) are hydrocarbons containing one or two C atoms bounded by Cl and usually H atoms and include TCE, TCA, trichloromethane [TCM], tetrachloromethane [TeCM] and tetrachloroethene. They are used in paint manufacture as well as for cleaning purposes in the electronics, engineering and dry cleaning industries. The physical and chemical properties of chlorinated hydrocarbon solvents are noteworthy. Specific densities greater than 1 allied with low viscosities, low sorption onto geological material and chemical stability mean that they migrate quickly on entering the subsurface. Resistance to microbiological degradation means that they are persistent pollutants (Eastwood et al., 1991). These are also the most widespread organic pollutants of urban groundwater in several countries including Australia (Benker et al., 1996), Denmark (Moller & Markussen, 1994) and the UK (Nazari et al., 1993; Lawrence et al., 1996).

Other synthetic organic chemicals of concern include chlorinated aromatic compounds (chlorinated benzenes, polychlorinated biphenols [PCBs]), dioxins and polynuclear aromatic hydrocarbons (PAHs) (Kiely, 1997). Methyl tertiary butyl ether (MTBE), used in the oxygenation of gasoline, is a groundwater contaminant of growing concern in the UK (Burgess et al., 1998).

25% of pesticides in the USA are of urban/industrial origin (USGS, 1999). Pesticide presence is widespread in groundwater wells in the USA (USEPA, 1990). These substances are toxic even at low concentrations (Freeze & Cherry, 1979). Therefore, pesticide residues have very low levels in the EC drinking water guideline: 0.5 µg/L for all pesticides and 0.1 µg/L for individual components (Robins, 1996).

Landfills contain chemically hazardous materials that can be mobilised by infiltrating waters to form a leachate (Howard et al., 1996). These contaminants have the ability to migrate off-site unless careful leachate control management is undertaken (Kiely, 1997). Modern landfills are often lined but older unlined landfills have the ability to release large amounts of contaminants in the subsurface. De Roche and Breen (1988) found evidence of leachate migration off-site with elevated levels of total dissolved solids, boron, ammonia, and iron in wells downgradient of a landfill in north-western Ohio. Jankowski and Acworth (1997) found leachate plumes downgradient from municipal landfills are characterised by anaerobic conditions associated with an absence of DO, high concentrations of both major and minor ions, the presence of hydrogen sulphide and methane and the absence of nitrate and sulphate. In poorer countries, the most serious risks to groundwater occur when uncontrolled tipping, as opposed to sanitary landfilling, is practised and where hazardous industrial wastes, including drums of liquid effluents, are disposed of at inappropriate sites (Morris et al., 1994). Other sources of contamination that threaten groundwater quality include underground storage tanks, surface impoundments, land application of waste, nuclear waste repositories and mining operations (Bedient et al., 1994).

Case study: Birmingham

The Birmingham aquifer (Fig. 4.3) is part of the greater Triassic Sherwood Sandstone aquifer underlying the region. Overlying the sandstone are Quaternary glacial deposits of thickness 0 to 40 m which confine the aquifer in west Birmingham (Ford & Tellam, 1994). The aquifer is confined by the Mercia Mudstone to the southeast of the Birmingham fault. This fault restricts groundwater flow between confined and unconfined aquifer sections, as can be seen by a difference in the piezometric surface on either side of it (Jackson & Lloyd, 1983). Groundwater velocities in the sandstone are 10 to 100 m/year (Hughes et al., 1999). The Birmingham plateau has the Trent and its tributaries to the north and east, and the Severn and its tributaries to the west and south.

Birmingham was a market town that grew rapidly in the twelfth and thirteenth centuries. By 1600 the local advantages for industry were apparent with coal, ironstone and wood all available. Major canal building in the 1700s opened up the surrounding Black Country region to industry (Freeman, 1966). Industry rapidly expanded along these and also along railway lines from around 1840 onwards (Ford & Tellam, 1994). The dominant industry in Birmingham

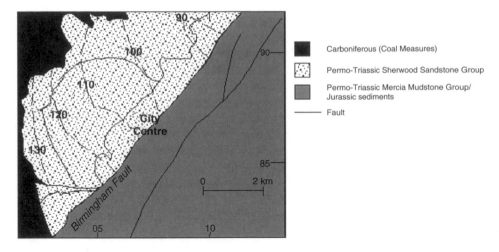

Figure 4.3.   Geology and groundwater of Birmingham (from Ford & Tellam, 1994).

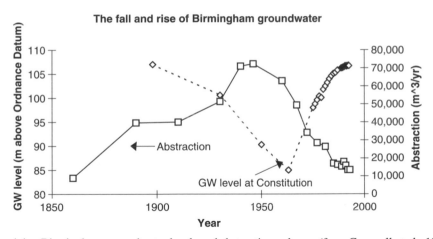

Figure 4.4.   Birmingham groundwater levels and abstraction volumes (from Greswell et al., 1994).

has been metal working with large amounts of sulphuric and other acids being used in the city since the early 1700s (Hughes et al., 1999). Birmingham reached its industrial peak with the post war economic boom of 1950–70 but since then there has been a steady decline in manufacturing industry. The entire outcrop of the aquifer is now urbanised while 70% was urbanised prior to 1900 (Rivett et al., 1990; Ford & Tellam, 1994). Birmingham Council records put 1997 population figures at 1 017 000 over an area of 254 km^2.

   The Triassic Sherwood Sandstone aquifer was used extensively for public supply up until around 1900 when Birmingham established reservoirs in Wales and since then almost all groundwater pumping has been for industrial usage (Hughes et al., 1999). Abstraction peaked in the early 1950s but water levels had been dropping since 1900, as can be seen from Figure 4.4. This clearly shows that groundwater levels have risen quickly since abstraction decreased after 1960. This has led to de-watering programmes to protect certain buildings.

As in Nottingham, recharge values were found not to vary greatly before and after urbanization as water supply losses compensated for the effects of impermeabilisation (Greswell et al., 1994). Predictive modelling of groundwater levels by the same study showed groundwater levels may rise by over 10 m in the city centre area.

Several regional surveys of groundwater geochemistry have been conducted (Jackson & Lloyd, 1983; Rivett et al., 1990; Ford & Tellam, 1994; Hughes et al., 1999). These surveys sampled abstraction boreholes, usually at industrial sites. The main natural sources for major ions include rainfall and fall out as well as the dissolution of aquifer carbonate and silicate minerals while anthropogenic sources included industrial wastes, processing chemicals, sewage, road de-icing salts, urea, domestic wastes and fertilisers (Ford & Tellam, 1994). Hydrochemical data (Cl^- and NO_3^-) and carbon isotope data suggest that recently recharged groundwaters exist in the aquifer to the west of the Birmingham fault whereas older groundwaters (correlating with lower chlorides and nitrates) occur to the east where the aquifer is overlain by mudstone (Jackson & Lloyd, 1983). Ford and Tellam (1994) also found that Cl^- and NO_3^- were significantly lower where the aquifer is confined east of the fault. Hughes et al. (1999) found by using sulphate isotope signatures that high sulphate values in the confined zone waters were from gypsum dissolution. Overall major ion quality is good with only nitrate median values approaching EC maximum admissible levels (Lerner & Tellam, 1992). High barium concentrations were postulated to be from natural sources such as barite (Ford & Tellam, 1994). Deep, non-industrial boreholes were usually uncontaminated, suggesting good quality water at depth in the aquifer.

Groundwater contamination appears to be site specific with the poorest groundwater quality under industries with large pollution potentials, such as electroplating and metal finishing works. Total dissolved solids under food processing sites were 470 mg/L, while metal industry boreholes had values of approximately 800 mg/L (Ford & Tellam, 1994). They showed boron, a common compound in metal-related industries, gave the most clearly defined correlation between land use and chemistry (Table 4.1).

Triassic sandstones have high sorption capacities (e.g. Spears, 1987), but metal loading was severe enough to cause over-loading at some metal industry sites (Ford & Tellam, 1994). Ford et al. (1992) have shown that pH levels are falling over time due to the combined effects of acid spillage, organic degradation and water levels rising into calcite-sparse regions in the upper aquifer. This will increase metal mobility and result in deterioration in groundwater quality over time. A possible solution where pH-dependent species are of concern is to deepen well casings to increase the flow through the deeper carbonate-rich parts of the aquifer; this could result in long-term immobilisation (Tellam, 1995). Hughes et al. (1999) also found evidence of metal-working industry contamination in the aquifer from heavy sulphate isotope signatures.

Significant organic contamination was found in the aquifer. Chlorinated solvents, in particular TCE, were the most prominent organics found by Rivett et al. (1990). They found

Table 4.1.   Average boron concentrations in Birmingham groundwater under varying land use (after Ford & Tellam, 1994).

Industry	Metal working	Gas works	Other industries	Service industries
B concentration (μg/L)	745	400	200	20

78% of the boreholes sampled had detectable levels of TCE with 40% above the WHO guideline value of 30 μg/L and 30% over 100 μg/L. Other chlorinated solvents were widespread but usually at low levels with the odd exception of TCA and PCE. This study related the contamination to land use with high TCE attributable to its use in metal cleaning. Drift and mudstone covering were found to have some influence in protecting the sandstone aquifer, as was the existence of an extensive unsaturated zone. Deep-cased boreholes and low abstraction boreholes were also found to have lower solvent levels. Concentrations of BTEX compounds (benzene, toluene, ethylbenzene and xylene), phenols, and C_6 to C_{30} n-alkanes were present at low levels, generally less than 2 μg/L (Rivett et al., 1990). Inorganic contamination was found to be much more diffuse than organic contamination because (a) inorganic compounds have much more numerous sources while solvents are more used within specialised industrial applications; (b) solvents have been in use since the 1950s while many inorganic contaminants have been in use since the 1700s; (c) organic solvents are less mobile than inorganic anions, but more mobile than heavy metals (Ford & Tellam, 1994).

Case study: Coventry

The city of Coventry is generally underlain by Permo-Carboniferous strata, with some Bromsgrove Sandstone and Mercia Mudstone to the east (Fig. 4.5). The Triassic strata have a gentle regional dip of 1 to 2° to the southeast. Glacial drift deposits (sand and gravel, boulder clay, laminated clay and silt) are intermittent in the region with clay deposits up to 20 m in places (Lerner et al., 1993a). The sequence is alternating sandstones and mudstones with occasional conglomorates.

Coventry's industrialisation began around the silk and ribbon manufacture in the early 17th century together with watch and clock making from the middle of that century. Proximity to Birmingham, local coalmining and good transportation links all stimulated growth (Freeman, 1966). Until the mid 1970s the economic prosperity of Coventry was based around its manufacturing industries of engineering works, car plants and machine tool factories. Economic recession from 1973 onwards led to large-scale job losses. Recent times have seen large old industrial sites in the inner city replaced by new high technology manufacturing.

Average rainfall is about 670 mm/y with potential evapo-transpiration of 510 mm/y. Direct recharge from October to April dominates in the surrounding rural area (Lerner et al., 1993a). Anthropogenic recharge sources in the city include water mains, foul and storm sewers. The discontinuous lenticular units form a complex multilayer aquifer system extending down 500 m in places. This gives rise to strong vertical gradients in recharge/discharge zones. There is a regional groundwater flow pattern from the natural recharge areas to the north and west to stream discharge to the centre and east (Lerner et al., 1993a). Groundwater makes up about 24% of Coventry's public water supply with the remainder of the supply from Oldbury Reservoir and the river Severn (Lerner & Tellam, 1992). Inorganic groundwater quality is good. Virtually all of the major ions and metals are at elevated levels in urban areas with respect to rural wells but levels are not excessive (Lerner & Tellam, 1992).

Chlorinated hydrocarbon solvents were expected to be major contaminants in the aquifer due to large industrial usage of the solvents in the Coventry area. A survey of 32 industries in the Coventry area in 1985, by the Coventry Pollution Prevention Panel, shows just how great their usage is in the region (Table 4.2).

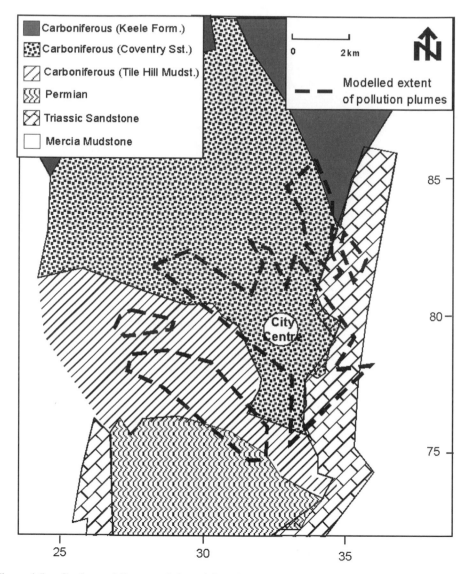

Figure 4.5.   Geology of Coventry (adapted from Lerner et al., 1993a) with thick dashed lines representing the modelling of CHS pollution in the urban area (adapted from Burston et al., 1993).

Table 4.2.   Summary of CHS use by 32 Coventry industries (1985).

CHS	Total quantity used per annum (L)	Max. quantity used by one industry (L)	% industries used by	% of total CHS use in the UK	% of total CHS use in Coventry
TCE	420 000	80 000	53	18	59
TCA	286 000	74 000	72	21	40
TeCM	207	200	9	29	<1

Table 4.3. Summary of CHS concentrations in Coventry groundwater (from Burston et al., 1993).

CHS	TCM	TCA	TeCM	TCE	TeCE
% Boreholes with CHS > 1 µg/L	43	52	31	72	40
% Over British legal limits	–	–	10	38	10
British limits µg/L	–	–	3	30	10
Mean concentration µg/L	2.1	11	4.9	67	4.1
Median value µg/L	0.7	1.2	0.1	9.8	0.4
Maximum value µg/L	21	270	80	1100	53

Samples were collected from 42 existing boreholes at 23 sites, and the results are sum-marised in Table 4.3. CHSs were found to be major pollutants in the Coventry aquifer (Burston et al., 1993).

TCE is the most common CHS, and Burston attributes this to its long history of use in Coventry in the metal-working, car and components manufacturing industries. Poor disposal practices were common in the past, and spillages are commonplace still. TCA has replaced TCE over the past 20 years and its presence at fairly low levels but throughout the urban area shows that CHSs are still causing problems despite improved disposal methods. Mains water in Coventry contains high TCM concentrations as a by-product of chlorination. TCM quan-tities were higher towards the city centre, possibly reflecting the higher densities of water supplies and sewage here. TeCE and TeCM are less widespread than the other CHSs but are still significant pollutants. TeCE contamination probably resulted from its use in the dry cleaning industry (Burston et al., 1993). TeCM was used in fire extinguishers and high levels were found at the Old Fire Station near the city centre (Eastwood et al., 1991). The fact that this site was disused for 16 years prior to sampling shows the persistence of CHSs in the subsurface.

Most of the boreholes showing CHS pollution were at industrial sites utilising CHSs. Sites underlain by thicker mudstone deposits had lower CHS concentrations. Groundwater flow modelling was used to determine possible present and future extent of CHS pollution (Fig. 4.5). However the complex spatial distribution of pollutants at one site examined in detail highlighted the difficulty in defining a 3-D picture of pollution in such a complex hydro-geological system with so many potential pollution sources (Bishop et al., 1993a,b). They conclude that the low organic content of the Sherwood Sandstone in the region means that sorption of the organic contaminants is not significant. In addition, the aerobic nature of the aquifer means that the degradation of the CHSs will be a slow process. Future aquifer pro-tection allied with groundwater treatment appear as the only viable options for the aquifer (Lerner et al., 1993b).

Higher levels of inorganic solutes, such as boron, were found not to correlate with the CHS distribution suggesting that sewage and not industrial pollution is the source of these elevated values (Burston et al., 1993). Heavy metal contamination was generally low and restricted to industrial wells (Nazari et al., 1993).

Case study: Belfast

The major Triassic Sherwood Sandstone aquifer in Northern Ireland is approximately 150 km² and underlies Belfast and the areas to the East and South of the city (Fig. 4.6). It is Northern Ireland's most important aquifer as it coincides with the most populated and indus-trialised region of the Province (Manning et al., 1970; Robins, 1995; Kalin & Roberts, 1997).

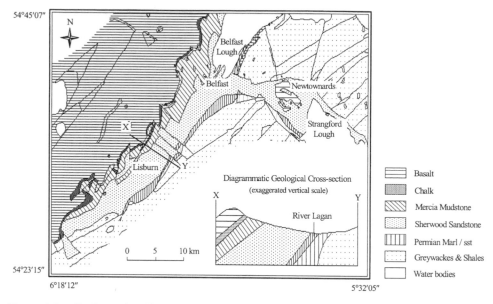

Figure 4.6.   Geology of Belfast area (from Cronin, 2000).

Groundwater provides about 18 000 m^3/d of the total public water supply in the Belfast/ Newtownards urban areas (Cronin, 2000). The majority of the public water supply for Belfast comes from the Mourne Mountains surface reservoir to the south east of the city. The Triassic Sherwood Sandstone has been supplying water to farmers for centuries (Bennett, 1976). It also provided water for the thriving mineral water trade of the 19th century, which exported water to every corner of the British Empire, but this ceased when industrial activity contaminated the groundwater beneath this area of Belfast (Kalin & Roberts, 1997). It was used substantially by local industries to feed steam boilers up to the 1930s. Abstraction declined after this due to the demise of steam power (Bennett, 1976) and also due to the fact that many of the linen manufacturers found the water to be too hard for their purposes (Manning et al., 1970). However Lisburn Borough Council continued using groundwater for public supply after constructing their first deep well in 1934 (Thompson, 1938). The Triassic Sherwood Sandstone Group has been in use as an aquifer in the Newtownards/Comber area since the end of the last century (Manning, 1972).

In the Belfast hills there are a large number of poorly constructed landfills from which solutes could leach. These sources pose a threat to the potential of good quality groundwater being available for utilisation. Investigations into the groundwater quality of the springs to the west of Belfast concluded that various landfills were causing localised pollution (HandES Ltd., 1997). The degree of contamination of these Chalk springs may reflect considerable dilution in the aquifer. The Department of the Environment (Northern Ireland) has collated data on the various activities in the area that pose a threat to groundwater pollution. These activities have been classified into high, medium or low risk categories (Table 4.4). The high-risk activities locations (Fig. 4.7) show a widespread distribution of potentially hazardous activities in the study area (Cronin, 2000). They overlie a diverse sequence of geological units so that the entire aquifer is designated a maximum vulnerability rating in the Groundwater Vulnerability Map (DoE NI, 1994).

Table 4.4.   Industrial activities around Belfast posing a threat to groundwater pollution.

Risk	Activity
High	Asphalt Works, Landfills, Chemical Works, Gas Works, Iron and Steel Works, Lead Works, Metal Finishing and Electroplating, Petrol Stations and Fuel Storage, Paper and Pulp Manufacturing, Plastics, Scrap Yards, Shipbuilding, Tanneries, Timber Treatment.
Medium	Mechanical Engineering, Sewage Works, Railway Land, Tobacco Works.
Low	Mineral and Printing Workings, Reclaimed Land, Textiles, Food Processing.

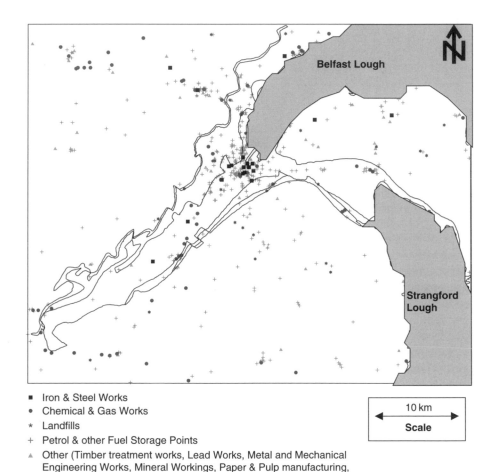

■   Iron & Steel Works
●   Chemical & Gas Works
∗   Landfills
+   Petrol & other Fuel Storage Points
▲   Other (Timber treatment works, Lead Works, Metal and Mechanical
    Engineering Works, Mineral Workings, Paper & Pulp manufacturing,
    Scrap Yards, Tanneries, Shipbuilding Yards, Plastics)

Figure 4.7.   Industrial activities in Belfast area that are a threat to groundwater quality (from Cronin, 2000).

The thick glacial drift covering of low permeability clay has proven to be a good natural defence against contamination of the deep Sherwood Sandstone aquifer. Groundwater flow modelling highlighted the importance of water from the Chalk formation as a recharge source in the Lagan Valley as well as inflow from greywacke strata in the Newtownards area

(Yang et al., 2000). Further groundwater flow modelling work determined recharge to be approximately 43 000 m^3/d to the Sherwood Sandstone aquifer in Lagan Valley (Cronin, 2001). It is a slow moving system with maximum velocities in the range of 15 to 75 m/year (Cronin, 2001).

Despite the widespread high-risk activities in the area, water quality analyses suggest good quality water, except for certain wells that are subject to mixing with formation waters and/or saline intrusion in the vicinity of Belfast Harbour (Cronin, 2000). Elevated nitrate values in the Newtownards area often approach the EU Drinking Water Directive guideline level of 25 mg/L. This is especially the case in public supply wells. High levels of nitrate and sulphate in the sandstone wells suggest that there are preferential pathways for contaminants through the low permeability till. Nevertheless, McConville (1999) did not find evidence of preferential pathways. It is more likely these variations are indicative of poorly constructed wells (Cronin, 2000). This allows surface runoff a short circuit route into the groundwater. Industrial contamination of overlying alluvial aquifers in Belfast has meant that parts of this natural resource are lost, probably forever.

Discussion

Birmingham, Coventry and Belfast are all mature urban centres with long histories of industrialisation. Both Birmingham's and Coventry's groundwater contamination are strongly linked to land-use with the poorest groundwater quality under sites with large polluting potential, such as metal industry sites. Widespread use of organic chemicals at such sites led to significant localised organic contamination of the aquifer. TCE was found in over 70% of boreholes sampled in industrial areas in both cities. Indeed, Nottingham too has organic chemical contamination: TeCE was detected regularly due to its widespread use in the textile processing and dry cleaning industries (Barrett et al., 1997). TCE is of less significance here due to lower proportion of metal working industries in Nottingham compared with Birmingham and Coventry. The aerobic nature of these aquifers mean degradation is low while sorption is not significant due to low organic contents. Sections of the aquifer beneath both Birmingham and Coventry have been well protected by overlying units of low permeability mudstones or drift. Indeed, this is probably the main reason why groundwater resources underneath Belfast are not more contaminated as, although it has a widespread and varied distribution of industries, the Sherwood Sandstone aquifer here is overlain by significant thicknesses of low permeability glacial tills. Inorganic contamination in these mature cities is more widespread on the whole due to a larger number of potential sources with a more widespread distribution. In addition, the long history of industrialisation of mature cities has meant that inorganics have a long history of contaminating groundwater resources in the area. Heavy metal levels are generally not excessive but a potential cause for concern is that rising pH levels may mobilise large amounts in the future. Therefore industrial pollutants have created serious contamination in aquifers underlying mature urban cities. However these problems have tended to be localised and have not rendered the regional groundwater resources unusable.

Saline intrusion

Under natural conditions, a seawater/freshwater interface exists in coastal aquifers. This interface is not sharply defined but tends to consist of a mixing of salt and fresh water in a zone of dispersion around the interface (Freeze & Cherry, 1979). With the development of groundwater supplies and subsequent lowering of the water table or piezometric surface, the dynamic balance between fresh and seawater is disturbed (Domenico & Schwartz, 1990). Seawater can then intrude into usable parts of a coastal aquifer. Hence heavily abstracting coastal cities are extremely vulnerable.

In order to improve coastal water management, much effort has been put into the modelling of saline intrusion (Kishi et al., 1988) and submarine groundwater discharge to the sea (Li et al., 1999). Movement of the salt/fresh water interface in confined coastal aquifers is dependent upon pumping rate and tide action (Kishi et al., 1988). Van Dam (1983) studied in detail the shape and position of the salt water wedge in the various types of coastal aquifers (confined, phreatic and semi-confined) through the measurement of piezometric levels of the saline water. He concluded that the shape and position of the saltwater wedge depend very much on the hydrogeological characteristics of the aquifer.

Case study: Barcelona (Custodio, 1997)

Barcelona existed prior to the arrival of the Romans. The plain of Barcelona consists of alluvial fans and eolian deposits that had been exploited for water since Roman times by means of dug wells and water galleries. These were superseded by wells installed in the river valleys from the middle of the 19th Century. Growing problems with well outputs led to the installation of recharge wells injecting river water into the aquifer and also to the direct exploitation of surface water. Intense groundwater exploitation for urban and industrial development has led to the lowering of piezometric levels below sea level and the intrusion of seawater into the aquifer. This has led to the pumping of deeper (and more expensive) groundwater as well as the partial drainage of former wetlands and shallow water table areas. Industry has abandoned many saline wells or used them only for cooling purposes. This, coupled with the introduction of treated surface water, has seen groundwater levels recover in recent years. This has led to the flooding of several underground carparks and tunnels. This problem has been added to by leakage from the water distribution network. Efforts have been made to achieve sustainable use by protecting recharge, controlling seawater intrusion and fighting pollution. Gravel and sand mining as well as waste disposal are now under strict control. Saline intrusion is being monitored and some wells used to extract seawater. Artificial recharge has continued and efforts are continuing to assess the impact of regional land use planning on groundwater resources.

Case study: Brighton

Brighton developed from the mid-eighteenth century as a holiday and residential centre on the south coast of England. Wells were drilled in the chalk aquifer underlying Brighton and Worthing in the 19th and early 20th centuries (Headworth & Fox, 1986). Uncontrolled abstraction led to saline intrusion. In 1957 a more sophisticated abstraction policy was introduced which alleviated the saline intrusion problem. Water was pumped near the coast in winter and further inland in summer in order to conserve winter storage. This has led to the stabilisation of water levels over the past 30 years despite abstraction increasing by one third. Winter pumping allows water that would otherwise leak to the sea to be intercepted while allowing inland recharge to occur (Lerner & Barrett, 1996).

Discussion

The degree of saline intrusion varies widely among hydrogeological settings and abstraction volumes. The traditional approach to reversing seaward migration into freshwater aquifers was to discontinue pumping from coastal wells and look for alternative sources of water.

However, population increases and lack of alternative resources means this often may not be a viable option. This has meant that alternative approaches are increasingly being looked at. These include such schemes as artificial-recharge systems, desalination plants and blending of waters of different quality. The Atlantic coastal communities of the USA have tackled saltwater intrusion under three different categories: scientific monitoring and assessment, engineering techniques and regulatory actions (USGS, 2000). Saline intrusion is a problem that can only be tackled with a concerted approach from all abstracters from the aquifers.

4.5   CONCLUSIONS AND FUTURE STRATEGIES

The many and varied sources of groundwater contamination have led to efforts in many countries to concentrate on preventing resources from being threatened (NRA, 1992; Aust & Sustrac, 1992). Hydrogeological and groundwater vulnerability and protection maps are the most common form for the graphical expression of data and information on aquifer systems at regional and national scales (Sotornikova & Vrba, 1987). Nitrate-vulnerable zones around supply wells are zones in which fertiliser application is carefully managed. This can help prevent elevated nitrate levels in the abstracted groundwater (GSI, 1999). However, the numerous point sources in urban areas make even the assessment of groundwater vulnerability a complex task. In combination with this are the number of abandoned boreholes and poorly designed wells which act as conduits for pollution to enter the aquifer. For example, compared to the 50 or so wells in use today, Birmingham has over 300 abandoned boreholes (*pers. comm.*, Greswell, 2000), some of which could possibly be compromising groundwater quality. Hence the protection of aquifers underlying urban areas is an extremely difficult, if not impossible task. Sustainable city development means that these problems must be addressed in order to minimise the overall threat to groundwater resources (Lerner, 1996).

Mature urban cities often have problems with rising groundwater levels, but this is a manageable problem with designated abstraction points. Artificial recharge may prove to be a beneficial resource policy in some cases but only if it is in a non-industrial area that would not result in a compromise of groundwater quality. Mature urban areas are generally sewered, meaning pathogens tend be localised issues except in urban areas served by a high density of septic tanks. Solute loads, however, especially nitrogen, can be high even in sewered areas. The education of both industrial operators and garden owners/park managers is essential to help combat rising solute concentrations. Authorities with responsibility for sewerage schemes need an objective evaluation scheme for determining a priority order in the restoration of leaky sewers (Eiswirth & Hotzl, 1994). This priority order is necessary because of poor financial resources available for sewer improvements. Better co-operation between municipal authorities, government bodies and other decision-making bodies is essential to look at financial resources available for groundwater protection.

Evidence from the case studies outlined, especially Birmingham and Coventry, suggest that virtually all heavy industry conducts activities that degrade the underlying groundwater quality. This may mean groundwater quality is unsuitable for potable supply without expensive treatment. However, this does not rule out groundwater being used by non-potable consumers. The localised nature of the contamination means areas outside the industrial zones generally do not have such serious contamination problems. Lerner and Tellam (1992) suggested that industries overlying aquifers could be required to abstract groundwater on site for their own use, resulting in a self-policing clean-up operation. This would encourage good, sustainable use of a resource with low costs and environmental benefits, whereby industry would take responsibility for the pollution it creates. Another advantage is that this would then help to solve the rising groundwater problems in mature urban cities, as groundwater abstraction would again be back to sizeable proportions. Such a policy would certainly create problems with issues like pollution arising from previous site use, adjacent site use, and pollution from other sources such as sewers or fertiliser/pesticide application. However, it is an example of a situation whereby potential polluters would be forced to take responsibility for the degradation they may cause. Better chemical handling practices must be adopted by industry in the future, but site lining may also be beneficial, as is now common practice in the construction of landfills (Tellam, 1995).

Lerner and Tellam (1992) state that the law has been ineffective in deterring polluters because there is both a lack of political will and a lack of resources for monitoring and enforcement. European Union driven policy has led to the UK Groundwater Protection Policy under which the degree of control on polluting activities is related to groundwater vulnerability and the proximity to public supply boreholes (NRA, 1992). Recent cases in the UK have seen companies prosecuted and fined for contaminating aquifer systems with CHSs and in the future companies will find preventing groundwater pollution increasingly in their own best interest (Lerner et al., 1993b). Lerner and Tellam (1992) suggest that engineers and scientists have a duty to address these shortcomings by pointing out the extent of pollution problems and by advising on the remediation, monitoring and enforcement issues.

Therefore, mature urban cities should eventually be able to provide sufficient supplies of ample quality groundwater. There are many advantages to exploiting this resource (outlined in more detail in Chapter 1). It can be used as baseflow for urban rivers to help alleviate flow problems in summer months. Use of urban groundwater for industrial purposes or public water supply has the added advantage of reducing current dependence on rural aquifers (and related wetland losses). If the water is of sufficient quality it can be added to potable drinking sources; if not it may be acceptable for amenity use in fountains, lakes etc. This still has associated quality implications. Processing waters for industrial purposes is another viable option. Blending of urban groundwater may be necessary to help solve these quality issues.

Evaluation of threats to urban groundwater requires the linking of many different types of data, from the fields of land use, hydrology, hydrogeology, chemistry, and microbiology. A Geographical Information System (GIS) is the ideal tool to do this combining of data. A GIS is a computer-based tool for mapping and analysis of data using statistics and database operations to produce visual displays (Maguire et al., 1993). Such a powerful tool can overlay new data on existing information and allow the incorporation of groundwater flow modelling results also. This can then facilitate the assessment of the vulnerability of water resources from pollution threats. Such tools will undoubtedly be needed increasingly in the future to help to manage the complex interplay of activities affecting urban aquifers.

CHAPTER 5

Rapidly-urbanising arid-zone cities

Ahmed Alderwish, Kamal Hefny and Steve Appleyard

5.1   INTRODUCTION

Water resources management in the world today is perhaps most challenging in the Middle East region, where a combination of low rainfall and generally scarce water resources, coupled with rapid population growth, demand for water for irrigation schemes, and a complex and difficult political environment makes the allocation and protection of water resources particularly difficult. The population growth in the region generally exceeds the world average of 2.5% per annum and, on average, more than half the population lives in urban centres (Marcoux, 1996).

In this chapter, the general groundwater hydrology of arid zone alluvial urban environments will be presented. Two case studies were selected to represent the alluvial urban environments, the first being the capital of Egypt, Cairo, and the second Sana'a, in Yemen.

5.2   HYDROGEOLOGICAL SETTING OF CAIRO, EGYPT

With a population of more than 13 million people, Cairo is easily the largest city in the region, and like many other Mega-Cities in the world it continues to experience rapid population growth and spatial expansion. Cairo is situated on the Nile upstream of the river's delta, about 250 km from the Mediterranean coastline and downstream of the Aswan Dam. The urban area extends a distance of about 50 km along the river and is between 5 and 35 km wide, covering an area of about 500 km². The Nile runs through Cairo from south to north at slope of about 20 cm/km. Along the eastern border of the Nile alluvial plain, the Mokkatam Plateau rises steeply to about 150 m above sea level. A similar plateau to the west, known as Pyramid's Plateau, rises to an elevation of about 100 m above mean sea level.

Cairo has an arid climate with an average annual rainfall of about 25 mm, which mostly falls in winter months. Due to its arid climate, the city relies on the Nile and its alluvial aquifer to provide fresh water for domestic use, for irrigated agriculture and for industry. The river is also the primary receptor of wastewater and drainage generated by a variety of urban and peri-urban land uses (Myllylä, 1995).

Cairo is located in the alluvial valley of the Nile. The valley is underlain by alluvial sediments consisting mainly of sand and gravel with interbedded silt and clay. The main aquifer beneath the Greater Cairo urban area consists of a sand and gravel member of

Late Pleistocene age. The aquifer is 20 to 140 m thick and underlies much of the Nile flood plain and portions of the adjacent slightly elevated plains. The sediments have a permeability of 20 to 70 m/d, and a transmissivity of 1 000 to 11 000 m^2/d. Clayey sediments of Pleistocene age form the base of this aquifer. This aquifer is generally semi-confined by overlying silty or clayey sediments that are 5 to 20 m thick, and is locally unconfined where the aquitard is absent. In urban areas there is also a large amount of man-made fill that overlies the aquifer.

The silty or clayey sediments that overlie the main aquifer have a maximum thickness in the vicinity of the Nile, and progressively thin with distance from the river. They generally have a low permeability, ranging between 0.005 and 0.2 m/d. The water table (known locally as the "subsoil water table") in these sediments varies in depth from a few centimetres to a maximum of about 2 m beneath the Cairo urban area. Recharge to these sediments takes place through the disposal of water on the ground surface, through leakage from water and sewerage distribution networks, and locally from leakage from the underlying aquifer where there are upward potentiometric heads.

5.3   GROUNDWATER MANAGEMENT ISSUES FOR CAIRO

Urban recharge

Recharge and groundwater flow in the Cairo region has been greatly changed by the construction of the Aswan Dam and by urban development. The Nile cuts partially into the aquifer, creating a hydraulic connection with the groundwater flow system. Prior to the construction of the Aswan Dam, river flow recharged the aquifer during the flood season each year (July to September), and drained groundwater during the low-stage season (January to February). Since the construction of the Aswan High Dam, water levels in the river are controlled throughout the year with an average rise of about 0.30 m higher than the pre-dam corresponding low levels, excluding the flood period. Since under the post-dam conditions the fluctuations in the river water levels are very limited, the interaction between the river and the aquifer is governed now by the changes in groundwater levels. Accordingly, the river acts as a drainage channel to the eastern bank in the area extending from the southern district of Helwan to the northern boundary of the city. Meanwhile, along most of the western bank, the river acts as a recharge boundary.

The Pleistocene aquifer in the city of Cairo and its environs is continuously recharged through the following sources:

- Downward infiltration from the subsoil water table in the semi-confining top member to the sand and gravel member.
- Lateral inflow from the Nile recharging areas of relatively lower groundwater levels on the western bank and north east of the city.
- Outflow from the Nile Valley aquifer at the southern boundary of the Greater Cairo area recharging the area west of the Nile.

On the other hand, discharge from the aquifer takes place by the following processes:

- Groundwater abstraction to the west of the Nile and north east of the city.
- Lateral outflow from the aquifer draining into the river along the eastern bank.
- Outflow leaving the north-eastern boundary of the city and entering the Nile Delta aquifer.

Infiltration from the subsoil water table is becoming an increasingly important source of recharge in urban areas because of population growth, and increasing leakage from the ageing water, sewerage and drainage infrastructure in the city. The additional recharge in urban areas has caused the water table to rise substantially, and has reached the ground surface in some parts of the city, causing extensive damage to buildings and roads. Shallow groundwater contains substantial amounts of dissolved sulphate (often more than 400 mg/L) and is aggressively attacking concrete and mortar, particularly in older buildings of historical value. The problem has increasingly become extensive since the 1960s; in 1962 the area where the water table had reached the ground surface was about 0.4 km^2, and this progressively increased to 1.3 km^2 in 1972, to 1.9 km^2 in 1982, and to 2.7 km^2 in 1990. Numerical modelling has suggested that the problem will continue to worsen unless active management is undertaken. The option of pumping and drainage to lower the water table locally is being explored, but there are limits to the extent this can take place because of the risk of land subsidence and differential settlement which can further damage buildings. Local dewatering may be necessary to protect buildings in historical precincts of the city that are of international significance and are important sources of tourist dollars. However, the immediate priority in dealing with the problem of the rising water table is to reduce the amount of water leaked to the subsurface in urban areas by replacing old water and sewerage infrastructure and increasing its capacity. This has commenced in some parts of Cairo, and is being assisted by aid money provided by the World Bank and the USA (Myllylä, 1995).

Potential sources of groundwater contamination

Groundwater beneath urban areas in Cairo is at risk from contamination from sewage, from agricultural land use and industrial wastewater disposal, although little work has been undertaken to determine the extent of the problem because groundwater only comprises 8% of the city's water supply. Potential sources of groundwater contamination are located both within urban areas of Cairo, and upstream of the city.

Although largely ignored as a source of groundwater contamination in urban environments, leakage of pollutants from river flow can be a significant source of contamination (Foster et al., 1998; see also the Boshan city case study in Chapter 7). The Nile Valley upstream from Cairo as far as the Aswan Dam is intensively settled, and in this area there are 43 towns with populations exceeding 50 000 and 1 500 villages discharging their wastes to the river. There are also 35 major factories discharging about 125×10^6 m^3 of largely untreated effluent to the river in the same area, and it is estimated that about 2.3×10^9 m^3 of drainage water contaminated with nutrients and pesticides from agricultural land is discharged to the river upstream of Cairo (Myllylä, 1995). Monitoring of the Nile since the 1970s has indicated that water quality has deteriorated due to industrial and domestic wastewater discharge (El Gohary, 1994). Concentrations of contaminants are likely to be diluted by the large volume of river flow, but there is a risk that some contaminants will accumulate in the aquifer with ongoing river recharge. There is also a risk that the load of contaminants discharged to the river will progressively increase with increasing population density along the Nile upstream of Cairo. It is estimated that by 2025 there will be about 86 million Egyptians living along the Nile, making it one of the most densely populated river basins in the world.

Within the urban areas of Cairo, the disposal of domestic and industrial wastewater is likely to be a significant source of groundwater contamination. Currently, there are at least six wastewater treatment plants servicing Greater Cairo, and none of the treated effluent is discharged to the Nile in the vicinity of the city. There are also plans to divert about $5 \times 10^6\,m^3/d$ of treated effluent as part of large desert reclamation projects. However, there are large problems with the operation of the network of sewers in the urban area that may be causing groundwater contamination. The sewerage system of Cairo has not been designed to cope with the growth of population in the region, and there is likely to be significant leakage from the system. About 25% of the population of the city is unsewered and so there is likely to be a significant amount of domestic wastewater disposed to ground, particularly in illegal shantytowns in parts of Cairo such as the "City of the Dead". Most of the wastewater collected in sewered areas is not adequately treated as many of the treatment plants are not functioning adequately, and it is estimated that only about 15% of the wastewater collected is fully treated, 25% is partially treated, and up to 60% is carried raw via open drains to the Mediterranean (Myllylä, 1995). Leakage from these drains is likely to also be a significant source of groundwater contamination. The problems with Cairo's sewerage system are gradually being upgraded with the assistance of international aid funds, but it is likely to take many years to eliminate current problems with the system.

More than half of Egypt's industrial capacity is located within Greater Cairo, and comprises chemicals, textile and tanneries, steel, food, engineering, and cement production operations. Although industry only occupies about 1% of the urban area, there are about 120 significant industrial sites in Cairo (Platenburg et al., 1997). The Mustorod-Shubra industrial area in the north of the city is located near an important groundwater recharge area (El Arabi, 1999). Industries within Cairo use in excess of $160 \times 10^6\,m^3/y$ of fresh water, and discharge more than $120 \times 10^6\,m^3/y$ of effluent, of which only about $56 \times 10^6\,m^3/y$ is discharged to sewers (mostly without pre-treatment). Little work has been carried out to determine the extent of groundwater contamination from wastewater disposal, but it is possible that a large range of contaminants could be present in groundwater beneath industrial areas in Cairo.

The increased use of treated sewage to irrigate agricultural land on the fringes on the city also poses a threat to groundwater quality. Much of the agricultural land is located on the fringes of the flood plain where the clay aquitard is absent and where a large amount of groundwater recharge takes place. Modelling (El Arabi et al., 1996, 1997) suggests that effluent irrigation in some areas will eventually affect existing water supply wells.

Groundwater quality monitoring in Cairo

The protection of groundwater resources has recently been identified as a key management issue in the Egyptian National Environmental Action Plan, but there are limited groundwater monitoring data to help manage the resource (El Arabi, 1999). A national groundwater quality monitoring network is being developed to characterise water quality within the Nile Basin and in desert areas, and 12 monitoring sites in the national network are located in the Greater Cairo region (El Arabi, 1999). Although monitoring wells are only analysed for a limited range of chemical parameters, there are sufficient data to indicate that there are likely to be significant groundwater contamination issues in Cairo.

Data from the monitoring wells indicate that ammonium, nitrate, iron, boron, manganese and Total Oils and Grease concentrations at a number of sites exceed drinking water criteria. Nitrate concentrations ranged between about 20 and 100 mg/L, and are probably derived from leaking sewers and industrial discharge in urban areas. The presence of ammonium in excess of drinking water criteria also suggests that sewage is the source of nitrogen contamination in groundwater. The highest nitrate concentration was detected in the Hellwan industrial area (El Arabi, 1999).

The Total Oil and Grease analysis indicated that hydrocarbons were detected at 60% of the monitoring sites. There are no data to indicate the chemical composition of hydrocarbons in groundwater, but they could include a wide range of petroleum hydrocarbons, and possibly a number of chlorinated solvents.

Management to reduce impacts of groundwater contamination

There are a number of societal, institutional and technical obstacles for protecting water resources from pollution in Cairo. Many of the problems are linked to a lack of resources for funding pollution control measures. Salaries are generally low in Cairo, making it difficult to recover costs for water resource and pollution management measures without the assistance of international aid. There is also a significant "brain-drain" from Egypt to the Gulf States and other countries of highly skilled and educated people capable of implementing and managing water pollution management measures. Additional problems in water resource management include (Myllylä, 1995):

- Poorly coordinated legislation, and inadequate enforcement of regulations.
- Little coordination between a large number of agencies involved in water resource management.
- Inadequate monitoring and reporting of water quality.
- Poor quality equipment.
- A very low awareness and understanding by the general community of environmental issues.

Legislation in Egypt is being progressively reformed. In particular, there has been a significant bolstering of environmental protection legislation, which is leading to a more unified approach to pollution control coordinated by the Egyptian Environmental Agency (EEAA). EEAA is trying to address the industrial wastewater disposal issue by tackling water pollution problems at their source. A number of strategies are currently implemented (Myllylä, 1995), including:

- Improving industrial housekeeping. In some industries, the amount of contaminated wastewater can be reduced by up to 70% and solid wastes minimised by the use of appropriate technology. Industries with heavily contaminated wastewater may require treatment plants to be installed to treat water prior to disposal.
- Identifying problems specific to particular industries. Particular industries in Egypt such as the food and textile industries have specific pollution problems that can be tackled in a coordinated way across the industry.
- Installing wastewater treatment plants. The total pollution load to the environment will be reduced by progressively installing wastewater treatment plants for processing industrial effluent.
- Encouraging local design of equipment and facilities. Local design of industrial equipment and facilities for wastewater treatment and better local training of its use can reduce the implementation costs of industry by up to 70% and free resources for other pollution control measures.

As well as improving pollution control measures, a new national environmental policy for managing water resources has been established which also sets national standards for protecting groundwater resources (El Arabi, 1999). The new policy has established processes for improving groundwater monitoring in the country, for aquifer vulnerability mapping and for implementing wellhead protection for water supply wells.

5.4   INTRODUCTION TO SANA'A, YEMEN

This section highlights the environmental impact of accelerated urbanisation on the groundwater ecosystem of Sana'a metropolis. The Sana'a basin ($3\,200\,km^2$) is an intermontane plain located in the central Yemen Highlands. It contains the flat alluvium covered Sana'a plain (some $500\,km^2$) surrounded by rocky outcrops of irregular topography rising to about $3\,000\,m$ above sea level (Fig. 5.1). Sana'a city (15° 21N, 44° 12E), the capital of Yemen, stands on the southern part of the Sana'a plain with elevation of about $2\,200\,m$. The population of Sana'a urban area has grown very rapidly, at average rates of 9 to 11% per year since 1975, reaching a total of 1.5 million in 2000. The urban population explosion has created major problems for government agencies responsible for providing adequate facilities to its citizens. As in other developing countries, urbanisation in Yemen is so rapid that planners have little time to make provision for this phenomenal growth. In addition political, institutional, and legal factors may exacerbate the problem. The rapid urban sprawl places demands on housing, transportation, water supply, and energy sources. Further, excavated land or old quarries are often turned into dumping grounds, which can pollute both water and the air. The population boom and industrialisation "urbanisation" can significantly affect the environment. With water resources for example,

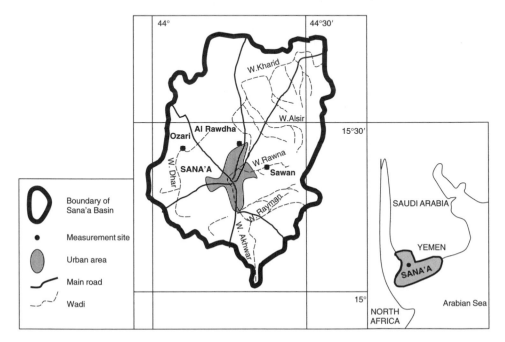

Figure 5.1.   Location map of the Sana'a plain.

there will be an increase in water demand and groundwater abstraction and a lowering of the water table and consequently, severe water shortage. Similarly trends can be traced for other resources. Sana'a is one of the fastest growing cities in Yemen. Rapid industrialisation, and consequent demographic changes, has created stress on civic amenities and outstripped the resources of the authorities to provide basic needs such as clean water. Thus, it is imperative that policy makers as well as donors should consider the basic infrastructure demands and environmental factors in urban planning.

Environmental impact is a product of population growth and consumption of resources, particularly land and water. The municipality's records show the existence of acute shortage in health services, educational facilities and other public services in Sana'a. The environmental situation in urban centres in Yemen is critical, reflected by: the high priority being given by the Government to environmental matters as evident from setting up of Environmental Protection laws, quality standards, protected areas, preservation of ecologically significant areas, archaeological sites etc. Waste disposal problems are also being addressed, focusing on urban sewage systems, refuse landfills and industrial and hazardous waste, and effluent from manufacturing industry.

The study assesses the current environmental aspects of urban groundwater and related basic services. Issues relating to urban hydrology, resource depletion, and water quality are outlined together with description of related basic services including storm drainage, domestic water supply, sanitation and handling of industrial and hazardous waste in regard to environmental aspects of groundwater.

5.5   HYDROMETEROLOGICAL ENVIRONMENT OF SANA'A

The climate in Sana'a is semi-arid with an average annual rainfall of 235 mm. Yemen's rainfall pattern is sparse and erratic in frequency, duration and distribution. Data from meteorological stations in Sana'a indicate that the monthly distribution of rainfall varies substantially from year to year at any location, and spatial distribution is characterised by high intensity rainfall events of short duration over limited areas. Two distinct rainfall seasons can be identified. The first season is a result of the Red Sea Convergence Zone (RSCZ) and occurs from March to May. During the month of June before the second rain season starts, there is usually a transition as the RSCZ retreats in front of a "monsoonal" Intertropical Convergence Zone (ITCZ), which dominates the Sana'a area between July and September. The moist air comes from the south (Indian Ocean) and converges with the north warm dry air from the north. During the dry season from October to March, the weather is characterised by a persistent dry easterly/north-easterly airstream. The Mediterranean effect reaches the Yemen once every few years.

Rainfall analysis over the period 1974–1993 indicated that there is a change in the contribution of each season to the annual rainfall. The earlier period was wetter and the second rainy season contributed more than during the recent period. The variability is higher for the second rainy season in the recent period, whereas the variability of the first season is becoming lower. The annual variation of rainfall is almost constant over the two periods (Alderwish, 1996).

The hottest month is July, with a mean monthly temperature of 22.5 °C and the coldest is December (14.1 °C). The highest observed daily air temperature was 34.5 °C on 4/7/87 and lowest was -9.6 on 20/1/87. Daily variation of air humidity is quite distinct with its maximum observed at night and minimum in the afternoon. The mean monthly relative humidity goes up to 51% in March and April, while during dry weather it drops to 35.8% (June). The mean monthly wind speed is small varying from 1.7 m/s in November to 2.3 m/s in July. The range of daily mean wind speed is 1.1–3.5 m/s. Due to the elevation of the Sana'a basin the mean monthly air pressure varies between 787.5 millibars (November) and 783.6 millibars (July). The mean sunshine hours observed monthly vary between 7.4 hours in August and 9.7 in January.

The Sana'a plain is drained by the Wadi As Syla, which is a tributary of Wadi Alkharid. Wadi As Syla runs from south to north through the centre of the city for 22 km. Smaller wadis from the mountains in the east and west flow into Wadi As Syla. The outflows of the Wadi As Syla are minimal and only after exceptional rainfall does it overflow into Wadi Alkharid.

A major feature of the natural environment are the mountains that surround the city. Despite the scenic attractions, the steep bare mountainside couples with intensive rainfalls of short duration, causing occasional serious damage due to sediment-laden flooding. These special geographic features impose certain difficulties on the design, operation and maintenance of road, storm, and foul drainage, and hence pose special environmental problems. The reliability of runoff data is poor for most of Yemen. Storm water drainage facilities are limited in Sana'a and where they exist, they often do not function properly. Hence during the rainy season, urban storm water causes flooding of the streets and wadis. The floods cause damage to the asphalt roads, shortening their useful life and requiring more maintenance. In addition to the suspended solids load swept into the town by the storm runoff from the surrounding mountains, the floodwater contains oil and domestic wastewater from overflowing cesspits. People open the covers to the sewers (where existing) and the resulting influx of storm water causes problems at the sewage treatment plant.

The environmental implication associated with poor quality of surface water runoff and road drainage is the infiltration of contaminants to the shallow groundwater aquifer along the Wadi As Syla. This is one of the main sources of urban recharge with annual average of $1.4 \times 10^6 \, m^3/y$ (Alderwish, 1996). The health implication is that the water related disease of malaria, which was of low prevalence, has become endemic due to more water bodies, increasing urbanisation, irrigation development and the mobility of residents and immigrants. There are malaria carriers in Sana'a and schistosomiasis is also present.

5.6   GROUNDWATER RESOURCES FOR SANA'A

The main source of groundwater in Sana'a is the highly transmissive Cretaceous Sandstone aquifer. The Cretaceous Sandstone is unconfined in the central part of Sana'a Basin, but overlain in the south and west by Tertiary and Quaternary Volcanics. Unconsolidated alluvium covers most of the urban area in the central plain, providing a shallow aquifer and acting as the main pathway for recharge to the deeper aquifers.

The dominant environmental problem not only in Sana'a but throughout Yemen is the overexploitation of groundwater resources. The Cretaceous Sandstone aquifer, where most of the present production wells for Sana'a public water supply are located, is heavily over pumped, with 80–90% of the pumped water being abstracted by the agricultural sector. The average pumped amount (over last 30 years) is in the order of $82 \times 10^6 \, m^3/y$ while the average recharge for the same period is $66 \times 10^6 \, m^3/y$. Groundwater flow models indicate that by 2005 both the eastern and western well fields will lose most of their present production capacity and that additional supplies will have to come from outside the Sana'a

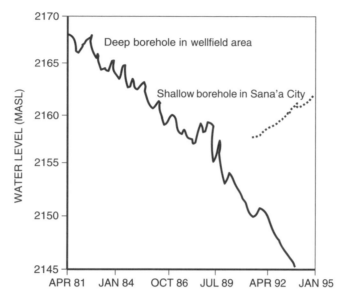

Figure 5.2. Hydrographs showing contrasting trends in water levels: regional decline and urban rise.

plain. Even if all uses of groundwater for irrigation were stopped immediately, the aquifers of the Sana'a plain could not sustain the water production to satisfy the basic needs of the growing population (Kruseman, 1996). Recently, there was a feasibility assessment for seven potential additional sources for Sana'a water supply (Alderwish & Al-Eryani, 1999). The main conclusions drawn from the study were: firstly, except for desalination, the development of any other water supply will be essentially a mining process; secondly, pricing, conservation policies, and comprehensive management in general are the keys to sustainability, and not the yield per se; and lastly, enforcement is a must for implementation, but it should also seriously consider the rural population's rights as users of water, rather than the owners, so that their economy can be sustained.

In contrast to the regional trend of declining water levels due to over-abstraction (4–5 m annually during the 1980s), groundwater levels in the shallow aquifer under the city are steady or rising (Fig. 5.2). This is a result of the continuous increase in urban recharge since the early 1970s following the rapid population growth and rising demand for water in Sana'a. In 1993, over 80% of the urban recharge infiltrated through cesspits ($13.6 \times 10^6 \, m^3$), with minor contributions from leaking pipes, wadi flows, industrial losses and irrigation return. Although infiltration from cesspits is a significant source of groundwater contamination, the potential impact (gains and losses) of that on the water resources management (including the pollution hazard) should be assessed while considering all factors, e.g. adequacy and efficiency of the existing sewerage system, limited resources, type of pollution and possible natural attenuation processes, and dilution within boreholes (Alderwish, 2000).

5.7   WATER SUPPLY FOR SANA'A

Historically, urban water supplies were obtained from dug wells and ghayls tapping the unconsolidated Quaternary deposits in the plain and minor wadis. The traditional way of life prevailing before the 1960s allowed a reasonable balance to exist between the amounts of water supplied from these sources and the city's domestic and agricultural requirements. Exploitation of deep aquifers for irrigation and municipal supplies started in the mid-1970s. The groundwater extraction has been largely uncontrolled.

Public water distribution and quality

To provide Yemen's urban areas with modern systems of central water supply and sanitation, the National Water and Sanitation Authority (NWSA) was established in 1973. At present the public water supply network has very strict quantitative limitations; it can only serve intermittently 40% of the households. The reasons for this are the limited availability of additional water sources, insufficient coverage of the distribution network, and insufficient financial means for expansion and maintenance. It is the intention of NWSA Sana'a branch to supply water to as many people as possible, but given the present state of the public network, the organisational/institutional conditions, and the expected urban growth, it is extremely doubtful whether the public network will ever reach 100% of the urban population.

The raw supply of water in Sana'a relies fully on the local groundwater resources. At the present, NWSA produces its water from the Eastern (10 wells) and Western (27 wells) wellfields plus a number of wells in Hadda (6 wells), Musaik (16 wells), and Assar (8 wells) areas. Only waters from the Western and Eastern well fields and from two wells in Musaik and two in Hadda areas are being chlorinated. Water from the other wells is pumped directly into the distribution network. The study by Dar Al Handasah (1999) showed that when the source of families' water supply or their method of sewage disposal were related to the incidence of diarrhoea, intestinal infections, and skin diseases over some areas in Sana'a, it was clear that pumping water without disinfecting had important health implications. Alderwish (2000) also showed that a relationship exists between the incidence of diarrhoea, intestinal infections and skin diseases, and public water supplies that have not been disinfected. The important implication of this conclusion is that as it is neither possible to cover the whole city by sewerage system (because of financial constraints) nor to abandon groundwater abstraction within the urban area (due to limited water resources), a feasible adaptive response is to disinfect all supplied water and hence avoid health problems due to biological pollution.

Although the distribution system is equipped with three reservoirs constructed as balancing reservoirs, the reservoirs cannot be filled because of the present imbalance between demand and supply. Consequently, this has led people to install tanks (below ground surface) to receive and store water. These tanks are not clean and usually have iron sedimentation at the bottom. Furthermore, some of these tanks have cockroaches, dead mice, algae, and other sources of pollution within.

At the source, water samples analysed from the well field in Sana'a indicate variable quality and even deterioration for some wells in the Western Well Field. However, the quality of the water provided by NWSA complies with WHO standards for safe drinking water. It is doubtful, though, that the quality at the tap can be considered safe. The reason for this is that the distribution network is operated on a rotational basis and consequently a deterioration of the water quality in the system is most probable. Al-Hamdi (1994) carried out research on groundwater pollution due to infiltration of domestic wastewater in the Sana'a city area. One of the objectives was to identify direct microbiological contamination of the distribution network. It was assumed that this was caused by the temporarily depressurisation of, and subsequent infiltration of domestic sewage into, the transport pipelines for drinking water.

No coliform bacteria were detected in any of the drinking water samples. This was attributed to the retention of bacteria by soil particles. Using data from NWSA files would provide a picture that is far from complete because monitoring is not done regularly or comprehensively. There are no NWSA standard requirements for the monitoring of public supply water. It would be a valuable step forward if some moves could be made in this direction as new public supplies are increasing in number and serving an increasing number of people – more investment in laboratory facilities is probably required. The ideal water quality assessment would be based upon a comprehensive set of samples taken simultaneously and correctly across the city, and analysed by the same laboratory by the same analyst.

Private water distribution and quality

Since 1980, as the public water supply system could not cover all of the urban population, a private water supply sector developed to fill the gap between the quantity of water supplied by NWSA and the city demand. All over the city, complementary water for domestic use is provided by the private sector, which has developed into an important economic activity.

Private water in Sana'a is provided via numerous small local distribution networks and a large fleet of water tankers (small trucks that can carry about $4\,m^3$). It is estimated that the private sector provides about 40–50% of the water used for domestic purposes in Sana'a (Kruseman, 1996). The public water supply is often available only once every week. Consequently the middle- and upper-income residents, who can pay the high purchase price, buy water from the private sector. So far, no comprehensive study of the private water supply sector has been carried out in Sana'a. The water for domestic use is distributed mainly via tankers and private house connections. It is estimated that about 2 800 truckloads are sold daily, and that 20 000 to 30 000 families are connected to the private distribution networks. Assuming that about 190 000 people are connected to private distribution networks, the average per capita consumption would be about 50 l/c/day.

Water samples analysed from private wells (Alderwish, 1992, 1996; Kruseman, 1996) indicate deteriorating water quality over a large part of Sana'a city (old city, Al-Qa'a, and Al-Hasaba) as reflected by the high values of EC (2 000 to 4 000 μS/cm). Cesspits that drain domestic wastewater to the shallow aquifer microbiologically pollute the Quaternary deposits. Depending on the subsoil conditions, the safe distance between the bottom of a cesspit and the intake level of a well should, for bacteria elimination, be between 75 and 285 m. Investigations carried out by SAWAS technical team revealed that distances as short as 7 m frequently occur (Foppen et al., 1996). Several of the samples from dug wells are polluted and have one or more constituents above the WHO permissible limit (Alderwish, 1993). As a consequence, about 85% of the private wells used for domestic water supplies provide water that is unsafe from a microbiological point of view, according to WHO standards.

Drinking water from private vendors

Private vendors provide "hygienic" drinking water in plastic containers of 10 and 20 litres capacity for about 60% of the urban population. Stations are equipped with various water treatment systems, which can include sand filtering, chlorination, reverse osmosis, etc. The quality of the water from these stations is very questionable. In 1997, a team from NWSA visited 14 stations and concluded that most of the stations are simply cheating the public, and the following was observed: chlorination was not being done or was done

Table 5.1. Public water supply and per capita consumption in Sana'a.

Year	Net public supply ($10^6\,m^3/y$)	Population	Population receiving public supply (%)	Per capita consumption (L/d)
1997	12 167 000	1 218 152	37	75
1998	13 500 000	1 322 061	35	80

Source: Statistical and planning department, NWSA.

improperly; most of the purification systems components were not working (left merely as decoration); the ground and pipes to fill containers were not clean; some stations had more than one raw water source; no expert technician was present in any stations to follow up the efficiency of the station and the quality of observed water; and finally, to date there is no comprehensive investigation nor regular and proper monitoring by any public authority.

Per capita water consumption

The capacity and type of the services provided by the public and private water supply systems indirectly control the present consumption of water. Per capita consumption for the population with public supply is estimated as 75 l/c/d in Sana'a (Table 5.1). These figures show one of the least per capita consumption in the world. The shortage of water supply is probably affecting social and hygienic conditions of the urban population and may cause several skin diseases. However, it could be argued that if the relatively cheap public water supply services are improved and if water became available on continuous basis, then the population may immediately start to use water at a rate that might not be sustainable in the long term. Consequently it may be justified to choose a policy aiming at a consolidation of the present low supply levels of the public network, but also to ensure that, even on an intermittent basis, water from the public system will become available to large portions of the urban population. This argument is left open for further future discussion.

The prevalence of private water vendors highlights the high and growing demand for water that is not currently being met by the municipalities, and the trend for the immediate future appears unchanged. Questions about water quality, both public and private, and the adverse health effects are cause for serious concern. The immediate introduction of standards for monitoring the quality of public supply of water is a priority. There should be monthly monitoring of NWSA wells, as well as regular monitoring of private vendors and private water supply sources. Given the adverse economic effect on the poor of their reliance on private distributors, it is suggested that a comprehensive study of the private water supply sector in Sana'a be undertaken. Priority should be given to remedial action programmes focusing on community health education, directed at women and children, regarding safe water-handling practices as well as sanitation and hygiene.

5.8   GROUNDWATER QUALITY IN SANA'A

The use of water depends upon its quality, which is defined by the physical, chemical and microbiological properties of the water. A perceptible or measurable increase in any component can be seen as contamination, provided quality remains within the limits for any particular application, whether for human consumption or some less demanding use. Once the accepted limits are exceeded, then the water can be regarded as polluted and has to be treated, or an alternative source found (Hem, 1985). Yemen's national water standards for drinking water quality and irrigation have been set by EPC and are mainly inspired by WHO 1984 guidelines in which there are tables relating to physical properties, microbiology, and inorganic and organic content.

Groundwater quality normally reflects water-rock interaction. That water represents the "baseline" water quality. However, even in undeveloped and sparsely populated areas where a natural baseline might be assumed to apply, there will be some man- and animal-derived additions, albeit at negligible concentration. Sana'a has been there for thousands of years and so some degree of local water quality degradation has, presumably, occurred for a very long time. Water resources are increasingly at risk of contamination from potentially polluting activities at the surface. Several previous studies showed that both surface water and groundwater have a large range of concentrations, indicating pollution mainly from urbanisation. Within the urban areas, two main sources of groundwater contamination can be identified; the existing sanitation and industrial waste handling. Several maps have been prepared to identify trends and anomalies for correlation (Italconsult, 1973; Mosgiprovodkhoz, 1986; Alderwish, 1992). Anomalous areas of high concentration are observed generally in zones that coincide with the city's main sewage route, densely populated areas and industrial areas. The plume area of contaminated groundwater is delineated by a zone of elevated conductance. The conductance of groundwater in the plume exceeds $1\,000\,\mu S/cm$ in a zone that is 2 to 3 km wide and 18 km long. It is widest at Sana'a city. The distribution of electrical conductivity indicates one main plume extending from the city toward the north (Bani Hawt) (Fig. 5.3). However, small local anomalies are also present along its longitudinal axis and coincide with the presence of additional sources of the contamination, namely, the textile factory, the maturation pond, and downstream flows of partially treated effluent at the discharge point at the end of the buried pipe near the corner of Sana'a airport (Alderwish & Dottridge, 1999). There are great risks of withdrawing contaminated groundwater, which is commonly found in highly populated areas, or the areas with water mains but no sewers within the Sana'a urban area. The problem is quite critical where extraction from the shallow aquifer is commonly the source for privately supplied water in un-serviced or poorly serviced areas of the city.

Environmental aspects of sources of contamination (current sewerage system and industrial waste collection and disposal) with regard to groundwater of Sana'a are described below.

Sanitation

Historically, solid human waste was collected in a "natal", a small room located below the bathroom. The solid waste was dried and used for either fertiliser or fuel in bathing-houses (hamam, traditional sauna). Liquid waste, washing water and cooking water were collected in cesspits with maximum depths of around three meters. The traditional forms of sanitation were well suited to the society with low water use that prevailed before the 1970s. The old forms of sanitation are rapidly giving way to water-flush sanitation as new water main supplies are developed in Sana'a. The interplay between sewage disposal and water mains supply is an important one, as domestic low water use is giving way to high water use. Where there is mains water supply but no sewers the traditional cesspits (despite becoming deeper than 15 m) become over-loaded hydraulically and are sources of

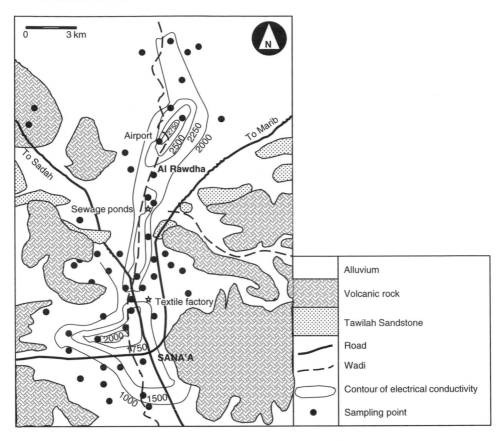

Figure 5.3.   Simplified geology and contours of electrical conductivity in the Sana'a basin.

Table 5.2.   Sewerage system coverage during 1997 to 1998 in Sana'a.

Year	Population connected to sewers (%)	Sewer-connected population also connected to public mains (%)
1997	9	25
1998	10	29

Source: Statistics and Planning Department, NWSA.

groundwater pollution. The sewerage system covers only part of the rapidly growing cities, and hence most of the population still uses the traditional on-site sanitation. Coverage of the sewerage systems in Sana'a during 1997 and 1998 is shown in Table 5.2.

In 1984, prior to installation of the treatment plant (sewer system), the return flow to the alluvial aquifer from domestic waste through cesspits was estimated to be $8.3 \times 10^6 \, m^3$ (Mosgiprovodhoz, 1986). In 1993, after partial provision of a sewerage system in the city, the return flow through cesspits was estimated to be $12.5 \times 10^6 \, m^3$ (Alderwish, 1996). In 1995, wastewater production in Sana'a was estimated as $23.5 \times 10^6 \, m^3$, of which $6.5 \times 10^6 \, m^3$ was collected and/or treated (Keetelaar, 1995). Eighty percent of the remaining production is believed to recharge the Quaternary deposits ($13.6 \times 10^6\,m^3$) by means of cesspits. These figures indicate that at least 2.5 times as much sewage is still being directed into the alluvial aquifer as is being discharged at Al Rawdha and that substantial pollution loads are being imposed on groundwater beneath Sana'a city. Not knowing the degree of interconnection between the alluvial aquifer, which is the receptacle for all the pollution the city can offer, and the underlying main aquifer strata, which supplies urban water, makes it difficult to assess the degree of seriousness of pollution in the main aquifers. Nevertheless, the continuously declining piezometric levels in the Cretaceous sandstone makes vertical penetration of pollutants more probable, induced by the downward hydraulic gradient, and this warrants more concern by NWSA who is responsible for supplying drinking water to the public.

Sewage treatment should remove most of the pathogens and suspended solids and add oxygen to the sewage, but does not significantly change salinity except as the result of evaporative concentration and production of nitrate by oxidation of ammonia and organic nitrogen. Recent analyses of sewage treatment outlets, (88 mg BOD/L, 532 mg COD/L, no data on microbiological quality), reflect the degree of treatment received by rudimentary facilities and overloaded and poorly managed maturation ponds. The effluent produced is hence not suitable for reuse except for limited agricultural purposes and is still a serious public health hazard.

A serious health and environmental problem exists today in the Nugom area with about 35 000 people, where cesspits overflow and drain their domestic waste to the streets. This situation is due to the existence of an impermeable bed at shallow depths that hinders vertical drainage of sewage water. The contamination of groundwater resources in these areas are further being aggravated by locals who pour acids into cesspits, after emptying, to kill bacteria which are believed to clog the porous wall of the cesspits.

With regard to waterborne microbiological contaminants, the access of wastewater to domestic water supplies is the main reason for the presence of waterborne diseases. Waterborne diseases occur when pathogenic bacteria and viruses are transported passively through domestic water distribution mains, followed by ingestion of the pathogens by humans and animals. Intestinal bacterial pathogens known to have occurred in contaminated drinking water supplies include strains of *Salmonella, Shigella*, pathogenic subspecies of *Escherichia coli, Vibrio cholerae, Yersina enterocolitica*, and *Campylobacter fetus*. These organisms can cause problems from mild gastro-enteritis to severe and sometimes fatal dysentery, cholera or typhoid. In addition, organisms naturally present in the environment that are not regarded as pathogens may occasionally cause diseases. Such organisms in drinking or even bathing water may cause a variety of infections including those of the skin, and also mucous membranes of the eye, ear, nose and throat among people whose defence mechanisms are impaired.

In a recent microbiological groundwater quality study conducted in Sana'a (Foppen et al., 1996), it was found that more than 80% of the well samples far exceeded the WHO guideline value for un-piped drinking water quality. Faecal coliforms were detected in 50% of the wells, ranging from 10^2 to 9.2×10^4 bacteria per 100 mL. In nearly 11% of the sampled wells *Salmonella* ssp. contaminated the water. Microbiological contamination is not likely to be limited to the Quaternary unconsolidated deposits, since drilled wells in the Tertiary Volcanic and Cretaceous Sandstone were also polluted. The study concluded that there is a rough correlation between the distance from cesspit to well and the number of bacteria. Considering the amount of microorganisms in domestic wastewater, the amount of infiltrating domestic sewage is likely to cause sanitary groundwater problems. At present around 250 wells abstract some $13-17 \times 10^6 \, m^3$ in the Sana'a city area (Alderwish & Dottridge, 1998; Siddiq et al., 1996).

Industrial and hazardous waste

The degree of hazard posed by waste depends mainly on the amount and nature of chemicals disposed or "emitted", while the effect on plants and animals (including humans) depends on the fate of a chemical in the environment. Industrial and hazardous waste can impact the environment in three principal forms; air pollution, water pollution and solid wastes. In the following, major hazardous waste producers in Sana'a together with the major media with which the contaminants are associated are briefly described. The available information indicates that wastewater has a number of generators (ACE, 1993).

Although in Sana'a there is some kind of zoning, it seems that it either came too late or it is not strictly enforced as reflected in the presence of industrial sites in the middle of residential areas. An old large industry, Yemen Textile Corporation Factory, was established at the boundary of what was then the city. Now however it is in the middle of the city. Other more recent smaller and medium-scale industries (e.g. small food and soft drink manufacturers) also exist within the city. It is understood that there is a plan for moving these small-medium industries out of city centre.

Currently only domestic waste is collected in Sana'a by the municipality. Some industries take care of their own waste and incinerate it in burning pits. Solid sludge and hazardous wastes from other industries are dumped together with other domestic waste on open dumpsites. Both private and municipal landfills have inadequate undersealing, drainage and controls on dumping. These sites present risks to the environment and public health, to the scavengers living and working among wastes, and for contamination of both watercourses and shallow aquifers. Liquid hazardous waste is disposed into the sewerage system or, as is the case for some industries, disposed with domestic wastewater, which is discharged into the surroundings without any treatment. There are almost certainly many other industry discharges which are connected illegally to sewers, some of which could badly affect biological treatment of ordinary sewage.

Previous studies conducted in Yemen (Haskoning, 1991; ACE, 1993) provide useful background information on hazardous waste management. Several hazardous waste producers and the nature of their waste have been described, however, without detailed quantitative information on many of the potential hazardous pollutants or the mitigation measures. Hazardous waste being produced within Sana'a area include: waste oil produced by commerce and vehicle servicing; halogenated solvents from dry cleaning services; cyanides from the production of polyurethane sponge; acids and alkalis from small laboratories within factories; batteries disposed of indiscriminately; sludge from factories. All of these are without any information on their composition or their quantities and hence it is difficult to assess how serious is their environmental threat. Nevertheless, pollution by hazardous chemicals is very harmful to humans and the environment, even at very low concentrations. Moreover, these chemicals are very long-lived and very hard to clean up from the environment. For example, if a hazardous contaminant reaches groundwater it will move in the direction of groundwater flow in the shallow aquifer and probably be intercepted at some time in an area where groundwater is used.

Hospital waste (hazardous and non-hazardous) is generally stored in drums, collected by the municipalities and dumped somewhere on landfills. Although public hospitals have their own furnaces to incinerate the waste, only the TB clinic in Sana'a is using its facilities. The number of hospital beds in Sana'a city in 1993 was estimated as 1 753. Using similar average solid waste production of 635 kg/y for every hospital bed (Haskoning, 1991) 1 100 tons/y of hospital waste is generated in Sana'a. These figures exclude wastes produced by private clinics and hospitals. It should be noted that, if separation of hazardous (pathological chemicals and radioactive waste) from non-hazardous waste is carried out at hospitals, the hazardous waste quantities could be reduced by 90%. The chemical waste from laboratories connected to hospitals, e.g. X-ray departments, blood banks and so on are disposed into the sewerage system (about 9 tons/y in Sana'a). Among these chemicals, there is cyanide, chromate, phenol and mercury. All are very toxic chemicals. Larger quantities of disposed chemicals are acids, alkalis and methyl blue. Major pollution control problems associated with hospital waste handling are: the potential spread of infections via contamination of workers using poor procedures to handle hospital wastes; and potential contamination of soils, surface waters, and groundwater due to inadequate storage and most probably inadequate disposal of hospital wastes.

The amount of *pharmaceutical waste* is difficult to assess due to illegally imported medicines. In Sana'a there are about 215 pharmacy shops. Some of the illegal operators indicate an average annual generation of 20 kg/y of expired medication. Attempts to estimate the amount of pharmaceutical waste in Sana'a indicate that about 30 tons/y is generated from suppliers of medicines and the pharmaceutical industry (YEDCO). Major pollution control problems associated with pharmaceutical industries are potentially soil and groundwater contamination brought on by disposal of expired medicines or pharmaceutical process residuals.

For *factory waste*, the textile factory in Sana'a is one of the worst, with hazardous pollutants released to the urban air, water and soil environment. The factory produces 15 million yards of finished textiles annually and produces several thousands m^3 of wastewater believed to be contaminated with a range of hazardous polyaromatic hydrocarbons, cyanide, heavy metals and paint solvents, together with mineral oil. This mixture is collected in three open drains and, after cursory settlement in four lagoons, is discharged at a point 400 meter north of the factory into a dry stream channel (Wadi As Syla). Some farmers use this water for irrigation (Alderwish, 1993). Further major pollution control problems associated with textile factories are concerned with their improper wastewater management. The Yemen Textile Corporation Factory and its environs need urgent attention with respect to its environmental acceptability, and immediate action could be taken by installing a properly designed and managed treatment plant to remove the hazardous chemicals before release from the factory.

Another factory producing liquid wastes of similar or less significance is a paint factory. It produces effluent containing oil and solvents that is discharged locally (infiltrated) after receiving little treatment. A further factory has its own treatment plant but unfortunately it is seriously overloaded and poorly managed. There is evidence of considerable infiltration in the vicinity of the plant. The effluent is acidic with a COD of over 12 000 mg/L but probably contains no non-biodegradable substances. The owner is aware that his present effluent disposal arrangements are inadequate. The major pollution control problems associated with plastics and chemical industry are threefold. They are: firstly, direct discharge of wastewater from plastics forming operations; secondly, soil and potential groundwater contamination caused by spills and unlined; and lastly, uncontrolled operating areas associated with paint manufacturing facilities. The contaminants may include lead, zinc, cyanide, oil and solvents among others. The pollution control problems associated with food and beverage industry include conventional organic contaminants (high BOD5 and TSS loading). The nature of the sludge produced is unknown, and using this sludge on food chain crops potentially impacts the health of farmers and food consumers. Uncontrolled onsite dumping grounds are used for the disposal of some industrial solid and semi-solid wastes, which also causes problems, although there is no information available on quantities.

Photographic waste includes photo laboratories' developing and printing chemicals, mainly solvents, silver and acids. The number of photo laboratories is still limited, but is growing. In Sana'a there are 10 laboratories using about 40 000 L/y (Haskoning, 1991). The chemicals are, without exception, disposed into the sewerage system. The major groundwater pollution control problem associated with photographic, printing and publishing industry is the potential contamination from discharge of untreated wastewater containing printing inks, solvents, silver, and acids.

For *petrol and diesel waste*, there are numerous petrol and diesel filling stations existing in Sana'a. Although the possibility of leakage of new tanks is small, they may cause environmental risk in the urban areas where slow, long-term leakage of fuel from buried tanks, sufficient to contaminate groundwater to the extent of only 0.1 mg/L, can render the resource unusable for a long time after the leak has been stopped (Edworthy, 1991). Major oil storage tanks, especially where they are located over an aquifer or adjacent to public water supply facilities, should be surrounded by a fuel-tight retaining wall or bund, impounding sufficient capacity to hold the contents of the tank in the event of spillage. The floor of the bunded areas should also be sealed to prevent infiltration of oil.

5.9   SUMMARY OF THE ISSUES FOR SANA'A

Information for Sana'a indicates the high vulnerability of urban groundwater resources and the serious dimensions of the water crisis in the country. This is primarily manifested by decrease and shift in rainfall distribution, scarcity of water resources, depletion of groundwater, low efficiency of water use, and low coverage of water and sanitation services.

The flushing, transport and infiltration of urban storm runoff is contaminating the shallow alluvial aquifer and will require attention in the long term. Particulate matter and oil can be readily intercepted in properly designed "traps" which can be constructed as the urban roads are paved. For most of the city, road drainage will be collected by the storm drainage system and discharged externally. The critical nature of the water resources issue in Sana'a is evident from:

- Uncontrolled pumping from private wells for irrigation; industrial water use and water sales to urban residents.
- Decreases in groundwater levels of 4 m/y in the Sana'a aquifer during the 1980–1990s period.
- Widespread occurrence of water-borne and water-related diseases resulting from poor quality and unmonitored private water supplies.
- Diminished access to traditional community supply sources, such as shallow wells in alluvial aquifers, and contamination of such sources through poor sanitation and waste disposal practices.
- Conflicts over water allocation both locally between users and regionally between water resource development sectors, particularly agriculture and water supply agencies.

It is obvious that water supply sustainability requires immediate political and technical attention. Groundwater resources planning and management plus supporting legislation and implementation of effective controls on land use, water extraction and waste disposal are essential if Yemen is to overcome its worsening water crisis.

Sanitation and waste disposal practices at the personal, community and officially organised levels leave much to be desired. The Yemen authorities responsible for waste disposal are well aware of the problems but funding and technical support to remedy them are presently lacking. The generally poor states of environmental health are integrally linked with existing water supply and community, household and personal sanitation practices. The high prevalence rates of water-borne and water-related diseases are direct manifestation of these poor sanitation and hygiene standards. Traditional community water supply and household waste disposal methods cannot be maintained at the urban centres without ensuring environmental health problems. The severity of the existing environmental health problems is recognised by both Yemeni and international health authorities. Remedial action programmes are needed, focusing on community health education and on improving water supply, sewerage systems and waste disposals at all community levels. From an environmental viewpoint, the prevailing health problems must be a first priority. Due recognition of this aspect must be included in all major development projects, particularly those involving upgrading or expansion of municipal water supply or sewerage systems, and reuse of sewage treatment plant effluent.

The environmental implications of industrial and hazardous waste include:

- Municipal sanitary landfill sites with inadequate undersealing, drainage and controls on dumping. These sites receive solid, sludge and hazardous wastes, often including hospital wastes, and represent risk of contamination to both watercourses and/or shallow aquifers.
- Several industrial discharges are connected illegally to sewers, and can badly affect biological treatment of ordinary sewage, as well as polluting the effluent from the treatment works.
- Poor waste disposal and chemical storage practices at industrial sites, which are poorly located relative to local drainage or recharge zones.

5.10 CONCLUSIONS

In rapidly urbanising cities in countries such as Yemen or Egypt, there is poor integration of environmental concerns with economic growth, particularly with regard to groundwater. It should occur at the planning stage and focus on improving the health and welfare of residents. Measures to protect the urban environment and improve public health must be primary objectives any infrastructure project within urban areas. This requires that environmental impacts are assessed early in the design, and that due regard is given to the impacts on groundwater. Additional runoff and infiltration, the disposal of wastewater, and the presence of land uses with significant volumes of polluting chemicals, all put groundwater at risk. The hazards must be actively removed at the design stage, rather than waiting for pollution to occur.

The present wastewater collection and treatment facilities in developing cities jeopardise the environment, public health and the quality of their groundwater. On-site facilities (cesspits) are overloaded by the trend to greater volumes of domestic water use and population increases. This may be even more important in specific areas where water logging due to excessive infiltration of wastewater in the past has weakened the subsurface and endangered the stability of buildings, often in areas with poor and limited income. Western-style sewage collection and treatment schemes are expensive and disruptive to install. They rarely expand fast enough to keep pace with the often unplanned growth in the city and its population. They require good maintenance and control of the quality of inflows if they are to be effective in treating wastewaters. Even so, they produce effluents which have to be disposed of, and which can effectively export scarce water resources from a city in an arid environment. A new paradigm of water supply and disposal is needed for rapidly urbanis-ing cities, especially in arid zones.

The reuse of domestic wastewater from urban centres and the recycling of other human wastes in agriculture can produce significant economic benefits and help defray the large costs of municipal waste management. Reuse of domestic wastewater could be an impor-tant strategy for conserving water resources in areas suffering from major water shortages. The recycling of human wastes to add nutrients to, and improve the physical quality of, the soil is an ancient practice. In its modern form, the reuse of wastewater effluents for irriga-tion of crops offers attractive benefits, such as increasing water supplies for productive agricultural use, adding valuable fertilisers and micro-nutrients to maintain soil fertility, and reducing pollution of water resources. However, the use of domestic effluent requires strict quality standards which should be adhered to before reusing in agriculture. In par-ticular, industrial inputs into sewage systems must be strictly controlled, a difficult task when factories may be unlicensed and when illegal sewer connections are easy to make and conceal.

The present hazardous solid and liquid waste disposal jeopardises the urban environ-ment, public health and quality of groundwater below cities. Spillage of industrial chem-icals is a worldwide problem. The growth of industry and the upgrading of hospitals and other urban infrastructures is likely to exacerbate these problems. Consequently, to achieve a sustainable urban development in parallel with the extension of the health ser-vices and industries, care should be taken now to monitor and improve handling and man-agement of hazardous waste and chemicals. Environmental impact assessments should be conducted for all new industry establishments regardless of their scale. Initiating manage-ment programmes for reducing or preventing the generation of waste during production processes or other operation would be a first step to an economically and environmentally sound way of dealing with hazardous wastes. As complete elimination cannot be realised, ways to recycle the wastes should be sought. When reduction or recycling of hazardous wastes prove unsuccessful, treatment of waste should be required and may be done either through incineration or proper storage.

Groundwater monitoring is critical to water resources planning and management. It is also integral to environmental protection, and provides feedback on the management of waste disposal sites. Similarly, groundwater monitoring plays an important role in deter-mining the continuing viability and effectiveness of wastewater reuse programmes, a tech-nique which could be an integral component of water conservation in any country in the long term in selected areas. A comprehensive groundwater monitoring programme is needed, which includes delineation of responsibilities for data collection, sample and data analyses, and reporting. This programme should cover standardisation of sample collec-tion, analysis and recording of monitoring data. It should also assign centralised responsi-bility for the monitoring database, including basin geology and land use, well locations, aquifer data, pumping records and water quality data.

Urban groundwater monitoring networks are specialised and should be rationalised to include:

- Shallow aquifers in areas serviced by water mains but relying on cesspits for waste-water disposal.
- Random sampling of wastewater flows in drains and sewers in proximity to industrial areas and from treatment plants.
- Additional analyses for major chemical ions, heavy metals, organic and biological and pathogenic contaminants.
- Additional sampling sites or networks providing feedback relating to wastewater reuse, both treated and untreated, and managed and unmanaged schemes.

The focus for any country's water resource development should be on environmental protection: that is, water quality, water conservation, catchment management, and resource sustainability. In arid environments, water supply sources are often groundwater and mainly used for irrigation, sometimes to the detriment of other users. Here detailed planning for groundwater protection should be an essential part of the environmental assessment of most projects. Mitigating measures relating to protection of water quality and improving public health must be included in all water resource projects and any other projects impacting on surface or ground water resources.

CHAPTER 6

Urban areas of sub-Saharan Africa: weathered crystalline aquifer systems

Richard G. Taylor, Mike H. Barrett and Callist Tindimugaya

6.1   INTRODUCTION

Urban populations in sub-Saharan Africa are growing at a faster rate than in any other region in the world. The supply of water to many rapidly urbanising areas of sub-Saharan Africa depends upon groundwater supplied by manually pumped wells, spring discharges, and production boreholes equipped with motorised pumps. Studies of the quality of urban groundwater in this region are few so that the impacts of urbanisation on groundwater quality remain poorly defined. Although it is clear that further research is required to establish current conditions, this chapter reviews the available evidence of groundwater contamination in urban areas of sub-Saharan Africa from the few existing studies (e.g. Longe et al., 1987; Faillat, 1990; Malomo et al., 1990; Uma, 1993; Nkotagu, 1996; Barrett et al., 1998) and draws from detailed hydrogeological investigations conducted in rural areas (e.g. Chilton & Smith-Carington, 1984; Houston & Lewis, 1988; Wright & Burgess, 1992; Taylor & Howard, 1994; Chilton & Foster, 1995; Taylor & Howard, 1998a, 2000). Recommendations for further study are also discussed.

A number of factors contribute to the lack of published data concerning urban groundwater. One of the most striking characteristics of sub-Saharan cities is their comparative newness. Mature urban centres in many high-income countries have histories of infrastructure (sewerage and water supply) and industrial development that extend well beyond the last century. Whilst there is a long history of urban settlement in sub-Saharan Africa, most large cities have a colonial origin, and rapid urbanisation is predominantly a late 20th century phenomenon. Therefore, chronic problems associated with long-term, urban groundwater usage observed in other parts of the world, such as over abstraction and contaminant loading, either have not occurred or are at an earlier stage (Foster et al., 1996). The demand for applied research into groundwater is, thus, more recent than in regions where urban development of groundwater has a longer history.

A lack of resources in sub-Saharan Africa is another key hindrance to the study of urban groundwater issues. When basic needs of citizens are not met, dedication of limited resources to areas that do not manifest in an immediate and concrete service requires careful argument. Osibanjo (1982) clearly articulates the situation in many low-income countries:

"The plight of the scientist receives no sympathy from government which sees no particular reason why special budget funds should be allocated to research that may not yet have a proven social value."

Donors and organisations assisting recipient governments in the development sector often share this view, that development and applied research are mutually exclusive. For example, groundwater development schemes routinely fail to collect and archive even basic data such as drilling logs or hydrochemistry, despite their profound utility and the fact that such activity is well recognised as intolerable in high-income countries. The lack of detailed information on groundwater quality in the urban regions of sub-Saharan Africa is particularly alarming in view of the rate of urban population growth and the reliance by many communities, often low-income, on untreated groundwater for domestic use.

This chapter is divided into five main sections. The background to urban development in sub-Saharan Africa, including a historical overview of urbanisation and the characteristics of urban areas, is presented (Section 6.2), followed by discussion of the risk posed to groundwater by urban development in sub-Saharan Africa (Section 6.3). An overview of the hydrogeological environment is then presented in Section 6.4 that draws heavily from research in Uganda. This section summarises current knowledge of aquifer vulnerability and the potential for contamination, and highlights gaps in our current knowledge. A series of recent case studies linking urban groundwater quality and sources of contamination is then reviewed (Section 6.5). The final section highlights the need for further research, particularly in relation to the competing interests of providing sanitation and developing groundwater.

6.2   URBANISATION IN SUB-SAHARAN AFRICA

Urban areas in sub-Saharan Africa are characterised by phenomenal rates of growth that have occurred primarily during the latter half of the 20th century and into this millennium. Urban settlement in sub-Saharan Africa has, however, a history that extends back to medieval times. In southern Africa, the city of Great Zimbabwe is believed to have had a population of around 10 000 people. Curiously, the demise of this settlement in the 15th century is thought to have resulted from over population (Garlake, 1982). In West Africa, archaeological evidence from the 11th century reveals the capital of an empire in the vicinity of present-day Ghana that was inhabited by between 15 000 and 20 000 people (Gugler & Flanagan, 1978). The empires of Mali and Songhay emerged during the 15th and 16th centuries and featured cities such as Timbuktu and Gao with populations estimated at 75 000 people. Dispersal of people following the fall of the Songhay Empire to the Moors led to the development of settlements like Kano in present-day northern Nigeria. By the 19th century, cities of 25 000 inhabitants had been established at Ile-Ife in Yorubaland (now southern Nigeria), at Abomey in Dahomey (now Benin) and at Kumasi in Ghana.

Colonisation of sub-Saharan Africa had a significant and lasting impact on the location and character of urban settlements. Colonial regimes introduced new foci of control, a more extensive economy, and drew (or forced) men to work in ports, mining towns, railway towns and factories (Keuper, 1965; O'Connor, 1983). Urban settlements became centres for commercial activity exporting raw materials and agricultural produce, and importing manufactured goods (Gugler & Flanagan, 1978). Consequently, ports have grown and become capitals (Fig. 6.1) while former inland capitals like Timbuktu have shrunk dramatically. In the interior of east Africa, Nairobi developed in the late 19th century as a result of the building of the Mombasa-Uganda railway and became the capital of the British East Africa Protectorate by the early 20th century. Despite rapid urbanisation during the latter half of the 20th century (discussed below), urban centres in sub-Saharan

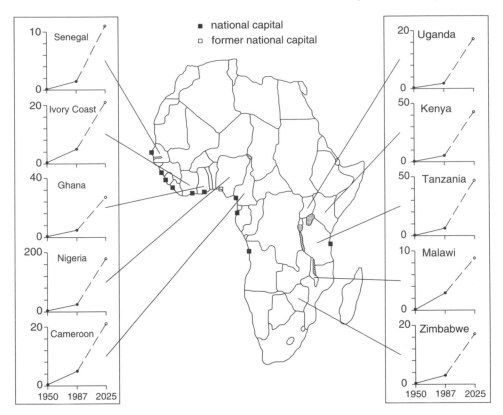

Figure 6.1. Urban population growth in sub-Saharan Africa. Plots of urban populations in millions versus time: 1850, 1987 and 2025 (project, dashed line) (United Nations, 1987 cited in Harris, 1990).

Africa continue to be largely administrative, commercial and transhipment centres. With the possible exception of Lagos, no industrial city has yet emerged.

Sub-Saharan Africa is one of the least urbanised regions in the world with less than one in three inhabitants living in an urban setting (Clark, 1998). Despite this current statistic, Africa is the only region where rates of population growth have not started to decline (Marter & Gordon, 1996). Urban growth, driven by both high rates of population growth and rural-urban migration in sub-Saharan Africa (White, 1989), is, at 8%, the most rapid in the world. In Figure 6.1, data from Harris (1990) are plotted which show this dynamic. In Nigeria, an urban population of 3.5 million in 1950 grew to 25 million in 1987 and is projected to be 179 million by the year 2025. An urban population in Cameroon of 0.4 million in 1950 rose to 5 million in 1987 and is forecasted to be 21 million by 2025. In Malawi, the urban population was 0.1 million in 1950, increased to 1 million by 1987, and is estimated to be 8.7 million in 2025. Globally, predictions of national population growth have since been revised and are generally lower than rates predicted in the early 1990s due, in part, to persistently high rates of infant mortality and the AIDS pandemic (Gardner-Outlaw & Engelman, 1997). Nevertheless, it is clear that stresses placed on urban environments in sub-Saharan Africa and their groundwater resources will be severe during the next few decades.

Rapid population growth and rural-urban migration have resulted in a proliferation of low-income, often illegal, settlements in many sub-Saharan African cities (e.g. Gugler &

Flanagan, 1978; Memon, 1982; El Sammani et al., 1989; Stren, 1989; Ebohon, 1996). Over the last few decades, the disparity between growth in the economic sector and the increase in the urban population has led to shortages in both conventional housing and employment opportunities in the "formal" sector of the city. The shortfall is made up, in part, by low-income settlements that are indifferent to environmental and construction guidelines but provide affordable shelter and an informal economy (Memon, 1982). A further characteristic of sub-Saharan African urban environments, and one that distinguishes them from mature cities in other parts of the world, is the presence of subsistence agriculture. Access to arable land in urban areas, no matter how small, provides the possibility of better living. In Kampala, Uganda, it has been estimated that 30% of all residents take part in urban agriculture. The largest urban agriculture sector is keeping livestock. In 1993, it was estimated that 70% of all chickens and eggs consumed in Kampala were produced within the city (IDRC, 1993). Features such as subsistence agriculture and a comparative lack of paved surfaces in relation to mature high-income cities increase the potential for rainfall-fed or direct groundwater recharge. This is particularly the case in the peri-urban areas that comprise the urban-rural interface.

6.3   URBAN GROUNDWATER ISSUES IN SUB-SAHARAN AFRICA

Urban development of groundwater in sub-Saharan Africa can be usefully divided into low-intensity abstraction ($<0.2\,L\,s^{-1}$) and more recent, high-intensity abstraction ($>2\,L\,s^{-1}$). High-intensity groundwater abstraction for piped supplies is less common but achieved using wellfields consisting of one or a series of high-capacity boreholes equipped with motorised pumps. Low-intensity groundwater use occurs via manually pumped boreholes and shallow wells, as well as dug wells and springs, and is widespread in humid, urban areas of sub-Saharan Africa due to the rapid pace of urbanisation and the inadequacy of centralised, reticulated water supplies. The pace of urbanisation has led to the incorporation of previously rural or peri-urban environments including their associated infrastructure into urban centres. There has also been a large migration into urban communities of rural dwellers, who are accustomed to using groundwater without charge and are, therefore, reluctant to pay for discharges from a centralised piped water supply. The inadequacies of urban, reticulated water supplies typically include the limited distribution of the pipe network, particularly in low-income settlements, and the unreliability of the service, which prompts the need for alternative sources.

A key, contemporary issue in the urban environment is the fact that, in addition to low-intensity abstraction, sewage disposal is also largely decentralised with disposal occurring on site, primarily through pit latrines. It is unclear whether this is a transitionary phase or whether decentralised groundwater abstraction and sewage discharge will remain a permanent feature of sub-Saharan cities. In the short term, conditions of rapid population growth mean continued reliance is placed upon the sub-surface environment as both a source of potable water and a receptacle of excreta, resulting in contamination of water points from sewage becoming an ever-increasing likelihood. The benefit to public health from the provision of on-site sanitation facilities is well demonstrated (Esrey et al., 1991) but a potential conflict remains between provision of on-site sanitation facilities and use of groundwater for potable water supplies (Foster, 1985; Gelinas et al., 1996).

Alternative forms of sanitation that involve containment and treatment can be problematic in sub-Saharan Africa where existing infrastructure is overtaxed and resources are limited.

In Dar es Salaam, Tanzania, Stren (1989) estimated that in 1985, only 10–15% of the population were connected to sewerage. Of the remaining population, septic tanks were used for liquid waste in mid to high-income areas, and sanitation in low-income areas consisted ostensibly of pit-latrines. Regular evacuation of septic tanks requires operational, emptying trucks and treatment facilities. In 1984, most trucks used by the city council in Dar es Salaam were no longer in working condition so that only 8% of the estimated daily production of 6 million litres of liquid waste could be removed (Stren, 1989). Uncollected and untreated sewage represents a threat, like pit latrines, to the quality of urban groundwater resources.

Refuse collection faces similar challenges. Without reliable refuse collection, households are resigned to depositing waste in open spaces. This presents an immediate risk to human health and enhances the possibility of contaminating underlying groundwater. Onibokun (1999) notes that landfills and dumps in Africa commonly occur in low-income areas where the population is less able to prevent landfill in their backyard, is likely not to benefit from waste collection, and invariably depends upon local, unprotected sources of water.

6.4   HYDROGEOLOGICAL REGIMES

Characteristics of weathered crystalline aquifers

In weathered crystalline rock environments, groundwater is transmitted by fractures in the bedrock and unconsolidated weathered materials that form the overlying mantle, or regolith (Fig. 6.2). Permeability within the bedrock and overlying weathered mantle derives from

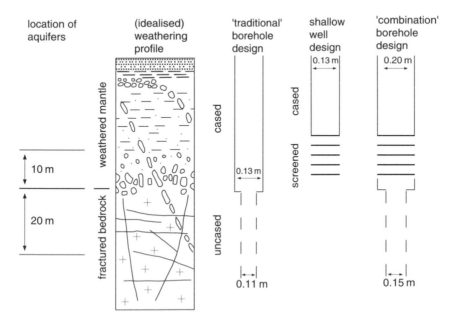

Figure 6.2.   Cross-sectional (vertical) representation of weathered crystalline aquifer system including the location of aquifers and well designs.

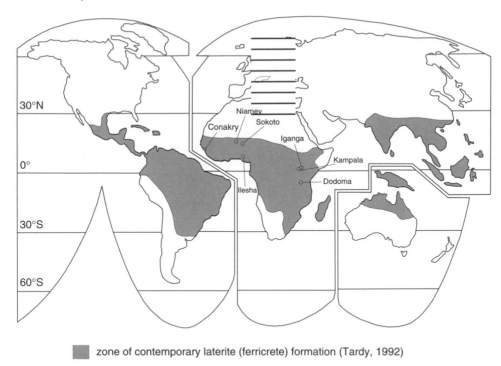

zone of contemporary laterite (ferricrete) formation (Tardy, 1992)

Figure 6.3.   Distribution of deep weathering environments (indicated by laterite development) and location of urban centres discussed in Section 6.5.

the prolonged, in situ decomposition of bedrock (McFarlane, 1991a; Nahon & Tardy, 1992; Thomas, 1994; Taylor & Howard, 1998b). As a result, weathered aquifer systems occur in areas where weathering has been undisturbed by Pleistocene glaciation or, significantly, by aeolian erosion. This includes large areas across equatorial South America, Africa and Asia (Fig. 6.3). The distribution of weathered crystalline rock environments is reflected in a map of contemporary iron-rich duricrust (laterite) formation prepared by Tardy (1992). In sub-Saharan Africa, the bedrock comprises primarily Precambrian "basement" rocks that include granite and gneiss (Key, 1992; Houston, 1995). Younger sedimentary terrains featuring limestone and sandstone occur in central and southern Africa, such as the Democratic Republic of Congo and Zambia, as well as coastal regions of west Africa, but are not the focus of discussion here.

Due to a primarily in situ development, mantles of weathered rock constitute the progressive degradation of bedrock materials and are therefore anisotropic. The thickness of the mantle is regularly in the order of tens of metres and depends upon a wide variety of factors which include tectonic setting (i.e. duration of weathering), bedrock lithology, and climate (i.e. humid versus arid conditions) (Ollier, 1984; Wright, 1992; Thomas, 1994). The lithology of a weathered mantle derived from gneissic bedrock, and prominent weathering reactions associated with its development, are summarised by Taylor and Howard (2000) in Figure 6.4. The summary is based upon detailed textural and mineralogical analyses of weathered profiles in Uganda. At the base of the profile, primary bedrock minerals such as mica and plagioclase are altered to mixed-layer (2:1) clays and kaolinite.

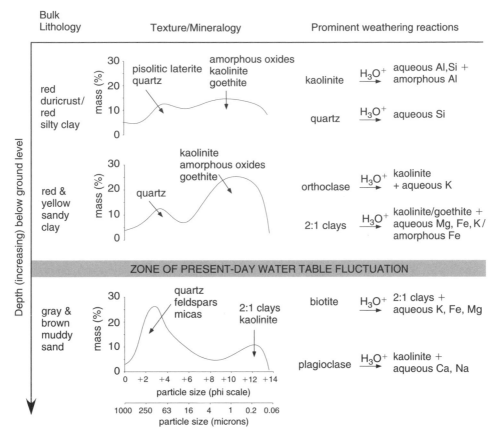

Figure 6.4.   Lithology, texture and prominent weathering reactions for a weathered mantle derived gneissic bedrock (adapted from Taylor & Howard, 1999a).

More aggressive weathering at shallower depths within the unsaturated zone transforms remnant primary minerals and mixed-layer clays into kaolinite and hydrous iron and aluminium oxides. Relatively stable constituents above the water table, quartz and kaolinite, are weathered towards the top of the profile and iron-rich duricrust (ferricrete or laterite) remains. The chemical aggression required to degrade kaolinite and quartz is not reflected in porewaters that typically exhibit a pH of between 5 and 7. Indeed, the weathering dynamics in the unsaturated zone of weathered mantles remain poorly understood. An improved understanding is likely to be found in the role of microorganisms, identified by several authors (Brimhall et al., 1991; Hiebert & Bennett, 1992; Vandevivere et al., 1994), in generating microenvironments of low pH and supplying complexing agents.

The lithological work in Uganda reveals three key characteristics about weathered mantles:

- Progressive leaching of metals is indicated from the base of the profile to its upper boundary. Hydrolytic processes play a central role in the development of weathered mantles. Furthermore, hydrochemical and geochemical evidence shows that development of iron or aluminium rich duricrusts does not depend upon absolute retention of these less mobile metal species (McFarlane & Bowden, 1992; Taylor & Howard, 1999a).

- The texture of the weathered mantle features a bimodal particle-size distribution as sand-sized, primary minerals are progressively weathered to clay-sized, secondary minerals at shallower depths. The poorly sorted texture of the weathered mantle has important implications: localised variations in the parent-rock matrix give rise to high degree of spatial heterogeneity in weathered mantle lithology; and, as recognised by McFarlane (1992), secondary structures of the bedrock such as quartz stringers are translated into the composition of the weathered mantle producing preferential pathways for subsurface flow and contaminant transport.

- Less aggressive weathering is associated with saturated conditions and the persistence of coarse-grained materials at the base of the weathered mantle. Lithological studies in Malawi (McFarlane, 1992) and Malaysia (Eswaran & Bin, 1978) support this observation and indicate similarly that the aquifer in the weathered mantle comprises a poorly sorted, muddy-sand.

Fractures in the underlying crystalline bedrock (Fig. 6.2) tend to be both subhorizontal and discontinuous. Groundwater exploration strategies in crystalline terrains (Boeckh, 1992; Greenbaum, 1992) commonly attempt to locate regional and hence, more continuous, fracture systems in the bedrock. Packer testing (Howard et al., 1992; Karundu, 1992; Taylor & Howard, 2000) and detailed geophysical studies (Houston & Lewis, 1988) indicate that fracture density generally increases towards the surface. This is consistent with the suggestion of other authors (Davis & Turk, 1964; Acworth, 1987; Wright, 1992) that fractures typically derive from decompression (i.e. sheeting) as a result of the removal of overlying rock in solution and erosion of pre-weathered (unconsolidated) material at the surface. Double (outflow) packer testing of fracture zones in the bedrock shows that transmissivities are typically less than $1\,m^2\,d^{-1}$ (Fig. 6.5) and that one or two highly productive zones account for the bulk transmissivity of the bedrock. The effective base of the bedrock aquifer typically rests between 60 and 80 m below ground level.

Low-intensity (predominantly rural) development of weathered crystalline aquifers

Development of groundwater resources within the weathered mantle and fractured bedrock of sub-Saharan Africa has, to date, occurred primarily for low-intensity, hand-pump supplies. Provision of potable water to the predominantly rural population of sub-Saharan Africa has utilised groundwater because, apart from its potability, it is more widely distributed than surface waters and is more reliable than rainfall collection schemes such as rooftop catchments. Traditionally, development schemes have targeted bedrock fractures because of the ease of well construction in consolidated formations and the belief that fractures underlying the unconsolidated weathered mantle are better protected from surface contamination. This approach also derives from the fact that water-well drilling has, in the distant past, been conducted in tandem with geochemical exploration of bedrock formations.

The bedrock aquifer is, however, characterised by highly variable, but generally low, well yields (Bannerman, 1973, 1975; Chilton & Smith-Carington, 1984; Houston & Lewis, 1988; Wright, 1992). As indicated by the results of pumping tests and packer testing in water-supply boreholes in eastern and southern Africa (Table 6.1), aggregate values for the transmissivity of bedrock fractures vary over three orders of magnitude with mean (geometric) values between 1 and $10\,m^2\,d^{-1}$. The thickness of the zone of fracturing varies

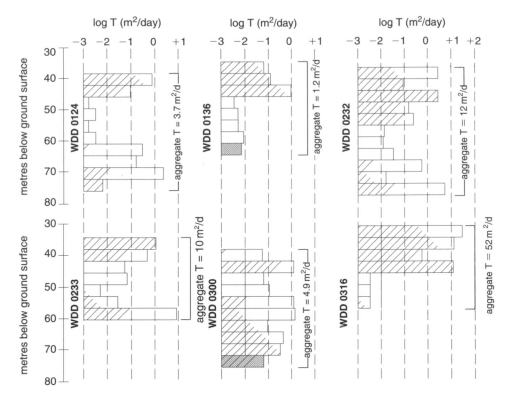

Figure 6.5.   Vertical, cross-sectional profiles of transmissivity (plot of $\log_{10}T$ versus depth) in the fractured-bedrock aquifer at 6 sites in central Uganda (Taylor & Howard, 2000).

Table 6.1.   Variability in the transmissivity of fractured crystalline rock (Taylor & Howard, 2000).

T range ($m^2 d^{-1}$)	T mean[1] ($m^2 d^{-1}$)	Number of sites	Method	Location	Reference
5–60	N/A	25	pumping tests	Botswana	Buckley and Zeil, 1984
0.8–80	5.5	60	pumping tests	Zimbabwe	Houston and Lewis, 1988
0.9–90	7.3	69	pumping tests	Zimbabwe	Houston and Lewis, 1988
0.07–250	1.1	22	packer testing	Uganda	Howard et al., 1992
0.3–32	3.3	17	pumping tests	Uganda	Tindimugaya, 2000

[1] geometric mean.

considerably but is typically between 20 and 40 m. Incidence of borehole failure is high and results, in part, from the poor storage characteristics of fractured crystalline rock.

Sustained abstraction from bedrock fractures has long been thought to depend upon leakage from the overlying, more porous weathered mantle (Chilton & Smith-Carington, 1984; Kafundu, 1986; Acworth, 1987; Houston & Lewis, 1988; Wright, 1992; Barker et al., 1992; Owoade, 1995). Recent studies (Sekhar et al., 1994; Taylor & Howard, 1998a, 2000; Tindimugaya, 2000) have noted that during prolonged pumping of groundwater from bedrock fractures, drawdown in the borehole was regularly observed to stabilise, implying the contribution of groundwater from an adjacent formation, (i.e. leaky-aquifer conditions). In Uganda, drawdown has also been detected in the weathered mantle following abstraction from bedrock fractures and observed to continue for more than an hour after abstraction ceased (Taylor & Howard, 1998a; 2000; Tindimugaya, 2000). This observation confirmed that leakage from the weathered mantle had occurred and demonstrates that the weathered mantle and fractured bedrock form an integrated aquifer system. Because of the discontinuous nature of bedrock fractures derived from decompression (sheeting), borehole failure can be expected under circumstances where there is no access to groundwater storage either in the weathered mantle or via the regional fracture network.

Hydrogeological studies of the weathered mantle in eastern and southern Africa yield estimates of transmissivity (Table 6.2) that generally vary less than the bedrock but give similar mean (geometric) values of between 1 and 10 $m^2 d^{-1}$. It is important to note that these estimates are a function of the tested depth interval (i.e. screened section) within the weathered mantle and not the entire thickness of the aquifer. The installation of shallow wells or piezometers is biased toward the upper (saturated) portion of the weathered mantle due to the problems of collapsing formations in the unconsolidated mantle during well construction, particularly at depths beneath the water table. Acknowledging that coarser, less weathered materials are found at greater depths, the actual transmissivity of the weathered mantle is expected to be greater than that indicated by tests conducted in shallow wells (Taylor & Howard, 2000). Nonetheless, field data fail to uphold the claim that the fractured bedrock represents a zone of high permeability and the weathered mantle constitutes a zone of low permeability (e.g. Rushton & Weller, 1985; Hazell et al., 1992), but rather support the assertion that the most productive horizon with the weathered aquifer system occurs at the base of the weathered mantle (Foster, 1984; Jones, 1985). Water balance calculations in Uganda (Howard & Karundu, 1992; Taylor & Howard, 1996) clearly indicate that the weathered mantle transmits the major portion of the recharge flux.

The preceding discussion regarding the characteristics of weathered aquifer systems has been based on studies of crystalline rocks such as gneiss, granite and amphibolite. It should be noted, however, that significantly more complex hydrogeological environments derive from the weathering of low-grade metasedimentary rocks such as phyllites and low-grade quartzites as well as volcanic rocks. For example, the north shore of Lake Victoria in east Africa is underlain by a suite of metasedimentary rocks known as the Buganda-Toro complex. Prolonged weathering of underlying rocks from the Triassic to mid-Cretaceous followed by stripping of much of the unconsolidated, weathered mantle from the mid-Cretaceous to early Miocene (Taylor & Howard, 1998b) gave rise to a group of

Table 6.2.   Variability in the transmissivity of the weathered mantle (Taylor & Howard, 2000).

T range ($m^2 d^{-1}$)	T mean[1] ($m^2 d^{-1}$)	Number of sites	Location	Reference
1–20	5.5	134	Livulezi, Malawi	Chilton and Foster, 1995[2]
0.2–5	2.1	81	Dowa, Malawi	Chilton and Foster, 1995[2]
1–60	5.2	64	Masvingo, Zimbabwe	Chilton and Foster, 1995[2]
2–10	3.4	27	Masvingo, Zimbabwe	Chilton and Foster, 1995[2]
0.2–40	4.6	6	Malawi/Zimbabwe	Chilton and Foster, 1995[2]
0.4–170	4.8	40	Mukono, Uganda	Taylor and Howard, 1995

[1] geometric mean; [2] references therein.

prominent mesas that are underlain by quartzitic material which is relatively more resistant to weathering than surrounding rock types. This geomorphic history has not only determined local relief, and its related effect on surface hydrology, but has also produced a subsurface that is characterised by alternating bands of highly weathered phyllite and fractured quartzite. In addition to posing a serious challenge to water-well drillers, this environment features an extraordinarily complex hydrogeology characterised by localised shallow flow systems and spring discharges. The presumption of consistent parent rock material, which may be applicable to cratonic areas and amenable to generalised discussion of hydrogeological characteristics, clearly does not apply in these terrains.

High-intensity (predominantly urban) development of weathered crystalline aquifers

High-intensity groundwater abstraction from weathered aquifer systems is a relatively recent phenomenon. At present, groundwater abstraction in the order of $5–30\,L\,s^{-1}$ is planned and developed in a number of regions of sub-Saharan Africa (e.g. Uganda, Ghana and Malawi) in an effort to supply potable water to urban communities without the high costs of treatment that are required for surface-water fed schemes. This is despite the fundamental uncertainty as to whether pumping of this magnitude is viable in the long-term (Taylor & Howard, 2000). The limited capacity of the weathered aquifer system is, to some extent, recognised and it is generally accepted that high-intensity abstraction will depend upon contributions from the weathered mantle. Indeed traditional borehole designs (Fig. 6.2) drawing from bedrock fractures rarely provide yields of greater than $0.5\,L\,s^{-1}$. Revised, combined high-capacity well designs do not leave a hydraulic connection to the overlying weathered to chance mantle but impose a connection within the open conduit of the borehole (Fig. 6.2). As such, they are referred to as "combination boreholes" as they attempt to exploit the hydraulic conductivity of bedrock fractures and the large storage capacity of the unconsolidated, weathered mantle. Combination boreholes have been demonstrated to be capable of providing yields of between 1 and $3\,L\,s^{-1}$ (Taylor, 1996, unpublished report). To meet anticipated urban demands of between 5 and $30\,L\,s^{-1}$, it is necessary to combine the discharge of several "combination boreholes" and invoke the concept of wellfields, such as the siting of several production boreholes in the same area.

Of fundamental importance to the planning of high-intensity groundwater abstraction is the finding of Taylor and Howard (2000) that the hydrogeological characteristics of weathered aquifer systems are a function of the dominant geomorphic process operating on the land surface. Work in Uganda (Taylor & Howard, 1998b) has demonstrated that the evolution of weathered environments can be resolved in terms of tectonically controlled cycles of deep weathering (in situ biogeochemical erosion) and stripping (colluvial and fluvial erosion). Deep weathering is dominant on tectonically stable land surfaces whereas on uplifted drainage basins, the land surface is primarily stripped. As shown in Figure 6.6, deep weathering is achieved by a recharge-dominated hydrology which serves to weather rocks more rapidly than weathered products can be removed by limited surface runoff. This hydrological regime supports a regional aquifer in the weathered mantle. Stripping results from a runoff-dominated hydrological system in which weathered products are removed more quickly than they are generated by intermittent recharge. An aquifer of highly localised extent subsequently occurs within a thinner weathered mantle. Because of the dependence of high-intensity abstraction on groundwater in the weathered mantle (see above), high-intensity abstraction is only possible in areas of deep weathering where a

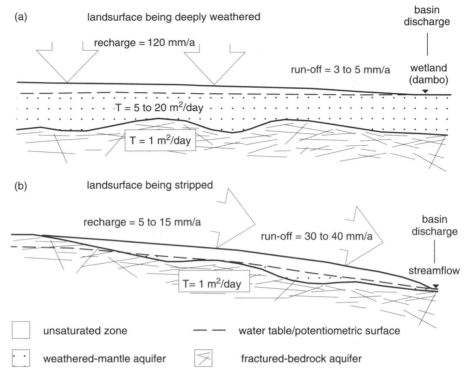

Figure 6.6.   Conceptual, cross-sectional model of the hydrological and hydrogeological character-istics of land surfaces being deeply weathered (a) and stripped (b) (Taylor & Howard, 2000).

regional aquifer in the weathered mantle enables high rates of pumping and favourable recharge characteristics can sustain the groundwater discharge.

The magnitude of estimated recharge across sub-Saharan Africa varies considerably (Table 6.3). A common characteristic of humid environments in sub-Saharan Africa is, however, the timing of recharge events. Evidence from stable-isotope tracers (Adanu, 1991; Geirnaert et al., 1984; Mbonu & Travi, 1994; Taylor & Howard, 1996, 1999b; Tindimugaya, 2000) and borehole hydrographs (Taylor & Howard, 1999b) shows that recharge events occur during periods of monsoonal rainfall when rainfall temporarily exceeds the evaporative flux. The dependence of the recharge flux on days of heavy rainfall (>10 mm day^{-1}) has been demonstrated by Taylor and Howard (1996) in Uganda. There is, consequently, a strong potential for surface or near-surface contaminants to be "flushed" into the weathered aquifer system along preferential pathways by monsoonally derived recharge events. The pattern of increased diarrhoeal illness during the rainy season is com-monly observed in sub-Saharan Africa. In the Gambia for example, Barrell and Rowland (1979) attribute massive increases in thermotolerant (faecal) coliform and faecal strepto-cocci concentrations in groundwater that coincide with the onset of the rainy season, to the flushing of faecal material of mixed human and animal origin into the subsurface.

Measures to protect groundwater resources typically involve the imposition of land-use restrictions in order to prevent the release of potential contaminants in the vicinity of

Table 6.3.   Estimates of groundwater recharge in sub-Saharan Africa.

Location	Estimate (mm yr^{-1})	Method(s)	Reference
Ghana	172	BS[1]	Asomaning, 1992
Kenya	50–60	Soil infiltration	Singh et al., 1984
Tanzania	10–16	Chloride tracers	Nkotagu, 1996
Tanzania	84–133	SMB[2]	Sandström, 1995
Uganda	17	SMB	Howard and Karundu, 1992
Uganda	200	SMB/flow modelling	Taylor and Howard, 1996
Uganda	120	BS/SMB/borehole hydrographs	Taylor and Howard, 1999b
Uganda	113–161	SMB/flow modelling	Tindimugaya, 2000
Zambia	80–281	SMB	Houston, 1982
Zimbabwe	12	Chloride tracers	Houston, 1990

[1] BS: baseflow separation; [2] SMB: soil-moisture balance.

groundwater-fed waterpoints. Definition of the area around a waterpoint that contributes to its discharge (i.e. capture zone) is, however, complicated by the preferential pathways along which recharge reaches the waterpoint. Subsequent delineation of wellhead protection areas (WHPAs), based upon an understanding of the waterpoint's capture zone, is compounded further by the contrasting mobility and survival of contaminants (e.g. nitrate, faecal streptococci, iron, organic solvents). Nevertheless, rough guidelines for protecting low-intensity (handpump) abstraction have been proposed (Lewis et al., 1980) and are the subject of active research (Taylor & Howard, 1995; Barrett et al., 1998; BGS, 2000).

No published information relating to the protection of high-intensity abstraction from weathered crystalline aquifers is currently available. High-intensity, simultaneous abstraction from the weathered mantle and fractured bedrock through combination boreholes will, nevertheless, feature complex capture zones. Drawdown within the bedrock generates significant, local hydraulic gradients that favour downward leakage from the weathered mantle, whereas drawdown in the more porous weathered mantle will induce gentle gradients over a larger area. Further study of the response of weathered aquifer systems to prolonged high-intensity abstraction is required before capture zones can be deduced with any precision. Delineation of WHPAs are necessary, however, to reduce the likelihood that borehole discharges are contaminated by local land-use, especially considering that rehabilitation of polluted aquifers is expensive and generally impractical (Foster, 1985). In the rapidly urbanising areas of sub-Saharan Africa, WHPAs are likely to be a source of future conflicts between urban resource management and the development of urban infrastructure, if imposed.

6.5   GROUNDWATER QUALITY

A significant volume of data has been gathered regarding groundwater quality in weathered crystalline aquifers in sub-Saharan Africa. Most data have, however, been collected in rural areas. This bias is understandable given the historic demography of sub-Saharan Africa, and the reliance of rural communities on untreated groundwater for domestic use. Mindful of the rate of urban growth in sub-Saharan Africa, urban groundwater quality is becoming an issue of concern due to the reliance on untreated groundwater in urban and urbanising areas, particularly in low-income settlements. It is also clear that many urban centres have little or no alternative to groundwater as a supply of water to support continued urban growth. Fortunately, natural groundwater in the weathered crystalline aquifers of sub-Saharan Africa is generally of potable quality (Clark, 1985; Taylor & Howard, 1994; Chilton & Foster, 1995). The principal threats to the potability of groundwater in this environment are discussed below and include undesirable levels of natural parameters, associated with the weathering of crystalline rock, and sewage contamination. In this section, discussion of rural groundwater quality is presented in order to provide a baseline for uncontaminated or natural levels of dissolved species, and to infer the effects of urbanisation based on the few case studies published to date.

Rural groundwater quality

Rural environments are relatively unaffected by anthropogenic inputs, and the hydrochemistry of groundwaters is largely determined by biogeochemical interactions that occur between infiltrating recharge and porous materials in the subsurface. In the crystalline terrains of sub-Saharan Africa, these materials commonly derive from granite and gneiss which are host to a wide range of mineral assemblages but typically include feldspars, micas and quartz. Prolonged weathering that is responsible for the development of weathered crystalline aquifer systems involves mineral alterations that are primarily hydrolytic and, so, are characterised by the leaching of metals from the substrate (Fig. 6.4). Hydrogeochemical data from Uganda (Taylor & Howard, 2000) presented in Table 6.4 and Figure 6.7 clearly demonstrate this dynamic. Table 6.4 provides a comparison of the concentration of dissolved metals in infiltrating waters, indicated by monsoonal rainfall (source of recharge), and groundwaters from two representative sites. For most parameters, an order of magnitude increase in aqueous concentration is associated with migration through the unsaturated zone to the water table. Figure 6.7 includes elemental geochemical data, plotted logarithmically, for the substrate (i.e. weathered crystalline rock) at the two sites where the groundwater samples were collected in Table 6.4. Leaching of Na, Ca, Mg and K from the weathered mantle is clearly demonstrated by depleted levels above the water table in the unsaturated zone. Although there is slight enrichment in the concentrations of Al and Fe in the weathered mantle, hydrochemical evidence in Table 6.4 indicates that these less mobile metals are not entirely retained. This is significant because despite relatively innocuous cations (Na^+, Ca^{2+}, Mg^{2+}, K^+) being the main aqueous products of weathering processes, it is less mobile trace metals such as iron, aluminium, chromium and arsenic that have been found in concentrations exceeding WHO (1993) drinking-water quality guidelines in groundwaters from weathered aquifer systems in Nigeria, Ghana, Malawi and Uganda (McFarlane, 1991b; McFarlane & Bowden, 1994; Taylor & Howard, 1994; Tijani, 1994; Smedley, 1996).

Table 6.4.   Hydrochemistry of meteoric waters in the central Uganda (from Taylor & Howard, 2000). All values are given in $mg\,L^{-1}$.

Meteoric water	Na	Ca	K	Mg	Al	Fe	NO$_3$	Cl	HCO$_3$	SO$_4$
Monsoonal rain[1]	0.2	0.9	0.6	0.4	0.09	0.09	0.6	0.9	1.2	0.6
Groundwater (Owang 6)	46	9.9	5.9	4.9	0.08	0.68	1.1	5.1	55	83
Groundwater (Apac MW)	6.7	30	12	7.5	0.30	0.12	4.4	1.8	89	36

[1] results derived from the median of 6 monsoonal rainfall samples.

Concern over significant concentrations of As, Cr and Al in water supplies arises from their toxigenic effects, which include a variety of physiological disorders that affect skin as well as cardiovascular and nervous systems. By contrast, high Fe concentrations ($>0.3\,\mathrm{mg\,L^{-1}}$) are an aesthetic concern as iron discolours water and gives it a "salty" taste.

metal concentration in the weathered mantle

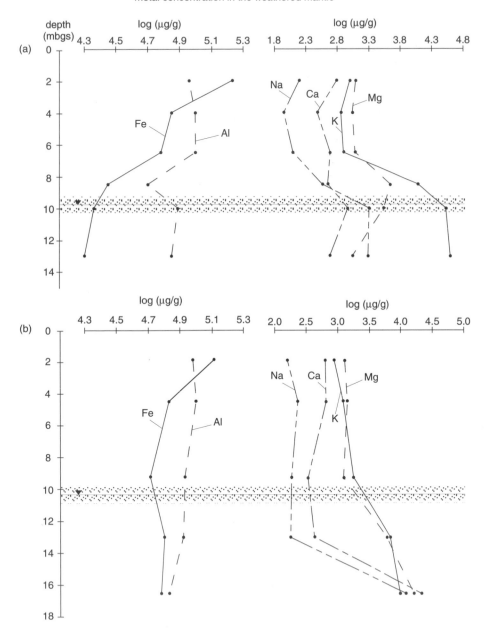

Figure 6.7.   Concentration of Na, Ca, K, Mg, Al, and Fe in the weathered mantle with depth at 2 sites, (a) Apac MW and (b) Owang 6, in central Uganda (Taylor & Howard, 1999a).

These should not be disregarded since the perception of consumers is critical to the successful adoption of safe water sources. The health benefits of consuming a safe source will be lost if users revert to less aesthetically offensive but biologically polluted surface-water sources. The mobility of these trace metals in the subsurface is not well explained by physicochemical (inorganic) models and there is growing evidence that microbial agents play a key role not only in creating low pH microenvironments to enable leaching, but also in providing complexing agents to enhance mobility. Indeed, McFarlane (1991b) demonstrates that standard sampling techniques, which include filtration and acidification of collected waters, remove Al and Cr from solution so that the quality of water being analysed differs from that being consumed. Filtration may also remove metals (and other species) that are bound to colloidal particles.

The possibility of abstracting groundwaters with undesirable, though natural, metal concentrations is clearly a factor that must be considered when developing groundwater from weathered crystalline aquifers. Hydrochemical evidence from Uganda (Taylor & Howard, 1994; Taylor, unpublished data), presented in Table 6.5 and Table 6.6, shows, furthermore, that there is little difference in the quality of groundwaters taken from the weathered mantle as compared with the fractured bedrock. Median values in Table 6.6 suggest that concentrations of iron are lower in the bedrock but the incidence of sites with iron in excess of WHO aesthetic guidelines is 35% and 50% for the bedrock and weathered mantle respectively. Targeting one of the aquifers in the weathered mantle or fractured bedrock (i.e. depth-specific abstraction) for development is, therefore, unlikely to reduce the chances of encountering undesirable trace metal concentrations.

Apart from undesirable levels of natural species, anthropogenic contamination of groundwater by domestic sewage has been identified in rural settings of sub-Saharan Africa (Langenegger, 1981; Foster, 1985; Taylor & Howard, 1994). In the absence of a significant mineral source of nitrogen in crystalline terrains, concentrations of nitrate exceeding WHO (1993) drinking-water quality guidelines ($45\,mg\,L^{-1}$) in weathered aquifer systems are presumed to arise from the mineralisation of organic nitrogen present in sewage. In each of the rural studies conducted in Botswana, Nigeria and Uganda, the source of nitrate in

Table 6.5.   Major ion hydrochemistry of rural groundwaters from central Uganda. Data are represented as medians and given in $mg\,L^{-1}$.

Aquifer	Sites	HCO_3	Br	Cl	NO_3	SO_4	Na	K	Ca	Mg
Weathered mantle	12	93	0.1	21	0.1	9	33	7.2	21	6
Fractured bedrock	50	182	0.2	24	2.2	17	32	5.6	42	12

Table 6.6.   Trace-element hydrochemistry of rural groundwaters from central Uganda. Data are represented as medians and given in $\mu g\,L^{-1}$.

Aquifer	Sites	Al	Ba	B	Cd	Cr	Co	Cu	Fe	Pb	Mn	Ni	Sr	U	Zn
Weathered mantle	12	101	120	6	<10	<10	<10	<5	345	<20	125	<10	140	<0.1	60
Fractured bedrock	43	<100	100	<5	<5	<10	<10	<5	<100	<20	<10	<10	190	<0.1	21

groundwater was unconfirmed either by the presence of microbiological indicators or the application of nitrogen isotopes. Each study did, however, identify a strong spatial correlation between denser human settlements and hence, an increased density of sewage sources, thereby linking elevated nitrate to domestic sewage. Concern over high nitrate concentrations in groundwater rests primarily in their risk to human health but also in its role in the loading of nutrients to surface waters (eutrophication). On its own, high concentrations nitrate and its less oxidised form, nitrite (NO_2^-) are linked to the development of stomach cancers and methaemoglobinaemia, a condition in which the ability of blood to absorb oxygen is impaired. In the case of methaemoglobinaemia, it is restricted to infants consuming contaminated water so the risk is effectively minimised if babies are breast-fed.

Impacts on groundwater quality aside, the dumping of domestic sewage into the subsurface via pit latrines is an effective, low-cost means of disposal in areas where water-borne sewage collection is not feasible. In a general sense, it is well recognised that improvements in sanitation facilities as well as the quality and quantity of water supplies reduce pathogen transmission. Indeed, urban sanitation has historically been neglected relative to improvements in water supply in low-income cities, yet improvements in excreta disposal may have a greater health impact than improvements in water quality and are the most pressing environmental and public health need in many developing countries (WHO, 1997). Clearly, there is a trade-off between the competing demands of providing sanitation and domestic water supplies from groundwater, as poorly designed or improperly located on-site sanitation facilities pose a risk to the quality of groundwater-fed sources.

In crystalline rock environments, the deeper, fractured bedrock aquifer is often assumed to be less susceptible to sewage contamination than the shallower aquifer in the weathered mantle (Acworth, 1987). Clearly, the weathered mantle aquifer is positioned more closely to surface sources of contamination, such as pit latrines. It should be recognised, however, that recharge to the fractured-bedrock aquifer occurs via the weathered mantle where preferential pathways such as remnant quartz stringers exist and can greatly facilitate subsurface flow and contaminant transport. Leakage from the weathered mantle to the fissured bedrock is, furthermore, enhanced by the development localised vertical gradients caused by groundwater abstraction from the bedrock. Recognition that weathered mantle and fissured bedrock form an integrated aquifer system entails that the bedrock aquifer is similarly susceptible to contamination.

Impacts of urbanisation

Many African capital cities, such as Lusaka, Dodoma and Conakry, use groundwater as their primary source of water. The demand for water of large urban centres requires a productive source such as the karst dolomite aquifer that underlies Lusaka. In contrast, weathered crystalline rocks that extend across much of sub-Saharan Africa, form a relatively weak aquifer system. Development of this aquifer system for piped water supplies is, therefore, primarily directed at smaller conurbations where water demands of $<30\,L\,s^{-1}$ may feasibly be achieved. Published studies of urban groundwater quality in sub-Saharan Africa are few and primarily feature investigations into sewage contamination of groundwater. There is an absence of data on other urban impacts on urban groundwater quality, such as the presence of industry. This situation has arisen from a combination of factors. Firstly, microbial contamination poses a greater short-term threat to health than industrial contaminants. Secondly, there is presently a lower level of industrial development in

sub-Saharan Africa than in other parts of the world, and facilities to analyse industrial contaminants such as volatile organic compounds and heavy metals in water samples have
typically not been established. One notable exception is Nigeria, where Nwankwoala and
Osibanjo (1992) demonstrate the widespread distribution of organochlorine pesticide
residues in surface waters. Nevertheless, current lack of comprehensive monitoring and
the perception that sources of contamination other than sewage are of low priority means
that there are no baseline data from which to gauge industrial contamination. Regional
summaries of available knowledge pertaining to urban groundwater quality in the weathered crystalline aquifers of sub-Saharan Africa are presented below.

West Africa

In the city of Ilesha in southwest Nigeria (population 600 000) where there is significant
dependence upon hand-dug wells for domestic water supplies, Malomo et al. (1990)
examined the relative influences of anthropogenic activities, such as agriculture and the
disposal of human and animal wastes, together with bedrock lithology on shallow groundwater chemistry. Ilesha is underlain by weathered crystalline rocks including biotite
gneiss, amphibolite, quartzite and quartz schists. 86 dug-wells penetrating the weathered
mantle were sampled and half of these exhibited nitrate concentrations exceeding $50\,mg\,L^{-1}$.
Plotted contours of major and minor ion cut across geological boundaries and were more
clearly associated with anthropogenic factors as highest solute concentrations occurred
beneath areas of highest population density. Data in Table 6.7 indicate elevated concentrations of nitrate and chloride in urban areas relative to rural sampling sites (Table 6.5 and
Table 6.7). The ratio of nitrate to chloride in shallow groundwater from urban sites suggests that recharge derives, in part, from sewage (Morris et al., 1994).

In Sokoto, Nigeria, Uma (1993) sampled 40 dug wells completed within the weathered
mantle. 12 of these were located in the urban area, and the depth of water ranged from 5–15 m
below ground level. Nitrate contamination is clearly demonstrated with higher concentrations observed at the end of the rainy season. This is similar to observations in The Gambia
(Barrell & Rowland, 1979), where elevated concentrations of sewage-derived bacteria were
detected in shallow wells during the rainy season. In Sokoto, however, Uma (1993) found no
correlation between depth to water table and nitrate concentration. Comparisons with historic data suggest that nitrate concentrations have risen since 1985. Nitrate concentrations
in rural wells outside Sokoto are high (mean: $75\,mg\,L^{-1}$) and may reflect the application of
fertilisers.

Gelinas et al. (1996) investigated the impact of well construction on groundwater quality from a shallow aquifer comprising detrital sand that underlies two districts of Conakry
in the Republic of Guinea. The districts, Bonfi and Hafia-Mosquee, are characterised by
high population densities, limited facilities for sanitation and waste collection, and inadequate piped water distribution systems. Under these conditions, shallow groundwater was
abstracted from open, unlined hand-dug wells and used for domestic water supplies. The
quality of this supply was, however, poor and attributed to the crude well design that was
unprotected from contamination at the ground surface. Wells were rebored and cased with
concrete rings. Metal covers and concrete plinths were added and an enclosure was set up
around the well, though pulley systems were still used to abstract water. Despite these
measures, the quality of shallow groundwater in the dug wells remained poor featuring
high nitrate concentrations and thermotolerant (faecal) coliform counts (Table 6.7).

Table 6.7. Urban case study groundwater quality data. All data are median values (mg L^{-1}).

Study area	NO$_3$	Cl	SO$_4$	HCO$_3$	δ^{15}N (‰)	TTC[1] (cfu/100 mL)	FS[2] (cfu/100 mL)	Reference
Ilesha, Nigeria	35	34	2.8	28	–	–	–	Malomo et al., 1990
Ilesha[3], Nigeria	1.5	15	1.8	12	–	–	–	Malomo et al., 1990
Sokoto, Nigeria	50	8.1	7.2	156	–	–	–	Uma, 1993
Ivory Coast	69	24	–	–	+13	–	–	Faillat, 1990
Niamey, Niger	42	–	–	–	+14	–	–	Girard and Hillaire-Marcel, 1997
Bonfi[4], Guinea	164	53	23	29	–	15 000	1 960	Gelinas et al., 1996
Bonfi[5], Guinea	140	60	23	37	–	7 200	3 195	Gelinas et al., 1996
Hafia-Mosquee[4] Guinea	90	34	19	18	–	9 200	1 385	Gelinas et al., 1996
Hafia-Mosquee[5] Guinea	91	41	14	21	–	20 950	2 850	Gelinas et al., 1996
Dodoma, Tanzania	41	238	83	296	–	–	–	Nkotagu, 1996
Iganga, Uganda	35	32	–	–	–	0	–	Barrett et al., 1998
Kampala[6] Uganda	67	59	–	–	–	544	–	Barrett et al., 1998
Kampala[7] Uganda	22	21	–	–	–	14	–	Barrett et al., 1998
Wobulenzi, Uganda	13	18	22	98	–	–	–	Tindimugaya, 2000

[1] thermotolerant coliforms; [2] faecal strepotococci; [3] rural area surrounding Ilesha; [4] modernised wells; [5] open wells; [6] high-density population; [7] low-density population.

The data suggest, therefore, that contamination of the shallow aquifer is diffuse. Nitrate concentrations exhibit a strong, positive correlation with population density and a negative correlation with well depth.

In Conakry, shallow groundwater was also investigated for evidence of industrial contamination (Gelinas et al., 1996). Analyses of heavy metals did not, however, provide evidence of extensive contamination, despite the weakly acidic nature of the groundwaters. Metal concentrations (Cr, Mn) exceeded WHO (1993) guideline values for drinking-water quality at two sites. Oils, greases and organic halogens (detection limit 0.05 mg L^{-1}) were analysed at five sites in a residential part of Conakry where it had been reported that inhabitants were using old wells to dump used oils and greases. None of these substances were, however, detected.

In the Ivory Coast, the source of high nitrate concentrations in fissured crystalline rock aquifers underlying five "concentrated settlements" was investigated using stable isotope ratios of nitrogen ($^{15}N/^{14}N$) (Faillat, 1990; Faillat & Rambaud, 1991). Nitrate concentrations of up to 200 mg L^{-1} were observed in wells that tap bedrock fissures at depths of between 40 and 70 m beneath a mantle of in situ weathered rock that varies from 5–40 m in thickness. Potential sources of nitrogen in these humid environments are considered to include domestic sewage and leaching of N from both soil and rotting plant debris following deforestation when villages were established. Although denitrification is suggested to have affected observed stable isotopic ratios of nitrogen and, hence the identification of nitrogen sources, this interpretation is not well supported by the oxidising conditions in groundwater (E_h: +303 to +573 mV) and isotopic ratios ($\delta^{15}N$: +6.9 to +15.3‰) that, with one exception (+25.5‰), show little evidence of isotopic enrichment in the heavy isotope expected from denitrification. Indeed, observed isotopic ratios (median $\delta^{15}N$: +12.6‰) suggest that nitrate is more likely to derive from a combination of sewage and soil sources.

In Niamey, Niger, Girard and Hillaire-Marcel (1997) also used stable isotope ratios ($^{15}N/^{14}N$) to trace the source of nitrate in the underlying fractured crystalline rock aquifer. They concluded that latrines were the major source of nitrate where isotopic ratios were enriched in the heavy isotope (i.e. $\delta^{15}N > 15‰$), but that a combination of sources, such as latrines and soil, resulted in isotopic ratios in nitrate that were less enriched (i.e. $\delta^{15}N <$ 15‰). In some cases, up to 85% of the groundwater nitrate load could be attributed to the soil. Here, groundwater pollution was thought to result from deforestation during urban expansion, and fertilisers from surrounding rice fields were not found to be a major contributor.

East Africa

In Dodoma, Tanzania, Nkotagu (1996) reported high nitrate concentrations in urban groundwaters that exceeded the national drinking-water quality guideline of 100 mg/l. Dodoma is underlain by in situ weathered crystalline rock (gneisses, granites) with an average thickness of 50 m that is succeeded with depth by fissured crystalline bedrock. Curiously, samples from deep boreholes in the fissured bedrock exhibited higher nitrate concentrations than samples collected from shallow wells in the weathered bedrock. In the absence of a significant mineral source of nitrogen, nitrate in groundwater necessarily results from the surface. It is, however, unclear whether this suggests that boreholes are more susceptible to on-site, point contamination via the well itself, or whether nitrogen-bearing compounds exist in shallow groundwaters but have not yet mineralised into nitrate. Urban sampling sites in Dodoma demonstrate considerably higher nitrate concentrations than rural areas and are considered to result primarily from sewage effluents.

In Uganda, Barrett et al. (1998) monitored nitrate and standard bacteriological indicators of sewage (i.e. thermotolerant coliforms) in two urban areas in order to assess the impact of on-site sanitation on groundwater quality under contrasting hydrogeological conditions and population densities. Study sites were located in Kampala and Iganga (Fig. 6.8). Iganga has a population of 38 000 (2002) and is situated in southeast Uganda (Fig. 6.3). Kampala is the country's capital with a population of over a million people (1 208 500 in 2002) and is located on the northern shore of Lake Victoria. Both urban areas are underlain by weathered crystalline basement. The nature of the groundwater flow regimes in the two areas differs significantly. Iganga is situated in an area of low relief area with a relatively thick layer of in situ weathered gneissic rock (generally around 30–35 m) with a water table that typically occurs between 10 and 15 m below ground level. Groundwater

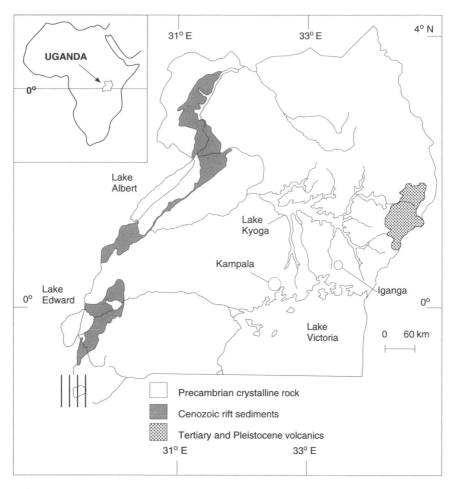

Figure 6.8. Location of study areas in Uganda, Kampala and Iganga, superimposed on a generalised geological map of Uganda.

samples were obtained from boreholes that were cased through the regolith and penetrated the top of the fractured basement. A groundwater-fed, reticulated piped water supply exists in Iganga but the distribution network is limited. Low-income and peri-urban populations that are not served by the piped supply rely upon hand-pumped boreholes.

Kampala has a variable relief that derives from long-term differential weathering of the underlying bedrock types that include alternating beds of phyllite, quartzite and low-grade schists. The thickness of individual layers of weathered and fissured rock (quartzite) is generally low and shallow, localised groundwater flow systems predominate. Sampling in Kampala was undertaken from protected springs that draw shallow groundwater primarily from fine sand at depths of 2–5 m below ground level. Parts of Kampala are served by a reticulated piped water supply that is fed by surface water (Lake Victoria). Like Iganga Town, the distribution network is limited, so that many low-income and peri-urban communities are often dependent upon springs and handpumped boreholes for household supplies (Howard & Luyima, 1999).

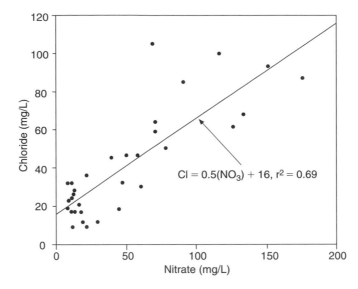

Figure 6.9.   Plot of chloride versus nitrate concentrations in groundwater from the town of Iganga in southeastern Uganda.

Several key findings from monitoring in Iganga and Kampala are described below:

- Concentrations of chloride and nitrate are higher in Iganga Town (Table 6.7) than those that have been observed under similar hydrogeological conditions in rural areas of Uganda (Table 6.5). Chloride concentrations are plotted against nitrate concentrations in Iganga in Figure 6.9. A significant correlation ($r = 0.83$) between the two species points to a single source of contamination. A plot of median chloride and nitrate values from Table 6.7 (Fig. 6.10) reveals a relatively consistent ratio of nitrate to chloride elevation is observed in urban weathered crystalline basement aquifers throughout sub-Saharan Africa. According to Morris et al. (1994), elevated ratios of nitrate to chloride concentrations ($>2:1$) are indicative of domestic sewage loading to groundwater. It is likely that high concentrations of nitrate throughout the regional-scale aquifer underlying Iganga Town result from an effectively diffuse input of sewage from on-site sanitation. Continued urbanisation is, therefore, expected to increase concentrations of inorganic contaminants in groundwater.
- In Iganga, thermotolerant (faecal) coliforms are generally not observed in the sampled boreholes that draw groundwater primarily from fissured bedrock. Presumably, thermotolerant coliforms that are discharged at the surface are largely attenuated by the thick mantle of weathered bedrock which separates the surface sewage sources from the well intake. It is, however, important to recognise that, if thermotolerant coliforms had been used alone as indicators of faecal contamination, there would be no indication of faecal contamination despite convincing evidence of sewage-derived inorganic contamination discussed above. The study highlights similar findings elsewhere (Foster, 1985; van Ryneveld & Fourie, 1997; Powell et al., 2003) that the absence of indicator bacteria does not necessarily indicate the absence of sewage-derived pathogens or other contaminants like nitrate.

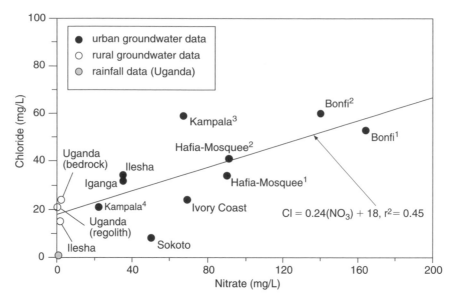

Figure 6.10.   Plot of chloride versus nitrate concentrations in groundwater from the weathered crystalline aquifers throughout sub-Saharan Africa.

- In Kampala, thermotolerant coliforms are detected in high concentrations in areas of high and low population densities, though concentrations are considerably greater in densely populated areas (Table 6.7). Median nitrate concentrations of 22 and 67 mg L^{-1} (Table 6.7) and high nitrate to chloride ratios (Fig. 6.10) suggest that nitrate has a faecal origin. Higher nitrate concentrations are generally associated with higher population densities, which provide a more continuous, dispersed source loading (van Ryneveld & Fourie, 1997). Interestingly, there is a poor correlation between concentrations of thermotolerant coliforms and nitrate. Several sites have high bacterial counts but low nitrate concentrations and vice versa. This may reflect different groundwater flow systems. Short, highly localised flowpaths from source to receptor (e.g. protected spring) will offer a limited period during which to mineralise organic nitrogen into nitrate but also a reduced capacity with which to attenuate sewage-derived bacteria. In contrast, longer flowpaths serve to promote mineralisation of organic nitrogen (into nitrate) and to increase attenuation of microbial contaminants.

6.6   CONCLUSIONS AND THE FUTURE

Urban populations are currently growing at a faster rate in sub-Saharan Africa than in any other region of the world. Many urban areas in this region are also comparatively new on a global basis. The existing infrastructure of piped water supply, sewerage and refuse removal is unable to meet demand in rapidly urbanising centres. Many urban areas, particularly low-income and peri-urban areas, rely upon localised abstraction of untreated groundwater through protected springs and wells for water supplies, and localised sanitary facilities such as pit latrines and surface dumps for the disposal of waste. Despite these ominous demographics, the quality of groundwater in urban areas of sub-Saharan Africa has, to date, been the subject of very few detailed studies. Limited data, presented in this chapter, demonstrate that, in terms of groundwater quality, there is an inevitable conflict between uncoordinated, localised waste disposal and groundwater abstraction in areas of increasing population density.

The most important aquifer systems in sub-Saharan Africa exist in the weathered crystalline rocks. These aquifer systems consist of discrete, subhorizontal fractures in the crystalline bedrock and an unconsolidated muddy-sand matrix in the overlying weathered mantle (overburden). Development of these groundwater resources has primarily been directed at low-intensity abstraction through hand-pumps. There is, however, growing interest in more intensive groundwater abstraction in order to supply urban piped systems without incurring the high costs of treatment that are required for surface-water fed schemes. Prolonged, intensive groundwater abstraction relies, however, upon contributions directly or indirectly through leakage into fractures, from the more porous and generally more transmissive aquifer in the weathered mantle. The sustainability of groundwater-fed supplies is uncertain but is more favourable in areas where tectonic stability has permitted the establishment of a regional flow system in the weathered mantle. Uplifted areas exhibit an aquifer of highly localised extent so that intensive groundwater abstraction is severely restricted. The in situ, weathered origin of the mantle serves to propagate heterogeneities within the bedrock into the unconsolidated mantle. This gives rise to highly variable aquifer characteristics that stem from broad variations in bedrock lithology and preferential pathways that derive from bedrock discontinuities such as quartz stringers.

Natural groundwater quality in weathered crystalline aquifers is generally good but undesirable concentrations of some metals (e.g. Al, Fe) are a necessary consequence of the weathered origin of the aquifers. Sewage is identified as the major source of contamination of urban groundwater in this region. Sewage is, however, typically the only source of contamination that has been investigated. Very few analyses have been undertaken for urban contaminants that are commonly identified in other parts of the world. These include petroleum products, chlorinated solvents, pesticides and heavy metals. Monitoring surveys are typically limited to nitrate, chloride and standard bacterial indicators of sewage contamination (thermotolerant coliforms, faecal streptococci). The limited body of available evidence suggests that shallow, urban groundwaters in sub-Saharan Africa are commonly contaminated by nitrate as a result of on-site sanitation (pit-latrines). Nitrate concentrations are strongly associated with population density and are, therefore, expected to rise in the near future. Standard bacterial indicators of sewage contamination are also detected in significant concentrations in shallow groundwater systems. It is, therefore, important recognise the vulnerability of shallow groundwater to contamination by sewage-derived pathogens (e.g. *Vibrio cholerae*).

Recommendations for further work

Two clear challenges for improved management of urban groundwater resources in sub-Saharan Africa emerge from this review:

- the need to develop groundwater and provide sanitation in an integrated manner that recognises the direct threat posed to the quality of groundwater resources by the disposal of sewage into the subsurface via pit latrines or leaky sewers; and

- the need to establish the impact of intensive abstraction from weathered crystalline basement aquifers and thereby enable delineation of wellhead protection areas to protect groundwater-fed, urban water supplies from domestic (sewage), agricultural or industrial contamination.

6.7   ACKNOWLEDGEMENTS

The views expressed in this chapter derive, in part, from field research supported by Department for International Development (U.K.), International Development Research Centre (Canada) and the Government of Uganda. Support in the form of a Post-Doctoral Fellowship and Canada-UK Millennium Fellowship from the Natural Sciences and Engineering Research Council of Canada to R. Taylor during the preparation of this manuscript is gratefully acknowledged. Discussion of the hydrogeological characteristics of weathered aquifer systems draws significantly from previous collaborative research with Ken Howard and Mark Hughes (University of Toronto), Guy Howard (WEDC, Loughborough), Mai Nalubega (Makerere University) and the Water Resources Management Department of Uganda. The assistance of Dr. Oliver Sililo (CSIRO, South Africa) who very sadly passed away in 2001, in improving the clarity and accuracy of this chapter is also gratefully acknowledged.

CHAPTER 7

Cities overlying karst and karst-like aquifers

Steve Appleyard

7.1   FORMATION OF KARST ENVIRONMENTS

Karst terrains are formed where there is extensive rock dissolution, which give rise to distinctive irregular landscape named after the Karst region in the Balkans. Karst features predominate in areas with carbonate rocks (limestone and dolomite), but may also form where underlying rocks contain large amounts of evaporite minerals, particularly halite and gypsum.

The large irregularities in the landscape in karst terrains are caused by the surface and subsurface removal of soluble minerals, particularly along lines of weakness in the rock mass, such as along bedding and joint planes. This dissolution can create interconnected voids that may range in size from small fissures a few millimetres in diameter, to large caves several tens of metres in diameter. In major karst regions, thousands of kilometres of caves exist, which extend to depths of more than 1 km in some areas. The underground voids eventually collapse, giving rise to large depressions at the land surface. The collapse of voids can be exacerbated by lowering of the water table caused by intensive groundwater abstraction, causing the rapid formation of sinkholes and other collapse features, and may cause extensive property damage in affected areas.

The erosion of karst collapse features may produce an irregular hummocky surface. In areas where erosion is particularly active, the erosion of karst features can give rise to a rugged landscape consisting of steep-sided valleys or steep isolated pinnacles of limestone characteristic of parts of southern China.

The formation of karst terrains requires both the presence of suitable rock types, and the presence of water with a suitable chemical composition to dissolve minerals within the rock mass. Limestone and, to a lesser extent, dolomite are soluble in water that is mildly acidic due to dissolved carbon dioxide and humic acids leached from soils. Carbonate minerals are dissolved by the water until it is about 99% saturated with respect to calcite, but the rate of dissolution is greatest when the water is less than about 65% saturated with respect to calcite (Palmer, 1984). Water undersaturated with respect to calcite will initially flow through fractures within the rock mass, progressively dissolving carbonate minerals. These become enlarged with respect to smaller fractures, and hence, even more water flows through the enlarged features due to the increased permeability. Dissolution mechanisms therefore favour the formation of large voids.

Cave systems can be formed above, at, or below the water table. They form when sufficient water flows through fractures in rock for non-Darcian flow to take place (Fetter,

1988). Vadose caves can form where a surface stream enters the ground in the unsaturated zone, phreatic caves can form where joint surfaces or bedding planes dip below the water table, and water table caves form at the water table itself. The orientation of cave passages is controlled by the pattern and density of fractures (Fig. 7.1). If fractures are widely spaced and steeply dipping, caves may form below the water table, whereas a dense or horizontal fracture pattern may favour the formation of caves at the water table (Ford & Ewers, 1978).

Although cave formation is favoured in humid climates, karst regions can be found in arid regions, either formed as the result of past wetter climates, or as the result of rare heavy rainfall events. Karst formation may also be accelerated where atmospheric pollution by sulphur and nitrogen oxides decreases the pH of rainwater significantly. The impact of large cities on karstic aquifers is independent of climate, as the presence of voids which may extend from the land surface to the water table makes these aquifers especially vulnerable to contamination (Aller et al., 1987), no matter what the average annual rainfall of the region.

Karst-like features can also develop in tropical or subtropical regions with lateritic soils. Voids can form in lateritic duricrusts due to the erosion of soft, poorly consolidated clays in an otherwise cemented ferruginous matrix. This can give rise to a spongy network of interconnected voids which may be at least several millimetres in diameter, and which can rapidly transmit water.

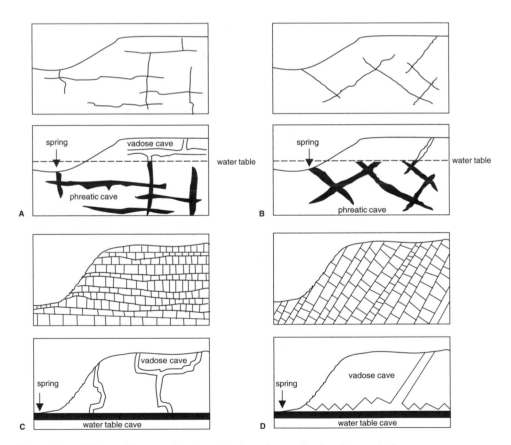

Figure 7.1.   Effects of fracture density and orientation on the development of caves.

In south-western Australia, the voids in duricrusts are often large enough to transmit fungal spores: in winter, the lateritic duricrusts form ephemeral perched aquifers that transmit the spores of the introduced fungus *Phytophthora cinnamomi* which is progressively killing the Jarrah forests in this region. In Brazil, voids in lateritic soils appear to be caused by termite activity (Mendonça et al., 1994). Where intensive urban development takes place, increased point sources of recharge such as stormwater soakwells and sewage disposal systems, coupled with intensive groundwater abstraction, can erode these voids to form sinkholes and caves which may be 5 m or more in diameter and several hundred metres long. These so called "pseudo-sinkholes" (Mendonça et al., 1994) may cause building subsidence problems, and can cause groundwater contamination problems of a similar nature to limestone karst aquifers. The formation of pseudo-sinkholes is a particular problem in the city of Brasilia in Brazil (Mendonça et al., 1994).

Karst-like features can also be formed as the result of processes other than by the dissolution or erosion of minerals. Large cave systems may also be formed in some volcanic terrains as a result of the exsolution and degassing of volatiles dissolved in molten lava flows, and the interface between individual lava flows may contain interconnected voids which may allow groundwater to transmit contaminants over long distances. Although not traditionally thought of as karst features, vuggy lava flow and lava cave aquifers show some of the hydrogeological characteristics of karst aquifers, and share the same extremely high vulnerability to groundwater contamination. It is for this reason that this aquifer type is included in this chapter.

7.2 HYDROGEOLOGICAL BEHAVIOUR OF KARST AQUIFERS

Karst aquifers show a wide range of hydrological characteristics. At one extreme are "underground rivers" where a surface stream may disappear underground and flow in an open channel or pipe underground, having non-Darcian behaviour. At the other extreme, some carbonate aquifers behave as homogenous, isotropic porous medium aquifers, and exhibit perfect Darcian behaviour. Most karst aquifers exhibit behaviour that lies between these two extremes, and groundwater flow may change from Darcian to channel flow over short distances. Many karst aquifers behave as a porous medium aquifer if treated at a regional rather than a local scale.

White (1969) proposed three conceptual models for karst aquifers in carbonate rocks: diffuse-flow carbonate aquifers which have limited dissolution features and behave like porous medium aquifers; free-flow carbonate aquifers, which have well developed dissolution features which transmit most of the groundwater in the aquifer; and confined-flow carbonate aquifers which have dissolution features, but where other confining units affect the direction of groundwater flow, and affect the response of the aquifer to external stresses.

Diffuse-flow aquifers are commonly found in dolomitic rocks that have a low solubility, and groundwater flow is mostly through fractures that have only limited dissolution features. Groundwater flow is commonly restricted to certain zones in the aquifer, and caves are generally small and not inter-connected. These aquifers generally have a well-defined water table.

By contrast, free-flow carbonate aquifers have substantial development of solution passages. Not only are many joints and bedding planes enlarged, but dissolution in some areas has created large conduits. Although all of the voids in the aquifer are saturated, the vast majority of groundwater flow takes place in large conduits, which have the hydraulic properties of underground pipes. Water flow rates in large conduits are similar to surface streams, the flow may be turbulent and carry a suspended sediment load. The water quality may be similar to surface streams, and discharge may take place at a limited number of large springs. Conduits drain rapidly in the absence of recharge, and springs may dry up. Conversely, flow increases rapidly in response to rainfall, and hydrographs of wells sunk in caves may resemble surface stream hydrographs. Because of the rapid drainage that takes place in these aquifers, the water table is nearly flat, and generally only has a small elevation above the regional base level.

7.3   GROUNDWATER CONTAMINATION IN KARST AND KARST-LIKE AQUIFERS

Groundwater in karst terrains is particularly vulnerable to contamination due to the presence of large voids that may connect the land surface to the water table, and due to the rapid rate that water may be transmitted through the aquifer. These features often limit the time available for pathogens introduced from the land surface to die-off, and reduce the ability of other contaminants to be removed by physical and chemical processes within the aquifer.

There are a number of possible sources of groundwater contamination from cities overlying karst and karst-like aquifers. One of the major sources of nitrate and microbiological contamination is on-site wastewater disposal in unsewered cities (e.g. BGS, 1994), and the disposal of stormwater into soakaways or constructed drainage wells can introduce a wide range of contaminants into karst aquifers (Smaill, 1994; Telfer & Emmet, 1994). Solid waste disposal is another major source of contamination, especially where large karst features like dolines are used as convenient landfill sites (Bodhankar & Chatterjee, 1994), or when landfill sites are located on major recharge areas (Zhu et al., 1997). Groundwater contamination by food processing wastes, heavy metals, petrochemicals, solvents, cyanide and pesticides from industrial and mining activities has also been reported from some cities overlying karst aquifers (e.g. Filipovic, 1988; BGS, 1994; Emmett & Telfer, 1994; Zhu et al., 1997).

Microbiological contamination

Unlike most other aquifer-types, karst aquifers are particularly susceptible to contamination by micro-organisms and water-borne diseases are often a major problem, each year killing or seriously disabling large numbers of people who drink untreated water drawn from these aquifers. Most porous medium aquifers have pore spaces which are fine enough to filter out most micro-organisms, and slow groundwater flow rates coupled with the large surface area of the aquifer matrix allow most bacteria, which generally have survival times of less than 70 days in soils (WHO, 1989), to die off within a few metres of contaminant sources. By contrast, pore spaces in karstic aquifers may allow large organisms like protozoa (1 μm to 1 mm in diameter) to be transmitted through the aquifer to water supply wells, even though survival times of these organisms are usually less than 20 days (WHO, 1989). Karst voids are so large in some regions that cave-dwelling invertebrates and even fish have been pumped from bores and wells!

Coliform and faecal coliform bacteria are commonly used as indicators of the possible presence of other pathogens in water (Clesceri et al., 1989), and these bacteria are usually considered to be the most important public health indicator for other infectious agents (other bacteria, viruses and protozoa) in water (McFeters & Stuart, 1972). Diseases that have been reported from the consumption of untreated water from karst aquifers beneath cities include cholera, dysentery, and a variety of other gastroenterital diseases (Bodhankar & Chatterjee, 1994). Coliform and other bacterial contamination of groundwater in karst areas is extremely widespread, even in rural areas in countries like the United States (Panno et al., 1996) which are considered to have land management practices minimising the risk of contamination taking place.

The microbiological quality of water in karst aquifers is extremely difficult to monitor, as numbers of microorganisms vary greatly with time. Studies of bacterial contamination in springs in a limestone terrain in Ireland (Thorn & Coxon, 1992) indicated that the number of coliform bacteria in spring water samples could vary from 0 to 300 colony forming units (cfu's) per 100 mL of water within a two hour period, with most of the bacteria being derived from dairy cattle. The bacterial quality of water is usually at its worst after heavy rainfall, and water from some springs was estimated to have travelled about 1 km from recharge areas within a 12 to 18 hour period after heavy rain. Similar results were obtained in a karstic area of Kentucky by Ryan and Meiman (1996). They indicated that it is often difficult to distinguish the source of contamination in karst springs because water from different parts of the catchment may arrive at the spring at the same time after rainfall.

Coliform levels in karst aquifers beneath unsewered cities can often exceed 1 000 cfu's per 100 mL of water (Morris et al., 1994), compared to drinking water guidelines of less than 1 cfu per 100 mL, and drinking untreated water is most likely to cause disease in susceptible individuals. Paradoxically, improving standards of living in some countries may increase the risk of exposure of people to waterborne disease, as the percentage of the population who are elderly and are immune-compromised increases. There is also the risk that antibiotic-resistant strains of bacteria will become endemic in groundwater systems due to the widespread use of antibiotics in agriculture in some countries. Other disease causing parasites such as *Cryptosporidium* that are resistant to many water treatment processes are also becoming more widespread in some urban centres.

Other contaminants

Karst and karst-like aquifer are contaminated by the same rage of land use activities that affect groundwater quality in cities overlying other aquifer types, but karst features may affect both the severity and distribution of contamination. Some contaminants may be detected in karstic aquifers which are not commonly found in groundwater because of filtering in the unsaturated zone. These include organic contaminants with large molecules such as lignosulphonate dyes (Motyka et al., 1994), pesticides (Zhu et al., 1997) and PAHs. Heavy metal contamination may also take place in karst aquifers, often because there is limited interaction of water with the aquifer matrix to allow adsorption of metals by the aquifer matrix to occur, or because bicarbonate and carbonate ions in groundwater form chemical complexes that enhance metal solubility.

7.4   ASSESSMENT OF CONTAMINATION IN KARST AND KARST-LIKE AQUIFERS

The presence of karst features can affect the distribution of contaminants in a groundwater contamination plume, because recharge and groundwater flow is commonly convergent towards karst conduits (Fig. 7.2). Therefore, contamination plumes may not progressively

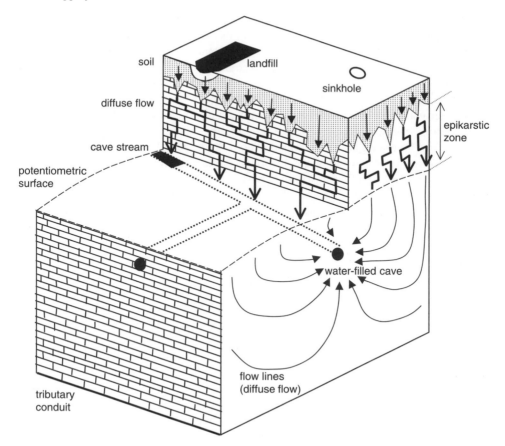

Figure 7.2. Distribution of groundwater contamination in a karstic aquifer (after Quinlan & Ewers, 1985).

spread and become more diffuse with distance from the source of contamination, but may converge towards cave streams, and then move rapidly to springs or other groundwater discharge features (Quinlan & Ewers, 1985). As the geometry of the network of small fractures and conduits controls the geometry of the groundwater contamination plume, the assessment of groundwater contamination in these aquifers is particularly difficult if there are extensive karst features. This is especially the case for dense non-aqueous phase liquids (DNAPLs) like chlorinated solvents, which may move under gravity in karst conduits and give rise to an extremely complex distribution of contamination (Crawford & Ulmer, 1994).

The first stage in any assessment of contamination in karst and karst-like aquifers is to compile existing geological data that may identify major conduits that will transmit contaminated groundwater (Crawford & Ulmer, 1994). This includes locating all springs, caves, cave streams, significant sinkholes, and lineaments (often a line of sinkholes). The use of a variety of geophysical techniques may also help locate structural features associated with karst conduits. Wherever possible, water level information should be gathered to give a regional sense of the groundwater flow direction. However, many karst aquifers have extremely high transmissivities, and there may be a negligible hydraulic gradient in some areas to identify the direction of groundwater flow. It is often necessary to use tracer techniques to identify the regional groundwater flow direction, to determine whether karst features are interconnected, and to measure groundwater flow rates.

The most commonly used tracers used in karst regions are a variety of dyes including fluorescein, tinopal 5BM GX, rhodamine WT and diphenyl brilliant flavine 7GFF ("direct yellow 96") (Crawford & Ulmer, 1994). A number of tracer tests are often carried out: preliminary tests with passive receptors are used to determine flow direction and interconnections between karst features, and more quantitative tests are used to determine groundwater flow rates.

Preliminary dye tracer-tests are carried out by placing receptors consisting of bundles of unbleached cotton or packages containing activated charcoal in all springs, cave streams, sinkholes, surface streams and wells in the vicinity of the test site. Dye is injected into a sinking stream or sinkhole, and receptors are checked and replaced at intervals of several days to weeks. The presence of dye is determined by eluting the activated charcoal with a mixture of propanol and ammonium chloride, and measuring the elutant with a fluorometer or spectophotometer. The unbleached cotton receptors are generally washed to remove mud, and tested for "direct yellow 96" under a long-wave ultraviolet lamp ("direct yellow 96" glows a pale yellow).

Groundwater flow rates are usually determined in quantitative dye tracer-tests. These usually involve collecting water samples at frequent intervals (often hourly) at some point distant from the dye injection site to accurately determine the breakthrough curve for dye arrival.

Groundwater monitoring in karst and karst-like aquifers can be relatively simple where karst features are small and localised. Under such conditions, these aquifers may behave like porous-medium aquifers, and monitoring bores and sampling programmes can be established in much the same way as for other porous-medium aquifers. However, when most of the groundwater flow is via karst conduits, groundwater monitoring is particularly difficult. Conventional monitoring bores may not detect groundwater contamination, because they may not be located in specific karst conduits which are carrying contaminated groundwater.

The most reliable groundwater monitoring sites in "mature" karst aquifers are springs and cave streams shown by dye tracing to receive discharge from the contamination source being monitored (Quinlan & Ewers, 1985; Quinlan, 1988). Monitoring bores are only useful when it can be demonstrated by tracing studies that they intersect the same karst conduits. Conversely, background samples should only be collected from springs and monitoring bores where it can be demonstrated by dye-tracing that there is no connection with the contamination source. The most representative groundwater samples are collected when bores in karstic aquifers are pumped at very low rates using peristaltic or similar pumps (McCarthy & Shevenell, 1998). If monitoring bores have a long screened interval, sampled water quality may vary depending on where the pump intake is located.

Periodic sampling may not give representative results in karst and karst-like aquifers, and generally the most effective time to collect groundwater samples is during and after major rainfall events rather than with a standard monthly, quarterly or six-monthly sampling regime (Quinlan, 1988). Contaminant concentrations in groundwater often peak during rainfall events, and very frequent sampling (one or more samples a day) is often required to determine peak concentrations and quantify the mass of contaminants discharged during the rainfall event. To reliably characterise the overall quality of karst springs, a monitoring programme must incorporate a flow-dependent, high frequency sampling strategy in addition to a periodic monitoring programme (Ryan & Meiman, 1996).

7.5   GROUNDWATER QUALITY PROTECTION POLICIES IN KARST ENVIRONMENTS

Karst aquifers are generally extremely vulnerable to contamination from surface land use, and so strong groundwater quality protection policies and enforcement measures are particularly important for protecting groundwater resources in cities overlying this aquifer type. The vulnerability to contamination is such that whole town water supplies in some areas like Croatia and Slovenia have been affected by pollution, causing epidemics of water-borne disease (Kresic et al., 1992).

Kresic et al. (1992) suggest that the first stage of groundwater protection in karst areas like the Balkan Region should be to investigate and understand the local hydrogeology, determine sources of contamination, and understand the interaction of pollutants with groundwater and aquifer rocks. Protection policies should be locked into planning measures, and should include pollution prevention measures such as either preventing in-ground disposal of liquid wastes or treating wastes prior to disposal; controlling usage of fertilisers and pesticides; and having adequate well-head protection zones around water supply bores (Kresic et al., 1992; Conservation Foundation, 1987). Kresic et al. (1992) stress that, wherever possible, urban planners should seek to prevent any development that may adversely impact on groundwater quality in karst areas, particularly where there is little or no soil cover. Kresic et al. (1992) indicate that the presence of a soil profile or sediment cover is probably the most important single physical factor that can ameliorate the impact of land use on groundwater quality in karst aquifers.

As carbonate rocks cover up to 35% of Europe and karst aquifers provide water supplies for millions of people on that continent, the European Commission Cooperation in Science and Technology Programme (COST Action Programme) established a working group to develop groundwater protection policies for these aquifers. The final report of the working group (European Commission, 1995) indicated that the guiding principle for karst aquifers should be pollution prevention due to their extremely high vulnerability to contamination. The protection programme should be based on:
- Establishing protection schemes.
- Developing adequate land use planning.
- Improving the handling of toxic substances.
- Operating a monitoring system.

Groundwater protection schemes should be based on a risk management approach considering the intensity of karst development, the hazard provided by a potentially polluting activity, the consequences of a pollution event, and the likelihood of pollution taking place. The main components of a protection scheme are land surface zoning maps, which encompass the hydrogeological elements of risk, together with codes of practice for those activities which encompass the pollutant loading and pollution control elements of risk. The COST report recommends that land use zoning in karst areas should include:
- Areas surrounding individual groundwater sources (source protection areas) often based on a 50 day groundwater travel time to provide protection from microbial contamination.
- Areas subdivided on the basis of the value of the groundwater resource (resource protection areas).
- Division of the entire land surface on the basis of the underlying groundwater to pollution (groundwater vulnerability maps).

The COST report recommends developing codes of practice for potentially polluting activities which include levels of response (R) to potentially polluting activities on the basis of four categories: R1 – acceptable; R2 – acceptable in principle subject to conditions being met; R3 – not acceptable in principle; and R4 – not acceptable. The report then recommends integrating groundwater protection zones and codes of practice in a matrix relating acceptable land uses to aquifer vulnerability, and prescribes different response levels to source protection and resource protection areas.

The COST report also strongly recommends that regulatory agencies develop public awareness and education programmes to "sell" the message of groundwater protection. Public awareness campaigns may take the form of regular newspaper articles, television slots and public seminars. Distribution of educational videos, leaflets and posters also contribute towards developing an awareness in the long term, and companies who may be willing to develop an "environmentally friendly" image may contribute to these campaigns. Another option is to include formal instruction on groundwater protection policies in primary and secondary schools as part of a syllabus on the protection of the environment. This could possibly be achieved through an international agreement to have a year devoted to the protection of groundwater, especially for countries where the understanding of environmental issues by the general public is poor.

These programmes are essential, as legislation and enforcement measures alone are likely to be ineffective unless the general community recognise the impact that land use can have on groundwater quality in karst aquifers, and accept the need for special measures to protect water quality. Even if regulatory agencies have defined hydrogeologically and economically rational policy for groundwater management for these aquifers, this does not mean it will be implemented. No matter how rational such policies may appear, they may not be considered politically attractive or acceptable, especially in the case of groundwater, which is "out of public sight" and therefore "out of public mind" (Foster et al., 1998). More powerful industrial or agricultural lobby groups may often interfere with the regulatory process.

In order to overcome these difficulties, regulatory agencies need to build social consensus to overcome resistance to policies to protect water quality and manage groundwater use. A key factor is the formation of well-informed water-user and environmental interest groups, together with other stakeholders interested in groundwater management (Foster et al., 1998). Such groups can assist the implementation of groundwater management policy if they are able to effectively lobby within the political process.

7.6 CITY CASE STUDIES

The city of Mérida, Mexico

Introduction
Mérida is the largest city in the south-eastern part of Mexico, and has grown rapidly over the last 30 years to its current population of about 550 000. The city is an important administrative centre located in the central part of the Yucatan peninsula (Fig. 7.3). The climate is tropical with a mean average rainfall of 990 mm, the highest rainfall occurring in the period June to October, and the period November to April being the driest months.

Mérida is totally dependent on a karstic aquifer for its water supply of more than 280 mL day^{-1}. The flat terrain and shallow water table make the provision of sewerage difficult and expensive, and consequently most of the wastewater and drainage disposal in the city is to the

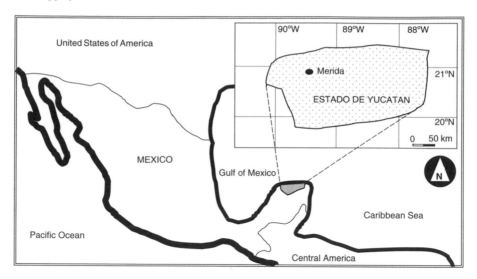

Figure 7.3.   Location of Mérida.

ground, increasing the risk of serious groundwater contamination. About 70–75% of the city is serviced by septic tanks, 20% by pit latrines, and only 5% of effluent is treated and disposed of in deep saline aquifers (Gonzales-Herrera, 1992). Septic tanks used in Mérida retain effluent for only a few hours, and this coupled with the shallow water table and the presence of karst features means that untreated effluent is likely to reach the water table very rapidly.

Hydrogeological setting
Mérida is located on a low-lying limestone plateau of Tertiary age. The plateau is a flat, karstic surface with little or no soil situated only a few metres above sea level, and has no surface runoff (BGS, 1994). Steep-sided sinkholes known as "cenotes" occur in the area, and these range in diameter from a few metres, to several hundred metres. The limestone forms an unconfined aquifer with a water table 5 to 9 m below ground surface and a very shallow hydraulic gradient (0.02 m km^{-1}). Groundwater is fresh to a depth of 40 m beneath the city and there is a gradational change to saline groundwater that occurs at a depth of 60 m. Geophysical logging has indicated that the aquifer is highly karstic (Buckley & McDonald, 1994), and this allows individual boreholes to have specific capacities exceeding 10 Ls^{-1}m^{-1} (Morris & Graniel, 1992).

Under natural conditions, the average recharge to the aquifer is estimated to be less than 100 mm y^{-1}. However, because of the high water usage (460 L day^{-1} capita^{-1}) in the city, coupled with in-ground disposal of wastewater and drainage, and leaks from the water distribution system, recharge beneath the city area is estimated to be about 600 mm y^{-1} (Morris et al., 1994). Despite the large increase in recharge due to urban development, there has only been minor mounding of the water table beneath the city due to the high permeability of the aquifer.

About 65% of the Mérida public water supply is pumped from two borefields outside the city limits, although a number of bores are still operational within the city. There are also a number of private groundwater users within the city area, and water for irrigation of public open space and plazas is usually pumped from shallow dug wells constructed for the purpose.

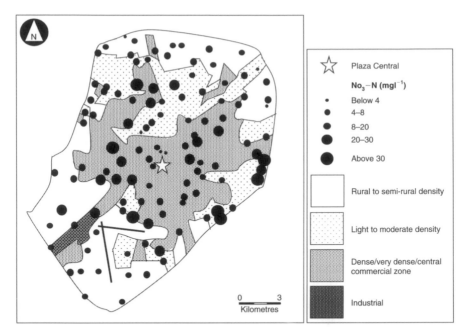

Figure 7.4.   Distribution of nitrate in groundwater in Mérida.

Groundwater contamination

The karstic aquifer beneath Mérida is highly vulnerable to groundwater contamination, both from microbiological and chemical pollutants.

The principal source of contamination is on-site effluent disposal, as septic tanks and soakaways and pit latrines are often constructed in karstic limestone within 1–3 m of the water table, ensuring that poorly treated effluent rapidly reaches groundwater. There is little attenuation of microbes within the unsaturated zone as karst fissures as these features are generally much larger than the size of the pathogenic microorganisms in effluent. Consequently, there is gross microbiological contamination at the water table beneath the city, with faecal coliform counts typically being in the range of 1 000 to 4 000 colony forming units per 100 mL (Morris et al., 1994) compared to normal drinking water requirements of less than 1 per 100 mL. Levels of microorganisms detected vary seasonally, with the lowest values generally being detected in the drier season, and the highest levels being measured at the wettest time of the year (July–August). Coliform levels in excess of 100 colony-forming units per 100 mL are commonly detected in deep bores, suggesting that microbes are transmitted thought the aquifer by vertical karst features.

Groundwater beneath Mérida is also contaminated by nitrogen as a result of wastewater disposal (Fig. 7.4). Groundwater beneath the city is aerobic, allowing ammonia in effluent to be oxidised to nitrate, and nitrate concentrations in groundwater are generally in the range of 4 to 30 mg L^{-1} (as nitrogen). Nitrate concentrations are lower than expected given the extent of in-ground wastewater disposal, probably as a result of the relatively low population density of the city (averaging 35 people per hectare), and the substantial dilution of nitrogen levels as a result of the high urban recharge and the large amount of groundwater throughflow in the highly permeable aquifer (Morris et al., 1994). Levels of Total Organic

Carbon (TOC) are also generally lower in the wet season due to the additional infiltration of water. TOC concentrations of up to $40\,mg\,L^{-1}$ occur in groundwater beneath industrial areas in Mérida, probably as a result of industrial wastewater disposal.

Sampling by the British Geological Survey has also indicated that there is also widespread contamination of groundwater beneath the city by chlorinated solvents including 1,1,1-trichloroethane, trichloroethene, tetrachloroethene and carbon tetrachloride (Gooddy et al., 1993). Measured concentrations of solvents were generally below drinking water guideline values, but it is suspected that these are anomalously low due to delays between sampling and subsequent chemical analysis in the United Kingdom. Solvent concentrations appear to be generally higher in groundwater in the rainy season than at other times of the year, possibly as a result of flushing of solvent residues from the unsaturated zone each rainy season (Gooddy et al., 1993).

Groundwater management
Although there is widespread microbiological and chemical contamination of groundwater beneath the urban areas of Mérida, the majority of the city's water supply is pumped from outside the city area and is relatively uncontaminated. Groundwater for public water supply is also chlorinated to minimise the risk of waterborne disease being spread.

Groundwater abstraction from large production bores in Mérida is licensed by the National Water Commission to control usage, but there are thousands of dug wells in the city which are not licensed. However, modelling carried out by the British Geological Survey (Morris et al., 1994) has shown that this groundwater management strategy is working well under current levels of groundwater abstraction, and there is a negligible risk of contaminated groundwater from the city being captured by peri-urban production bores. However, the BGS suggested that the risk of contamination would increase significantly if urban areas extended towards the borefields, or if groundwater abstraction increased significantly. The BGS recommended that wherever possible, production bores for drinking supply should be shut down within urban areas.

Although Mérida has a high per capita consumption of groundwater, over 90% of water pumped is not used for drinking supply, but is either used for non-potable uses or is lost by leakage from supply mains. There is scope for replacing at least a proportion of the higher quality water supplied from peri-urban borefields with lower quality groundwater for non-potable uses. In general, water quality in deep (15–40 m) boreholes beneath the city is better than water pumped from shallow wells, and therefore selective pumping of deep groundwater from beneath Mérida for non-potable use would have the advantages of helping extend the life of the peri-urban borefields, and would also improve shallow groundwater quality beneath the city by diluting contamination due to wastewater disposal. Improved water quality beneath the city would in turn reduce potential impacts on downgradient groundwater users.

Boshan, Shandong Province, China

Introduction
Boshan is located in the centre of Shandong Province in China (Fig. 7.5), and is an important mining, industrial and manufacturing centre which is particularly well known for the production of ceramics. Although the urban centre of Boshan only has a population of a few hundred thousand people, the surrounding Boshan District is densely populated and industrialised, and is in close proximity to the major industrial centre of Zibo City, which has a population in excess of 3.5 million. Like many towns and cities in densely populated

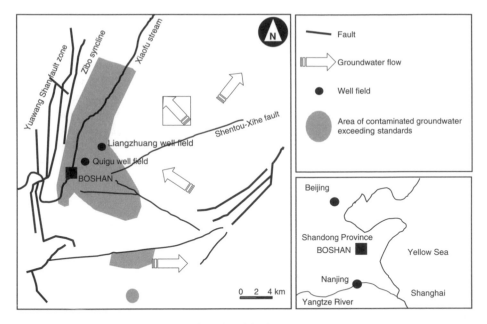

Figure 7.5. Hydrogeological setting of Boshan District.

regions of the world, it is difficult to distinguish a clear boundary between the urban centre of Boshan and the surrounding rural areas, which in effect form an extensive peri-urban region.

The district has a rugged relief with topographic elevations ranging between 150 and 1 000 m above sea level. The average annual rainfall for the district is about 670 mm, of which about 60% falls during the summer wet season in the period June to September.

Boshan and Boshan District are totally dependent on groundwater for water supply, because surface streams have become seriously contaminated through the disposal of mine water, sewage and industrial wastes. Currently, more than $80\,000\,\mathrm{m^3\,day^{-1}}$ of groundwater is pumped for urban and industrial use, of which $25\,000\,\mathrm{m^3\,day^{-1}}$ is pumped from the Tianjinwan wellfield located to the east of Boshan. Another well field, the Liangzhuang well field, was abandoned in 1986 due to contamination problems which made groundwater unsuitable for potable use (Zhu et al., 1997).

Hydrogeological setting
Boshan is located in the southern part of the Zibo syncline and is immediately underlain by a variety of sediments of Palaeozoic to Mesozoic age. Limestone, dolomitic limestone and dolomitic marl of Ordovician and Carboniferous age are the major aquifers in the area, and these are generally karstic. The transmissivity of the karstic aquifer ranges from 200 to $5\,000\,\mathrm{m^2\,day^{-1}}$, and groundwater yields from individual bores range from $2\,000$ to $5\,000\,\mathrm{m^3\,day^{-1}}$. The region has been intensively faulted and has broken up the Boshan District into a number of fault blocks (Fig. 7.5). Large faults appear to act as groundwater barriers, and have largely restricted groundwater flow between fault blocks. Each fault block with an area of 100 to $120\,\mathrm{km^2}$ can be considered to be a relatively independent hydrogeological unit.

Under natural conditions, recharge to the karst aquifer is by the infiltration of rainfall through solution features and jointing surfaces where limestone is exposed in each fault block, and is estimated to be about 200 mm y^{-1} (Zhu et al., 1997). However, over-exploitation of the groundwater resource, the regional water table has been lowered below the base level of many of the streams in the area, and leakage from streamflow now forms a significant part of groundwater recharge in the district. Intensive pumping of groundwater has also made a number of springs disappear: the Xiafou, Quigu and Luangzhuang springs were important tourist attractions in Boshan District and had a combined discharge of more than 50 000 m^3 day^{-1} before large scale groundwater abstraction commenced in 1970 (Zhu et al., 1997).

Groundwater contamination
There is widespread contamination of the limestone aquifers near Boshan, and this has greatly reduced the amount of groundwater available for potable supply.

Possibly the single largest source of contamination in the district is acid drainage containing high concentrations of dissolved iron and sulphate from coalmines. Currently, about 40 000 m^3 day^{-1} of acid mine water is drained from mines, and this has caused progressive increases in the sulphate concentration in groundwater, with concentrations commonly exceeding 500 mg L^{-1}. The disposal of industrial wastewater is another major source of contamination in the district, with in excess of 20 000 m^3 day^{-1} of effluent containing sulphate, petroleum hydrocarbons, phenols, cyanide, arsenic and heavy metals being discharged to streams or recharge basins near Boshan. Other sources of contamination include domestic wastewater, unlined landfill sites, and the widespread use of fertilisers and pesticides in the district. Groundwater beneath urban areas of Boshan typically has sulphate concentrations in excess of 500 mg L^{-1}, nitrate concentrations in excess of 10 mg L^{-1} (as nitrogen), and a Total Dissolved Solids (TDS) content in excess of 1 200 mg L^{-1}.

Seepage from surface streams is a major pathway for contamination reaching groundwater, particularly seepage from Xiaofu Stream that passes through urban areas of Boshan and flows to the north-east (Fig. 7.5). This stream has been severely contaminated by the disposal of industrial wastewater and landfill leachate, and leakage from the stream has degraded groundwater quality over a large area near Boshan.

Direct discharge of industrial wastewater to the ground is another important contamination pathway, particularly as many industries are located on outcrop areas of dolomite and limestone. An important historical source of groundwater contamination in the district is the Boshan Pesticide Factory, which is located on limestone bedrock in the recharge area of the Liangzhuang borefield. Prior to 1986, the plant used to produce the pesticide hexachlorocyclohexin (BHC), and the manufactured product was stored on bare earth. Wastewater was discharged directly into pits in the limestone. Contamination caused by the pesticide plant contributed to the closure of the Liangzhuang borefield, and BHC could still be detected in monitoring bores within the borefield until 1992.

Leachate from unlined landfill sites has also contributed to groundwater contamination in the district. The largest landfill in the Boshan district, the Ayu Landfill, is located in close proximity of the Liangzhuang borefield, and has been used for the disposal of municipal and industrial waste. Groundwater near the landfill has become contaminated with a number of toxic chemicals including chromium (VI), mercury, phenols, cyanide and nitrite that resulted in the closure of a number of bores within the Liangzhuang borefield.

Groundwater management

Widespread contamination and local over-abstraction have restricted groundwater use in Boshan. This has resulted in water supply borefields being pushed further and further from the urban centre of Boshan, with the new Tianjinwan borefield being located more than 10 km to the east of the city.

There are currently limited controls on land use in the district to protect water quality, and so new borefields are also vulnerable to contamination. Zhu et al. (1997) recommended that a water resource protection area be established around the Tianjinwan borefield. They also indicated that a number of industries and waste disposal sites should be relocated to areas underlain by sediments with a low permeability, because the risk of contamination in recharge areas of the limestone is so great. New landfill sites should be lined and covered, and be equipped with leachate collection systems to minimise the risk of contamination taking place. Zhu et al. (1997) also indicated that as seepage from Xiaofu stream and its tributaries was such a major source of contamination near Boshan, stream beds should be lined with low permeability materials to prevent further degradation of groundwater quality.

The combined properties of low storage capacity of the limestone aquifers (specific yields are only 0.003 to 0.01), local high recharge rates and limited soil cover allow water in these aquifers to be exchanged rapidly, and there is the potential to improve groundwater quality if polluting activities are removed and additional pumping is carried out. Groundwater contamination by chromium (VI) caused by two electroplating plants was reduced over a period of a year from concentrations of about $100\,\mu g\,L^{-1}$ to less than $10\,\mu g\,L^{-1}$ after their closure and a pump-and-treatment programme (Zhu et al., 1997). Monitoring over a three-year period has indicated that the cleanup programme has been successful, but is unlikely to be a cost-effective way of restoring groundwater quality for water supply.

Mount Gambier, South Australia

Introduction

Mount Gambier is located in the south-eastern corner of the state of South Australia (Fig. 7.6) and currently has a population of about 23 000. It is an important regional

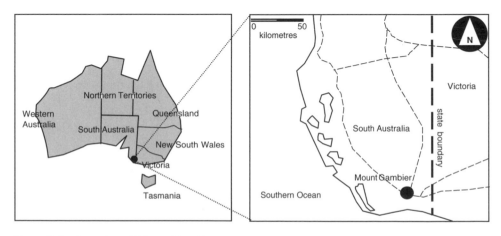

Figure 7.6.   Location of Mount Gambier.

centre for tourism and agriculture, and a number of food processing industries occur near the town. Mount Gambier has a Mediterranean-type climate with an average annual rainfall of about 700 to 800 mm that mostly falls during the period May to October.

About 4 000 mL y^{-1} is pumped from the groundwater-fed Blue Lake in Mount Gambier for water supply. Blue Lake receives discharge from a karstic limestone aquifer that underlies Mount Gambier, and the aquifer is extremely vulnerable to contamination from urban, agricultural and industrial land use in the region. Blue Lake is also the principal tourist attraction for the city, principally due to its unique habit of changing colour from a steel grey-green colour in winter to an intense blue in summer (Telfer, 1993). It is also picturesque, being contained in an extinct volcanic crater within a volcanic complex. The lake supports a wide diversity of aquatic life, although there are only small populations of some individual species.

Blue Lake has a surface area of about 60 ha, is up to 77 m deep, and contains about 37 000 mL of water. Rainfall adds only about 1% of the total volume of the lake, and evaporation consumes about 2% of the total volume. Groundwater inflow to the lake is approximately the same as abstraction (Turner, 1979; Ramamurthy, 1983) and there is no longer significant throughflow as was the case before the lake was utilised for water supply.

Hydrogeological setting

Mount Gambier is immediately underlain by limestone of Eocene age that forms a regional unconfined aquifer with a thickness of up to 150 m. The aquifer has a transmissivity of about 20 000 m^2 day^{-1} near Mount Gambier due to interconnected horizontal and vertical karst conduits that have developed near volcanic features (Telfer, 1993). There is no surface runoff in the area despite the relatively high rainfall in the area. The hydraulic gradient in the unconfined aquifer is very flat due to the high transmissivity, but the regional groundwater flow direction is to the south-west, with groundwater flowing from urban areas towards Blue Lake (Emmett & Telfer, 1994). The water table is generally more than 20 m deep beneath the city area.

Under natural conditions, recharge to the unconfined aquifer is by infiltration through soils and through karst features. However, natural recharge has been modified for the last 100 years by stormwater runoff in urban areas of Mount Gambier. A total of between 3 600 and 6 200 mL y^{-1} of urban runoff in Mount Gambier is recharged through between 300 and 500 constructed drainage bores and modified karst features within the city (Telfer & Emmett, 1994). In many of the bores, drainage of water takes place from a small interval (often only 0.5 m thick) tens of metres below the water table. Drainage bores are mostly equipped with silt traps, but these are not regularly cleaned to remove detritus, reducing their effectiveness in removing soil and debris washed into the bores.

Monitoring of Blue Lake has shown that stormwater is having an impact on water quality in the lake, as the lake salinity has declined by about 10–12% since the early 1980s. Mass balance calculations indicate that approximately 1 540 mL per year of groundwater currently recharging the lake is from urban stormwater (i.e. about 35% to 55% of the total groundwater input to the lake).

Groundwater contamination

Groundwater near Mount Gambier has been polluted by both industrial and urban point sources, and by diffuse sources of contamination. Telfer and Emmett (1994) suggest that because of the dual-porosity nature of the unconfined limestone aquifer, point sources of recharge will mostly be input into karst conduits and water will move rapidly towards Blue

Lake, whereas diffuse recharge will affect the aquifer matrix and water movement will be much slower. Consequently, diffuse sources of contamination are likely to take much longer to have an impact on lake water quality than point sources of contamination.

The most immediate potential point source of contamination in the capture zone of Blue Lake is the recharge of stormwater in urban areas of Mount Gambier (Fig. 7.7) which is probably delivered both into karst conduits and the aquifer matrix, with some interchange of water between these components. The quality of the stormwater probably varies considerably with time and location within the city, with recharge bores in industrial areas posing a greater threat to groundwater quality than bores in residential areas. However, sampling of water draining into several recharge bores indicated low levels of the pesticides atrazine, simazine, aldrin and dieldrin, and lead levels in some bores were above potable limits. Other bores showed traces of tetrachloroethene and pentachlorophenol, although at levels below potable limits.

The karst features in the unconfined aquifer near Mount Gambier have historically been a convenient solution to industrial waste disposal problems, and several cheese factories were located specifically in the area for this reason. Effluent including whey, blood and offal from abattoirs, and chemicals from wood treatment processes (copper, chromium, arsenic and creosote) has been discharged into karst features, creating a legacy of groundwater contamination problems, including a 1.5 km long contamination plume caused by whey disposal. These practices ceased after 1976 with the introduction of the South Australian Water Resources Act, and there have since been large improvements in waste disposal practices.

The largest impact on water quality in Blue Lake has been from nitrate contamination from a variety of diffuse sources, causing nitrate levels in the lake to increase from about 2.6 mg L^{-1} (as nitrogen) in the 1970s to current levels of about 3.9 mg L^{-1} (as nitrogen), and continuing increases in nitrate concentration could affect the long-term viability of the lake as a water supply. Contaminants from industrial sources have never been detected in water from the lake despite their presence in groundwater. This is possibly because Blue Lake has the physical and chemical characteristics of an efficient settling basin (Stumm & Morgan, 1981), and many pollutants may be precipitated out of the water column in the

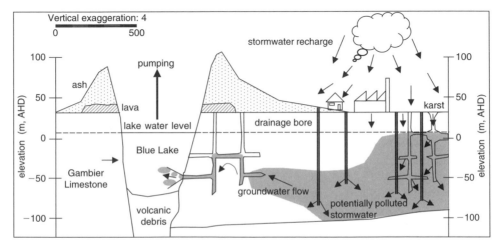

Figure 7.7.   Impact of stormwater recharge on Blue Lake.

lake. It is possible that changes in nitrate concentration in the lake are in part due to the enhanced capture of contaminated groundwater near the lake caused by increasing pumping rates from the lake with population growth in Mount Gambier.

Historically, one of the most significant sources of nitrate contamination was from septic tank drainage and the disposal of sewage direct to karst features when Mount Gambier was unsewered prior to the mid 1960s. These disposal practices have caused widespread contamination of the upper part of the unconfined aquifer (Waterhouse, 1977). Current sources of nitrate contamination of groundwater include the widespread use of fertilisers on gardens, sporting fields and parks within urban areas, and intensive grazing in rural areas surrounding the city (Dillon, 1989).

Groundwater management
The particular sensitivity of Mount Gambier's water supply to pollution was first recognised in a government review in 1973 (EWS, 1973), which recommended that:
- Mount Gambier should be proclaimed as a defined groundwater protection area under existing South Australian legislation.
- Below ground disposal of trade wastes should be phased out.
- Wastes being disposed of below ground should be given treatment to meet approved quality standards.
- Septic tanks should be replaced with connections to sewers.
- Government agencies maintain effective communication with primary producers and industry management on matters that may affect groundwater quality.

Many of these recommendations were subsequently acted upon in the following decades, but in the 1990s it became clear that the management of the groundwater resource was lagging behind the understanding of groundwater processes, and that the community also wanted environmental and aesthetic aspects of water quality managed in the lake rather than the lake being considered only as providing a water supply.

In response to these concerns, the South Australian Department of Environment and Natural Resources (DENR) undertook extensive community consultation, and together with community, industry and other government representatives, a management plan was developed for the area. The aim of the Blue Lake Management Plan (DENR, 1994) is to maintain water quality, aesthetic and tourism values, and to maintain the environmental values of Blue Lake. This aim has been addressed by defining groundwater management zones in the vicinity of Mount Gambier and Blue Lake based on the lake's groundwater capture zone; by setting up measures to prevent further groundwater contamination from existing land use in the area; by setting up measures to prevent historical groundwater contamination from affecting the lake; and by preventing vehicles and surface contamination entering the lake. A major focus of the plan is to prevent further pollution occurring from urban and rural land use, and this is being addressed by a number of strategies including:
- **Planning measures.**   It is important to ensure local government development plans incorporate criteria to minimise the risk of groundwater pollution occurring, particularly by preventing new polluting activities to be established with 2 km of the lake (Zone 1 groundwater protection area), and by either encouraging existing polluting activities in this area to relocate or take measures to minimise the risk to groundwater quality. By integrating local government development plans with groundwater management zones,

groundwater quality protection can be considered in planning of future development within Mount Gambier and its environs.

- **Stormwater management measures.** Stormwater disposal in urban areas is probably the most significant risk to water quality in Blue Lake, particularly in industrial areas. Stormwater disposal by the Mount Gambier City Council and all sites greater than 1 ha in the city area must be licensed. The Blue Lake Management Plan incorporates plans to identify and locate where potential pollutants are stored in Mount Gambier, and ensure they are stored in appropriately bunded areas to prevent spillages reaching drainage bores, and that industrial wastes are disposed of in a prescribed manner. There are also provisions in the Management Plan to ensure that drainage bores are properly constructed, that silt traps are properly maintained and cleaned, and to ensure that drainage bores can be sealed rapidly in an emergency to prevent spillages of pollutants contaminating groundwater.

- **Enforcement measures.** It is illegal to pollute groundwater in South Australia under provisions of water resource and environmental protection legislation. The Blue Management Plan recommends strategies to develop regulations and policies utilising water resource, environmental and planning legislation in a co-ordinated way to improve groundwater protection in Mount Gambier and surrounding areas.

- **Education and communication measures.** Increasing general awareness of the vulnerability of the karst aquifer to contamination and of the interaction of groundwater and Blue Lake is an extremely important part of the Management Plan. It is proposed to disseminate information through public displays, resource materials for schools, videos, brochures for tourists and residents, and through media releases. It is also proposed to improve lines of communication between government agencies, industry and community groups that have an influence on groundwater management in the area.

- **Research measures.** It is recognised in the Management Plan that ongoing research is required to improve groundwater management in the area. Specific research needs identified are determining the relative risks posed by various land uses on groundwater quality, determining the potential mobility of potential pollutants in stormwater and groundwater, and improving the design of silt traps and drainage bores used in the area.

An implementation group comprising representatives of government, industry and community groups has been established to implement these strategies. The implementation group is currently recommending that a new industrial area be established outside of the groundwater capture zone for Blue Lake, and is undertaking a review of diffuse contamination sources near Blue Lake.

Auckland, New Zealand

Introduction

Auckland is located on a narrow isthmus on the North Island of New Zealand, and is the largest city in that country, having a population in excess of 1 million (Fig. 7.8). The city has a temperate coastal maritime climate with an average rainfall of between 1 000 and 1 500 mm y^{-1} depending on local topographic effects, and is highly variable depending on the El Nino Southern Oscillation in the Pacific Ocean.

Auckland is an important commercial, industrial and administrative centre, and groundwater in the city is utilised by industry and for irrigation. Groundwater also provides about

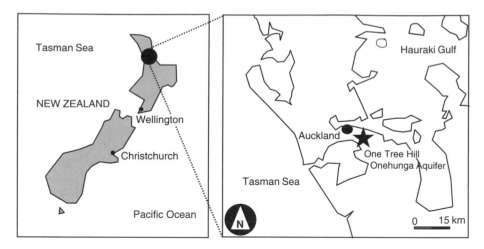

Figure 7.8.   Location of Auckland.

$19\,000\,\text{m}^3\text{day}^{-1}$ of water for the city's municipal supply, out of a possible sustainable abstraction rate of $23\,000\,\text{m}^3\text{day}^{-1}$ (Smaill, 1994). Moves by the city's municipal water supply company to increase its withdrawal to about 95% of the sustainable resource have been met by legal challenges by other industrial users to prevent a virtual monopoly of groundwater use (Smaill, 1994).

The high demand from Auckland's volcanic aquifers, coupled with the high vulnerability of these aquifers to contamination, creates complex groundwater management problems in this city.

Hydrogeological setting
Auckland is a city located in a complex volcanic landscape, and is dominated by extensive lava flows, scoria cones and tuff deposits of Tertiary to Quaternary age which collectively form the "Auckland Volcanic Field" (Viljevac, 1997a). Up to 48 volcanic cones have been recorded in the greater Auckland area, and tuff deposits and lava flows several kilometres long extend from volcanic centres, filling valleys incised in basement rocks comprised of interbedded sandstone and siltstone (Fig. 7.9).

The volcanic deposits in palaeovalleys form an extensive, mostly unconfined, aquifer. Groundwater flow is through fractures, joints, large cavities, and rubbly or scoriaceous zones in the volcanic rocks forming shallow aquifers with a maximum thickness of 35 to 45 m. The aquifers are recharged by the direct infiltration of rainfall, and through drainage bores and soakage pits, which are widely used in Auckland for stormwater disposal. The hydraulic properties of the aquifers vary considerably over short distances, and yields from individual bores can vary greatly depending on fracture density from very low yields, to yields greater than $1\,000\,\text{m}^3\text{day}^{-1}$ (Smaill, 1994). The hydraulic conductivity of the aquifers vary from 0.5 to $320\,\text{m}\,\text{day}^{-1}$, and the transmissivities vary from 68 to $4\,800\,\text{m}^2\text{day}^{-1}$ (Viljevac, 1997a). In general, groundwater flow is strongly controlled by the palaeotopography of basement rocks where this surface is above sea level, and groundwater flows down palaeovalleys towards the ocean. Near sea level, groundwater flow is along preferential paths through high permeability zones in basalt.

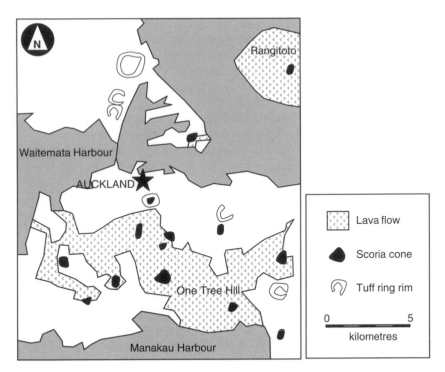

Figure 7.9. Volcanic deposits in Auckland.

Urban development has generally increased groundwater recharge in the Auckland area, and has greatly changed the spatial distribution of recharge. Under natural conditions, recharge to the volcanic aquifers is typically about 30% of average annual rainfall, or about 300 to 375 mm y^{-1} (Viljevac, 1997a). Recharge in urban areas is highest where stormwater is disposed of in soakaways or drainage bores, and may exceed 1 000 mm y^{-1} in some commercial or industrial areas where in-ground disposal is widely used, whereas recharge in residential areas is typically about 800 mm y^{-1}. The lowest rates of groundwater recharge occur in sewered urban areas of Auckland where stormwater is piped to the sea, and recharge in these areas is commonly less than 200 mm y^{-1}.

Ground soakage systems are the main stormwater disposal method in areas underlain by basalt. This is largely due to the high cost of constructing a piped disposal system through hard rock, and the high capacity of the volcanic rocks to take up stormwater. As the relative costs of constructing reticulated systems has increased and industry has become more densely developed, in-ground disposal has become the only economically viable method of disposing the large volumes of runoff produced in these areas.

Groundwater contamination
Groundwater in the volcanic aquifers is generally of high quality because of the short residence time of water in the aquifer, and the limited reaction of groundwater with aquifer materials. However, groundwater quality has deteriorated in some parts of Auckland, particularly in industrial areas. Groundwater in the most heavily industrial areas has abnormally high conductivity and alkalinity levels, and elevated concentrations of chloride, sulphate and some heavy metals.

The most severe occurrences of groundwater contamination have resulted from a number of point-sources including petroleum hydrocarbon contamination from a number of petrol service stations, leachate contamination from a number of closed landfill sites, and heavy metal contamination from a former fertiliser manufacturing plant. Contamination from the former fertiliser plant is particularly severe, as groundwater beneath the site has a pH of 3.5 and a copper concentration of $25\,mg\,L^{-1}$ (Smaill, 1994). An extensive groundwater contamination plume from the site has affected a number of private groundwater users, but fortunately has not affected water quality in nearby municipal water supply bores.

Despite the presence of a number of significant point sources of contamination in Auckland, the main source of contaminants entering the city's volcanic aquifers is stormwater disposal through soakaways and disposal bores. Runoff from roads, roofs and paved surfaces in Auckland is a source of chemical contaminants including petroleum products, pesticides, solvents, heavy metals and other inorganic compounds. Groundwater in some areas is also contaminated by faecal bacteria (Viljevac, 1997b) from road runoff and leaks from sewer mains, and there is a risk that pathogenic microorganisms can be transmitted large distances in groundwater due to the rapid rate of groundwater flow and large pore spaces in the volcanic aquifers. The contamination from these sources is currently not severe, but with pressure to increase in-ground stormwater disposal in Auckland, there is a risk of groundwater quality progressively deteriorating unless stormwater disposal is properly managed.

Runoff may also contain a large amount of suspended solids, which can clog fractures in disposal boreholes, greatly reducing their soakage capacity. Although cleaning of the disposal bores increases soakage in the short term, further clogging often makes the bores unusable. In the past, the solution to this problem has been to drill new disposal bores, and this has resulted in an increasing area of the volcanic aquifers being affected by clogging, reducing their capacity to transmit groundwater.

Groundwater management
Management of water resources near Auckland is carried out by the Auckland Regional Council (ARC), a government agency which is also responsible for environmental management, planning, transport and education in the region. Water resources are managed as part of an integrated environmental and natural resource management process prescribed by the New Zealand Resource Management Act, 1991. This allows many of the factors that may affect water quality, such as land use planning and waste management, to be handled by a single agency through the same legislative process.

The Auckland region currently has a population of about 1 million, a third of New Zealand's population. It is the fastest growing region in the country, with the population forecast to exceed 1.6 million over the next twenty years. This population growth is putting a strain on the region's water resources, particularly groundwater.

One of the major difficulties in managing groundwater use in Auckland is that abstraction has increased to the point where the city is reliant on the additional recharge produced by stormwater disposal to maintain sustainable yields, and yet stormwater disposal is also the greatest risk to groundwater quality in the city. The removal of stormwater disposal could reduce the available recharge in some areas by up to 80%, and could all but eliminate some groundwater resources. The cost of obtaining this water from other sources would be prohibitively expensive.

In order to manage this problem, the Auckland Regional Council has put in place policies that require all stormwater to be treated prior to aquifer disposal, and these are written into groundwater management plans for specific aquifers in the region. The type of treatment required varies with land use, but typically involves the removal of suspended sediment, which also removes a large proportion of the heavy metal load. Petrol service stations, and other activities that store or handle petroleum hydrocarbons, may also require stormwater to be treated with oil separators before disposal. The ARC have also developed regulations for the handling, storage and disposal of potentially hazardous materials on industrial premises to reduce the risk of contaminants entering the volcanic aquifers by stormwater disposal. All premises in groundwater areas with an area greater than $1\,000\,m^2$ that dispose of water to ground require a stormwater disposal permit to ensure compliance with these measures.

Poorly constructed and abandoned bores and boreholes are also potential conduits for contaminants to be introduced into groundwater in Auckland. Consequently, permits are required for drilling, altering the construction and abandoning all water bores in the region to ensure that they are of a suitable standard to prevent contaminants being introduced into groundwater, and to prevent undesirable mixing of water between different aquifers with a different water quality. An additional water abstraction permit is required for all bores, other than those used for small domestic or stock supplies or for fire-fighting purposes, to regulate water usage from volcanic aquifers in the region. Allocation limits for aquifers are established in groundwater management plans, and licensing ensures that groundwater usage is managed on a sustainable basis and that there is sufficient groundwater to maintain flows of natural springs and to prevent seawater intrusion.

7.7   FINAL COMMENTS

The protection of groundwater quality in cities overlying karst and karst-like aquifers is very difficult due to the often extremely high vulnerability of these aquifers to contamination by both pathogens and chemical contaminants. The rapid rate of groundwater movement and the presence of preferred flow-paths often make it difficult to establish wellhead protection zones around individual water-supply boreholes and wells, and therefore good land use planning and management of land use activities is particularly important for providing regional protection of groundwater quality for these aquifers.

In rapidly growing urban centres in developing countries it is generally not possible to implement an orderly planning framework that will protect water quality in karst and karst-like aquifers under an entire city. However, it is important that areas with good quality groundwater on the fringes of urban areas are identified and secured by planning measures before they are covered and contaminated by urban sprawl. Such planning can only be implemented with the support of a community who understand the importance of land use management for protecting groundwater quality. Education and public awareness programmes should therefore be an important component of any groundwater protection initiative developed for these aquifers.

CHAPTER 8

Groundwater management in urban alluvial aquifer systems: case studies from three continents: Agadir, Lima and Los Angeles

Gino Bianchi-Mosquera, Craig Stewart, Bob Kent and John M. Sharp Jr

8.1 INTRODUCTION

Freshwater supplies represent a very small percentage of available water throughout the world. Depending on the geographical location, water resources may be more or less plentiful, but availability is not always directly related to the wise use of the resource. Despite the many cultural, political, and economic differences that exist among countries in diverse regions of the world, a unifying theme for all locales is the need for freshwater to subsist and prosper.

This chapter evaluates three cities in terms of how water rights and water management issues will likely impact their future growth and quality of life. Agadir, Morocco; Lima, Peru; and Los Angeles, California represent three coastal cities at different stages of development, but in similar semi-arid geographical settings that overlay alluvial aquifer systems (Fig. 8.1). The cultural, political, and economic development of these cities has been both a result and a cause for current groundwater issues and concerns.

Los Angeles in southern California is a megalopolis that secured rights to import surface water in the early 20th century. The area was under Spanish colonial rule in earlier times and the rights of Los Angeles to certain local groundwater and surface water can be traced back to that era. The area's growth was supported initially by local groundwater and surface water supplies and subsequently by water imported from watersheds several hundreds of kilometres away. Despite access to supplies of surface water and groundwater from other basins, there is continued concern regarding the availability of water and demand on groundwater supplies as the area's population is projected to continue to grow, and the use by other rights-holders limit its surface water imports. As a result, groundwater basin management occurs in this region to a greater degree than in most places in the world.

A contrasting view is offered by Lima, another large city. Also subjected to Spanish colonial influence, it has a very different approach to water rights. According to the Peruvian constitution, all natural resources, including groundwater, are patrimony of the country and can only be assigned by the government. Thus, the key issues are similar to those of Los Angeles in that there is not sufficient local water for the city's ever-increasing population, but differ in that legal issues have been much less a factor in driving the water management equation.

Figure 8.1.   Geographic location of Agadir, Lima and Los Angeles.

Finally, Agadir represents a city that has grown significantly in recent years, but has not and probably will not reach a level of population comparable to Lima or Los Angeles. However, like those two cities, Agadir is also the main city in its region. Similarly, Agadir is dealing with the issues of insufficient water supply and in some cases deteriorating water quality, but may have more time to use the information gleaned from other cities' experiences. In Morocco, water rights and water management issues have historically been handled at the ministerial level. Water is considered a public resource and government has a total claim to water in the country. However, in 1995 Morocco passed a water law authorising the management of the country's water at the basin or watershed level, which, when fully implemented, will result in a more decentralised and participatory process for water management.

8.2   OVERVIEW OF STUDY AREAS

Agadir

The city of Agadir is located on the Atlantic Ocean approximately 400 km south of Casablanca. Agadir is a relatively modern city, completely rebuilt after a major earthquake in 1960. Greater Agadir, with a population of approximately 500 000 is the largest city in the Souss-Massa River basin (Fig. 8.2). The population of the Souss-Massa River Basin, with an area of approximately 27 000 km^2 has more than doubled from 865 000 in 1971 to 1.9 million in 1997 (Padco, 1998).

Agriculture and tourism dominate the region's economy. Agadir is also an important commercial and fishing port. The rebuilding of Agadir after the 1960 earthquake resulted in a large injection of capital and created a need for workers. This was partly responsible for the increase in migration from the rural areas of the Souss-Massa River Basin to the Agadir area. The planned and controlled redevelopment of Agadir resulted in high land

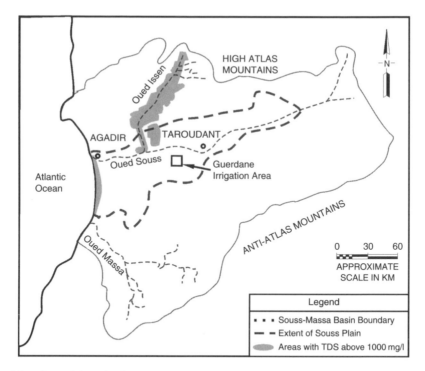

Figure 8.2.   Souss-Massa basin.

and housing cost, and most low-income workers settled in areas on the periphery of Agadir in shantytowns (bidonvilles). Many of these areas are without sewers or piped water. Even in the greater Agadir area, the effluents are discharged directly into estuaries without any prior treatment, although there is a project underway to upgrade the sewage system and build a wastewater treatment plant.

Lima

Lima is located along the coast of central Perú, within the Coastal desert physiographic region (Karakouzian et al., 1997). As Lima's population has grown from 500 000 in the early 1940s to over 8 million in the 1990s, the city's expansion has been constrained by the Andes mountains. These rise to over 5000 m above sea level to the east of the city, within 120 km of the coast. As a result, Lima has grown beyond its original siting along the banks of the Rímac river and has reached the Chillón to the north and Lurín rivers in the south (Fig. 8.3).

According to Ventura Napa (1995), the water resources of the Rímac Valley have been in continuous use since the first irrigation channels built by Lima's earliest settlers around 2000 B.C. The Spanish founded Lima at its current location in 1535 because of its proximity to the sea and a natural harbour, and for the ample water supply derived from the Rímac River. As the population grew, the ratio of urban to agricultural areas increased as well. Karakouzian et al. (1997) note that in 1910 there was approximately 600 km^2 of agricultural land to 12 km^2 of urbanised areas, but in 1995 the areas were 105 km^2 and 507 km^2, respectively.

The rapid and uncontrolled population growth observed in Lima has resulted in a transfer of water from agricultural use to residential, commercial, and industrial uses. The irregular

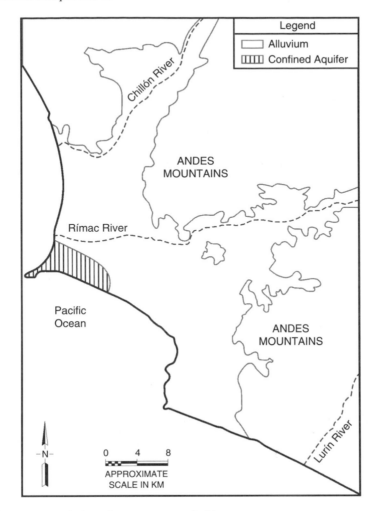

Figure 8.3.   Conceptual view of water resources in Lima.

flow of the Rímac, Chillón, and Lurín rivers throughout the year, together with the increasing demand by the growing population, continue to have great impact on the groundwater resources of the area. These are managed by SEDAPAL, the city's water supply and sewerage company. Lima's water availability is also reduced by the unusual climate that results from a combination of the cool Humboldt current in the Pacific Ocean and the nearby Andes mountains. Although it is located at about 12° south latitude, it has an arid climate with average annual temperatures that range from approximately 15 °C in July to 21 °C in January. Mean annual precipitation ranges from 10 to 26 mm, resulting in arid to semi-arid climate and conditions.

Los Angeles

The greater Los Angeles metropolitan area falls largely within a group of alluvial basins lying generally along the Pacific Ocean coast of southern California, USA. The major basins are

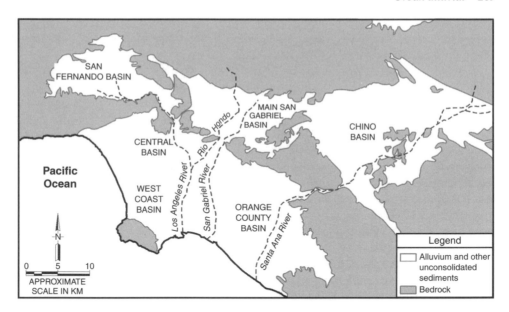

Figure 8.4.   Groundwater basins in the Greater Los Angeles area.

connected via narrows incised through hills, and together cover an area of more than $2\,600\,\text{km}^2$. The southwestern most of the basins border the Pacific Ocean, with tributary basins extending distances of approximately 30 to 80 km inland. The inland basins are flanked by mountain ranges that rise to maximum elevations ranging from about 900 to over $3\,000\,\text{m}$ above sea level. The general configuration of these basins is shown in Figure 8.4. Climate in the area is semi-arid, with about 250 to 400 mm precipitation on the basins and up to approximately 800 mm in the flanking mountains. Precipitation occurs mostly between November and April as a result of moisture-laden storms moving inland from the Pacific Ocean.

Prior to 1769, the Los Angeles Basin was occupied by Native American peoples, who relied on surface water features within the basin and surrounding upland areas. Spanish settlement of the basin began in 1769. The area was under Spanish colonial rule from 1769 to 1822, under Mexican rule from 1822 to 1848, and has since been part of the United States. Population growth was relatively slow throughout the mid-1800s, and has been relatively rapid but somewhat episodic since. From the late 1800s until the 1940s, land use in much of the coastal plain basin and most of the inland basins was either agricultural or ranching. Rapid expansion of industrial, commercial, and residential development into the inland basins began in the 1940s and 1950s. Currently, the population of the greater Los Angeles metropolitan area (Los Angeles and Orange Counties and parts of Riverside and San Bernardino Counties) is about 15 000 000.

8.3   HISTORICAL GROUNDWATER ISSUES

Agadir

Prior to the French colonial period (1912 to 1956), water management and allocation in Morocco was based upon complex ethnic, social and religious factors (Ilahiane, 1996).

Before the intensive irrigation practices which were implemented by the French and continued after independence by Hassan II, there was fierce competition for the available and limited water resources used for irrigation.

Rainfall is highly variable in the Souss valley and averages approximately 280 mm/y. Many of the major streams in Morocco, including the Oued Souss, the major river in the Souss-Massa basin, have irregular flows and are frequently dry in the summer. In the last 80 years, Morocco has experienced approximately 27 years of drought and the interval between droughts is only about 3 years (Swearingen, 1996). Prior to the construction of major dams, groundwater was the only storage available for irrigation water during dry months. Before the French expansion into north Africa, the technology and money to construct large scale dams did not exist in Morocco and a significant portion of the available water resources of the Souss-Massa basin was lost to the Atlantic during the wet months (Swearingen, 1987). During the French administration, the better farming land in Morocco which included the fertile Souss-Massa valley was expropriated by European and elite Moroccans. Today, much of this irrigated land is still concentrated in the hands of large landowners (Swearingen, 1987), and more than 80% of farmers own less than 5 ha each.

Major agriculture in Morocco began in the mid-1930s. This was referred to as the "California Policy" and was based on the irrigation and export of citrus fruits and vegetables. In 1985, the area under irrigation in Morocco was approximately 650 000 ha. Of these, approximately 100 000 ha or 15% of all irrigated land was located near Agadir in the Oued Souss valley.

Lima

During its first 17 years of existence after its founding in 1535, Lima's population obtained its drinking water directly from the Rímac river (SEDAPAL, 1997). In 1552, the municipal authorities recognised the health concerns that arose out of that practice and attempted to bring water from the springs in the La Atarjea area, several kilometers upriver of the city. A royal decree from the Spanish king in 1555 required that certain hygiene practices be followed in the provision of drinking water. As a result, the first city water authority was formed in 1556 to oversee the use and distribution of water from the city's canals and fountains. By 1578, city water was transported in clay pipes. The water distribution system at that time was generally implemented to provide water to key sectors of the city: churches, government buildings, and the homes of influential people.

As the city and the water demand grew, additional piping was installed and the La Atarjea spring source area expanded. By 1858, the water company was forced to dig infiltration galleries down to approximately 10 m below ground surface. The first public groundwater wells were drilled in the same area in 1872 and centrifugal pumps were used to bring the water to the surface. Water quality was a public concern until 1917, when the city's first chlorination plant was installed (SEDAPAL, 1997).

Lima's exponential population and geographic growth quickly outgrew the supply of water that could be obtained from the city's distribution network. This led to a significant increase in the number of wells being drilled throughout the city, as it was cheaper to install a well than to expand the water distribution system. Groundwater extraction for public use in Lima increased from about 0.5 m^3/s in 1955 to almost 8 m^3/s in 1996. Ventura Napa (1995) estimated that there were about 1 000 wells in use in the city. A subsequent study by Watkins et al. (1997) cited that there were about 2 000 wells registered in the city. Rojas et al.

(1995) state that there were 340 deep wells in the city, but only 280 were operating at a combined extraction rate of 7.2 m³/s. The overexploitation of water caused a decrease of about 20 to 30 m in the city's water table between 1969 and 1992 (La Touche, 1997). The water table is now typically found at about 95 m below ground surface throughout most of the city. Near the port of Callao, however, an unpublished study conducted by the authors noted shallow groundwater of low to moderate salinity at 15 m below ground surface 500 m inland from the shoreline.

Los Angeles

As with other states in the western United States, the historical philosophy of water use in California was to put every available quantity to its "highest beneficial use"; that is, to use it to "make the desert bloom." This approach involved development and use of groundwater beyond sustainable levels, resulting in conflicts and subsequent legal actions that allocated production rights and established management frameworks for the area's basins.

Under pre-development conditions, groundwater was relatively shallow within much of the basin area. Groundwater development began in the late 1800s, with numerous wells installed for domestic and agricultural use by the early 1900s. By 1905, about 8 000 wells were present on the Coastal Plain basins. An estimated 2 500 of these were flowing artesian wells. Fewer than 1 000 of these were equipped with pumping plants, and the rest were equipped with windmills or were used for domestic purposes (Mendenhall, 1905a,b,c). By the early to mid 1900s groundwater extraction from most of the Los Angeles area groundwater basins had reached a rate that was not sustainable under existing recharge conditions. Overdraft conditions had resulted in increased pumping lifts, reductions or loss in production from some wells, and seawater intrusion along the coast.

As the Los Angeles area has grown and its water demands have increased beyond the level which local supplies could support, engineering projects were constructed to convey water from other parts of the western United States to Los Angeles. These projects were monumental in scope, political complexity, and controversy. The first of these was the Los Angeles Aqueduct. During the very early 1900s the City of Los Angeles acquired much of the property and most of the water rights in the Owens Valley along the eastern side of the Sierra Nevada mountains, and constructed the Los Angeles Aqueduct to convey diverted surface water and extracted groundwater south to Los Angeles. The Los Angeles Aqueduct was completed in 1913 and provided water that supported tremendous growth of the city's area and population. In 1940 the Aqueduct was extended upstream to convey water from the Mono Basin. Currently, about 400 000 m³/y of water are delivered to Los Angeles via this aqueduct (Los Angeles Department of Water and Power, 2000). In 1941, the Colorado Aqueduct was completed between the Colorado River and coastal areas of southern California. The State Water Project was constructed in the 1960s and 1970s. This project consists of storage, diversion, and conveyance facilities to transfer water from northern California to the agricultural Central Valley and the urban areas of southern California. During recent years, imports of water by the Metropolitan Water District of Southern California (MWD; a regional wholesaler of water in southern California) via the State Water Project and Colorado River combined have been about 2.5×10^9 m³/y (MWD, 2000).

At the time groundwater development began, the chemical quality of groundwater in the Los Angeles area basins was generally good, and suitable for potable use. In most areas, TDS concentrations were less than 350 mg/l. Higher TDS concentrations generally corresponded to areas where local aquifer sediments were derived from marine sedimentary bedrock (such as along the southeastern margin of the Orange County coastal plain basin), or were along the coast in areas of tidal influence (Mendenhall, 1905c; Poland et al., 1959). Although the aquifer systems within the Los Angeles area basins still contain vast quantities of high-quality water, widespread degradation of groundwater quality has occurred during the past century. Primary causes of degradation have included seawater intrusion, discharge or unintentional release of industrial chemicals or wastes, and agricultural activities. Seawater intrusion was occurring in the West Coast basin by 1912 (Blomquist, 1992). The potential loss of groundwater resource in the rapidly developing coastal plain groundwater basins triggered comprehensive investigations and ultimately the construction of four major seawater intrusion barrier injection well systems. In the late 1970s and 1980s, groundwater in widespread areas of the San Fernando and San Gabriel basins was found to contain chlorinated solvents at concentrations exceeding drinking water standards, and numerous municipal water supply wells were taken out of use. In general, these impacts were greatest in areas where extensive industrial development had occurred in inland basins with predominantly coarse-grained alluvial sediments.

8.4   LEGAL, POLITICAL, AND ENVIRONMENTAL FACTORS

Agadir

The first water legislation was introduced in Morocco in 1914 (World Bank, 1995). Water is considered a public resource and the government has virtually a complete claim upon water. Exceptions are water rights acquired prior to 1914 together with some historical patterns of water use, typically traditional small-scale irrigated agriculture. Four government ministries are primarily responsible for water issues in Morocco: the Ministry of Interior (MI), the Ministry of Public Works (MTP), the Ministry of Agriculture and Agricultural Development (MAMVA) and the Ministry of Public Health (MSP). The Office National de l'Eau Potable (ONEP), which is under the MSP, and sixteen municipal water utilities (Regies) under the MI, are responsible for urban and some rural public water supply distribution. The Direction Generale de l'Hydraulique (DGH) under the MTP is responsible for providing bulk water in the river basins and also manages the construction and maintenance of surface reservoirs and dams. In addition, DGH is responsible for studying, planning, development, and allocation of all water resources, including groundwater (World Bank, 1995).

There are three major regional agencies in the Souss-Massa basin that are responsible for distribution of water. The Direction Regionale de l'Hydraulique (DRH) is in charge of hydraulic water resources management at the regional level, and coordinates water management committee at the river basin level. The Offices Regionaux de Mise en Valeur Agricole (ORMVA) is responsible for constructing canals or pipelines to transport irrigation water from dams to farmers. ORMVA has also drilled numerous irrigation wells which are also pumped into regional irrigation distribution systems. The actual network of canals and pipelines are maintained by ORMVA together with a complex system of irrigation user groups. Where associations (user groups) have taken over operation and maintenance responsibilities, they charge each farmer for water which is used to maintain the distribution systems. Where user groups have not been formed, ORMVA collects money directly from the farmers for operation and maintenance of the distribution system, but capital costs are not recovered.

ONEP is responsible for providing wholesale potable water to the Regie Autonome Multiservice d'Agadir (RAMSA) and other municipal users. In some cases standpipes have been provided for areas not served by a central piped supply. Approximately 60% of the potable water supply in the Agadir area comes from 31 groundwater wells, and 40% from surface water reservoirs. According to World Bank (1995), the average urban domestic water consumption in Morocco is approximately 100 l/c/d. The average urban domestic water consumption in Agadir is about 75 l/c/d. However, the urban water consumption is expected to rise to between 120 and 160 l/c/d by the year 2020, due to the increased standard of living and the installation of additional water lines, sewage collection and treatment systems by the Moroccan government or donor countries. There are two major population centres in the Souss-Massa basin which obtain a portion of their water supply from the Souss alluvial aquifer, and these are Agadir and Taroudant. In addition, there are approximately 80 rural communities, many of which have standpipes or wells to serve the communities that do not have piped systems.

For over 30 years, the Moroccan government has emphasised the need to develop Morocco's water resources. In 1967, a national goal was established to have one million hectares of land under irrigation before the year 2000 and in 1986 the government stated its goal to build one dam per year. The goals were both consistent with Morocco's long term strategy of "Not one Drop to the Sea" (World Bank, 1995).

Water resource planning in Morocco has historically been a centralised function. However, in 1995 Morocco passed a law which authorised the creation of watershed authorities to manage the water resources of the country's river basins (Water Law no 10-95). Each river basin authority can have between 24 and 48 members, comprising representatives of government, public water supply organizations, agriculture sector, industry chambers, provincial councils, ethnic groups, and irrigation user groups. The three major regional agencies, DRH, ORMVA, and ONEP, will probably play a major role in this new agency once it is created. The law has not yet been completely implemented; however, it is seen as a major step toward developing a decentralised participatory approach to discussions regarding water management (FORWARD, 1999). As part of this participatory process, the government expects the end users of the water to absorb a larger share in the cost of water.

The average urban charge for water in Agadir is US $0.03/m^3 and is seventeen times greater than the average charge of US $0.015/m^3 for gravity irrigation water, and about six times greater than the charge of US $0.056/m^3 for irrigation water under pressure. After adjustment for treatment and conveyance costs, irrigation water costs are estimated to be less than 20 percent of the cost of urban water. Because neither urban or irrigation costs includes the cost of construction of dams, urban charges frequently cover only 50% of the full cost of water, and irrigation charges less than 10% of the full cost of water. Where piped water is not available, water is frequently purchased from vendors at a cost between US $2.50 to US $5.00/m^3 (World Bank, 1995). Private well owners do not currently pay for water. When the 1995 water law becomes fully implemented, it will allow the Souss-Massa River Basin agency to charge approximately US $0.002/m^3 of groundwater.

Lima

Ever since Spanish colonial times, the central government has played a key role in the use and distribution of water. Typically, the Spanish settlers used the water during the day while the native communities used it at night and during non-working days. The Water

Code of 1902 assigned water rights to the owners of the lands through which the water flowed, but this was changed in the national constitutions of 1933, 1979, and 1993. These three documents concurred in assigning to the central government the right to all natural resources, including surface and groundwater.

The government's role in the management of water resources is currently defined by the General Law of Waters. Because Peru's government framework follows a sectorial approach, the various sectors such as mining, hydrocarbon, fisheries, and others have acted independently in addressing water quality issues. Fernandez (1998) noted that it is not uncommon to encounter contradictory legislation related to water issues. Because large portions of the water resources in Peru are used by the agricultural sector, water resources issues are managed by the National Institute for Natural Resources (INRENA). However, due to its political and strategic importance, the responsibility to manage the water supply and sewerage in Lima and its surroundings has been assigned to SEDAPAL, an independent company created in 1981 and wholly owned by the central government. Because it is not addressed in the General Law of Waters, there is no regulation that requires payment for use of groundwater, except in Lima. A private entity or individual is only required to pay a small fee to process a permit application to install a well. Currently, SEDAPAL will only issue such permits when it is assured through hydrogeological studies that the proposed extraction will not have a significant impact on the groundwater supply. SEDAPAL charges approximately US $0.33/m^3 of potable water for residential use, but residents of areas not served by the utility pay significantly more for water brought to them by private water trucks. Industries pay approximately US $0.19/m^3 for untreated water extracted from their on-site wells.

A new proposed Law of Waters has been awaiting congressional approval for several years. The proposed law would clarify the issue of water rights and how to obtain them. It would also establish hydrographic basin agencies to assign surface water rights, but would explicitly state that groundwater belongs to the owner of the land from where it is being extracted. The proposed law would only stipulate that the government has a right to restrict the use of groundwater if it impacts the water rights held by others.

Los Angeles

The legal framework controlling groundwater use and quality in Los Angeles as well as the rest of California is complex. The state has statutory authority for managing surface water. California statutes and laws controlling surface water rights have evolved over the past 150 years and include components of several disparate doctrines, including appropriative rights, riparian rights, and pueblo rights. However, the state does not have similar authority to manage groundwater. Instead, landowners in California generally have correlative rights to extract as much groundwater from beneath their property as they can put to beneficial use. In addition, groundwater beyond that claimed by overlying landowners can be appropriated for use by others, and prescriptive rights may be established through a period of uncontested use.

In the mid to late 1900s, widespread overdraft conditions and conflicting interests in groundwater production and use led to litigation involving water rights in each of the principal Los Angeles area groundwater basins. Groundwater rights in the San Fernando, San Gabriel, West Coast, Central, and Chino basins, as well as in the nearby Raymond, Puente, Cucamonga, and San Bernardino basins, were allocated through adjudication judgements, which established the framework for management of groundwater in each of these basins (CDWR, 1996). Although there are differences among the judgements for individual basins, each defines quantities of groundwater that can be pumped by parties to the judgement or other users, and establishes a watermaster with responsibility for overseeing management of the basin and collecting and reporting data to document basin conditions (Blomquist, 1992). Groundwater rights for the Orange County portion of the coastal basin were not adjudicated, but were resolved through a settlement that involved a special act of the California legislature approving formation of the Orange County Water District (OCWD) (OCWD, 1999). The adjudication judgement or settlement for each of the basins ties groundwater rights and extraction or replenishment fees to an identified basin safe yield or operating yield.

The result of this history has been a system of groundwater management that involves many diverse public, private, and judicial entities. Public water districts, court-appointed watermasters, flood control districts, the US Army Corps of Engineers, the US Geological Survey, private water companies, the California Department of Water Resources, the California Department of Health Services, the California Regional Water Quality Control Boards, and others are all involved in various management, recharge, data collection, or regulatory activities. Among these, the City of Los Angeles Department of Water and Power, which constructed and operates the Los Angeles Aqueduct and provides water to the City of Los Angeles, is the largest municipally-owned utility in the US. The Metropolitan Water District of Southern California (MWD) is a regional wholesaler of water that currently includes 27 member agencies and provides water for an area stretching from northwest of Los Angeles southward through the greater Los Angeles area to San Diego. MWD constructed and operates the Colorado River Aqueduct and holds contracts for the greatest amount of water imported from northern California via the State Water Project.

The groundwater management policies developed for each of the adjudicated basins and Orange County were highly dependent on availability of imported surface water. Although a variety of environmental factors can be tied to these policies, the greatest environmental impacts are probably those that have occurred in the areas from which water is imported. These impacts have included severe losses of surface water, riparian, and wetlands habitat in the Owens Valley and Mono Basin, reduction of flow and aggravation of water quality problems in the Colorado River, and water quality degradation and habitat loss along the river systems and San Joaquin/Sacramento River delta areas in Central and Northern California. It is noted, however, that a high percentage of the water transferred from northern California and the Colorado River is used for agricultural purposes in California's Central, Coachella, and Imperial valleys, and that these negative environmental impacts are therefore only partially associated with urban water use.

The cost for production of groundwater in Los Angeles area basins ranges from less than US \$0.05 to over US \$0.15/m^3. Current costs to wholesalers for imported water range from about US \$0.20 to US \$0.40/m^3, depending on season and level of treatment. In some basins, a "pump tax" collected on pumped groundwater is used by the water master or other entity for purchase of imported water for replenishment. Such a pump tax may be applied to all water produced or only to water produced in excess of a pumper's allocation. Recent pump tax charges for the Orange County, West and Central Basins are approximately US \$0.08 and US \$0.13/m^3, respectively (OCWD, 1999; Water Replenishment District of Southern California, 1996). In addition, pumpers in the adjudicated basins may be assessed a fee for support of watermaster activities. Together, these costs result in a total cost to the residential consumer that currently ranges from about US $0.40 to US $0.80/m^3 of potable water at the tap.

8.5   CONCEPTUAL REPRESENTATION OF THE AQUIFER SYSTEMS

Agadir

The Souss-Massa River Basin covers approximately 27 000 km^2 and is bounded on the north by the High Atlas Mountains; on the South by the Anti-Atlas Mountains; on the east by the joining of these two mountain ranges; and on the west by the Atlantic Ocean. The High Atlas and Anti-Atlas are composed primarily of limestone, dolomite, and shale, but they are also extensively intruded by igneous rocks. The valley floor consist of alluvium, with an occasional outcrop of Cretaceous strata (DGH, 1999). The thickness of the alluvium varies and in some locations exceeds 250 m. Groundwater is obtained primarily from the alluvium. The recharge rate is approximately 450 million m^3/y, but varies significantly with rainfall. More than 13 000 wells withdraw water for domestic, industrial and agricultural uses. The total estimated water supply (surface water and groundwater) available on a sustainable basis ranges from 794 to 1 085 million m^3/y. Total water consumption is approximately 965 million m^3, comprising 315 million m^3 from surface water and 650 million m^3 from groundwater. The demand for water in the Souss-Massa Basin exceeds the sustainable supply and the deficit is made up by mining groundwater. A ground water balance for 1976, 1979, 1985 and 1993 is presented in Table 8.1.

Approximately 94% of the water resources in the Souss-Massa basin are used for irrigation as compared to approximately 85–90% for Morocco as a whole (DGH, 1999). The remaining 5–6% are used for public and industrial supply. The major areas of irrigation are restricted to the Massa zone and the Souss Valley. The most intensive irrigation is centred near the town of Guerdane, approximately 50 km west of Agadir and near the Souss River

Table 8.1.   Water balance of the Souss aquifer (million cubic meters).

	1976	1979	1985	1993
Infiltration from rain and surface runoff	66.2	62.8	57.8	37
Infiltration in the base of the wadi	88.7	208.5	50.2	166
Return of surface irrigation waters	14.3	13.7	8	15.2
Upward flow from deep groundwater	3	3	3	3
Contribution from exposed aquifers	48	48.8	43.7	46.2
Total recharge	220.2	336.8	162.7	267.4
Underground discharge to the ocean	22	19.9	15	19
Drainage through the Souss alluvium	8.2	60.5	0	0
Pumping for irrigation by traditional farmers	116	73.7	11.1	83
Pumping for irrigation for public and private sectors	250.6	278.1	365.4	515
Pumping for drinking water	8.1	9.8	16.8	30
Total discharge	404.9	442	408.3	647
Groundwater balance (recharge-discharge)	−184.7	−105.2	−245.6	−379.6

Source: DGH, 1999.

in the centre of the valley. In the Guerdane area, the water table decreased by as much as 40 m between 1968 and 1986.

In the Souss-Massa basin, the rate of water usage is not sustainable even under the best planning conditions. Reuse of treated wastewater can play a role, but it will be insufficient to meet the projected needs of the area. The major impact will be on agriculture in the Souss-Massa region. Morocco's agricultural gross product contributes approximately 20% of the GDP and agriculture employment represents approximately 41% of total employment, two of the highest ratios for Mediterranean countries (Mohamed Ait Kadi, undated). The Souss-Massa Basin produces almost two-thirds of Morocco's exported fruit and vegetables (USAID, 1999). The over-pumping of the Souss aquifer has resulted in significant water level declines and by the year 2010 groundwater levels in some locations will be sufficiently low to make it uneconomical to pump. Some land now under cultivation may have to be abandoned with loss of agricultural employment, which will contribute to the continued migration from the rural areas to the Greater Agadir area.

Generally, groundwater in the alluvium contains water with TDS less than 1 000 mg/L, but some wells along the coast show the effects of seawater intrusion. Another area of poor groundwater quality exists east-northeast of Agadir, near the High Atlas Mountains. The source of this poor quality water (TDS > 2 000 mg/L) is infiltration of poor quality water from the Oued Issen (Issen River, Fig. 8.2) that drains a watershed underlain by Triassic evaporites (Boutaeb et al., 2000).

The major recharge areas for Agadir's water supply are rural in nature, and intensive industrial development has not occurred. Groundwater contamination by chlorinated organics such as PCE/TCE is therefore not a major threat to the water supply, but there is concern for the impact of agricultural chemicals used in the area. Most of the sewage from Agadir is untreated and discharges to the ocean, but the potential for contamination of the water supply by untreated sewage is low because the majority of the public water supply wells are not along the coast.

Lima

Lima is located on the alluvial fans created by the Chillón, Rímac, Lurín, and paleo-Canto Grande rivers. The extent of the predominantly unconfined aquifer, shown in Figure 8.3, is approximately 400 km^2 (Wild & Ruiz, 1995). Rojas et al. (1994) describe the Lima aquifer as consisting mostly of unconsolidated alluvial sediments that can be divided into two main sections. The upper 100 m or so of the aquifer is dominated by gravels and coarse-grained sediments within a sand and clay matrix. The lower 150 m or so of the aquifer is dominated by finer sands, silts, and clay. Interbedded silts and clays increase towards the port of Callao, where the aquifer is confined.

The aquifer boundaries are roughly defined by the Pacific Ocean to the West and the Andean batholith to the East, and by the Andes on either side of the Chillón and Rímac rivers as they flow through the higher elevation valleys on their way to the ocean. The base of the aquifer consists of low permeability Mesozoic sediments at the coast and the Andes batholith inland (Karakouzian et al., 1997). Lerner et al. (1982) note that minor ion chemistry studies suggest that there is very little recharge from the basement rocks into the aquifer.

Early studies of the Peruvian coastal aquifers conducted by Gilboa (1971) suggested that recharge to the aquifer occurred mainly through infiltration from the Rímac river and irrigation canals. Lerner (1986) added that leaking water supply piping was another significant component of recharge. He estimated that about 40% of the average water supply in 1978 recharged the aquifer, approximately $6 m^3/s$. Foster (1996) indicated that recharge rates in the Lima area have increased from about 5 mm/y to over 500 mm/y from utility system losses. Other main recharge contributions included irrigation losses of about $4.6 m^3/s$ and river recharge of about $3.4 m^3/s$. Precipitation does not contribute to recharging the aquifer. Karakouzian et al. (1997) cite pump test data from a 1994 study conducted by Binnie & Partners that indicated average transmissivities ranging from $1\,500$ to $3\,000 m^2/d$ across most of the aquifer, but increasing to $5\,000 m^2/d$ in the middle portion of the Rímac cone. Lerner et al. (1982) used pump test data to estimate a specific yield of 0.06 for the aquifer.

Few data are readily available to assess Lima's groundwater quality. Karakouzian et al. (1997) state that groundwater in Lima is of the calcium sulphate type and relatively hard, but meets international standards at most wells. At the time of their study, ammonia had not been reported in the wells, but high values of nitrate were observed in wells in the Chillón valley, probably due to agricultural activities. Rojas et al. (1995) note that one of the key concerns with groundwater quality in Lima was the presence of faecal coliform bacteria in the drinking water supply. The authors attributed this to the poor chlorination systems that exist at the various groundwater distribution points in the city, where groundwater is disinfected via chlorination before joining the main distribution network. Iturregui (1996) noted that a study conducted by the SEDAPAL privatisation committee found four wells that had been contaminated with gasoline. Typically, gasoline contamination in SEDAPAL wells has been identified by the smell of the water. The impacted wells were reportedly removed from the distribution network and pumped until the smell could no longer be detected.

Los Angeles

The Los Angeles area geologic basins were formed through Tertiary and Quaternary structural deformation. The maximum thickness of unconsolidated freshwater bearing sediments in the coastal plain basins is greater than $1\,000 m$. Alluvial deposits in the San Fernando, San Gabriel, and Chino basins range from several hundred to over 300 m thick. Because of their proximity to mountainous areas of crystalline bedrock, the flanks of the inland basins contain alluvial-fan deposits with a high percentage of coarse-grained sediment. These sediments are generally poorly sorted on a bulk scale, are relatively permeable, and comprise only crudely-developed aquifer/aquitard packages. The more distal parts of the inland basins contain sediments associated with a wider range of alluvial facies, and some lacustrine deposits. The coastal plain basins contain similar alluvial and minor lacustrine deposits and also unconsolidated marine deposits. Although relatively coarse-grained alluvial sediments occur where deposited along the ancestral channel areas of the major drainage systems, the coastal plain basins overall contain a relatively higher percentage of finer-grained sediment, and more clearly-developed aquifer and aquitard zones.

Estimated groundwater balances have been developed for each of the major groundwater basins in the Los Angeles area. Each of the basins is managed with consideration of these estimates. Together, total groundwater storage capacities in the basins is estimated to be on the order of $5 \times 10^{10} m^3$ (Blomquist, 1992), although some of this capacity is occupied by brackish water or is otherwise not readily useable. Under current conditions, which include enhanced recharge of storm flows, injection or percolation of some treated wastewater, and replenishment with imported water, cumulative operational safe yield for the basins is estimated to be in the order of $1.3 \times 10^9 m^3/y$. For comparison, annual imports of surface water into southern California by MWD are approximately twice this amount (MWD, 2000).

As land use in the area has changed and surface water imports have increased, the hydrogeologic conditions in the area also have changed. Estimates of areal groundwater recharge from precipitation range from about 40 to 75 mm/y (Geomatrix, 1997a). Initially, diversion of water from the area's rivers for irrigation may have reduced total groundwater recharge but increased recharge locally in irrigated areas. Further increase in areal recharge probably occurred in the early to mid 1900s as imported water was used extensively for agricultural irrigation. For example, in a part of the Chino Basin that is still agricultural, recharge from delivered or applied water (agricultural and urban) has been estimated to be about 250 to 300 mm/y (Geomatrix, 1997b). In more urbanised areas, including the San Fernando Basin and a part of the Santa Monica Basin (immediately north of the of the West Coast Basin), total areal recharge is estimated between about 175 and 350 mm/y, with at least 75% of this coming from delivered water (Geomatrix, 1997b). As noted by Foster (1996) and Geomatrix (1997a), leaking water supply distribution main and lateral pipelines, landscape over-irrigation, and leaking sewer lines contribute to urban recharge.

8.6   CURRENT GROUNDWATER MANAGEMENT ISSUES

Agadir

The Moroccan government has prepared several reports predicting water consumption in the Souss-Massa Basin for the next 20 years. Even under the most optimistic forecast which includes maximum collection of surface water, conversion to drip irrigation and reuse of treated sewage effluent, there will be a groundwater deficit of 50 million m^3/y at the end of the planning period. Surface water in the basin is collected and stored behind three major dams which were constructed between 1972 and 1991, but additional dam sites are limited. The total surface water storage capacity is approximately 650 million m^3. In addition to its use of the surface water for irrigation and potable supplies, water from the Aoulouz reservoir located near the head of the Souss Valley is also used to recharge the alluvial aquifer.

The most critical groundwater management issue in Agadir and the surrounding region will be the ability of the New River Basin Management Authority to develop a demand side water management policy. In addition, the declining water levels in the aquifer must be reversed or slowed to prevent the cost of pumping water becoming too excessive for economical crop production. Unless policies are implemented to conserve water, the export economy of the Souss-Massa Basin may decline with severe economic impacts to the Agadir area.

Lima

There are two main groundwater management issues that could represent a significant financial drain to SEDAPAL and have an impact on the city's population. The first issue is over-exploitation of groundwater to meet the demand, whilst the second is groundwater quality. SEDAPAL (1997) recognised that groundwater management was necessary to maintain the same level of service throughout the city in the long-term. SEDAPAL is using groundwater models discussed by Lerner et al. (1982), Wild and Ruiz (1987), and Watkins et al. (1997) to manage groundwater extraction in the city so that aquifer depletion is minimised or at least reduced until additional reservoirs planned in Andean areas are completed.

The concern over water quality is also an issue, both from a public health standpoint and from a financial perspective. Rojas et al. (1995) proposed that much of the faecal contamination observed in Lima's groundwater was a result of improperly located, constructed, or abandoned wells. They noted the existence of unprotected dug wells and open well heads throughout the city. In addition, there is no monitoring programme to adequately assess well conditions or water quality. Typically, groundwater is only tested for bacteriological parameters, conductivity, pH, salinity, and turbidity. Thus, it is not yet feasible to make a comprehensive assessment of current groundwater quality conditions in Lima.

Los Angeles

As summarised above, current groundwater management in the Los Angeles area occurs at a basin or watershed level. Each basin is managed somewhat differently and by a different combination of entities. However, there is a relatively high level of common ground and information sharing among the basins, through legal entitlements of water from upstream basins to downstream, common memberships in regional agencies, agencies having roles in multiple basins, and professional associations. Management strategies tried in one basin are watched by managers in other basins, and may be adopted if found effective. Examples of this include use of treated wastewater for recharge in seawater intrusion barriers or spreading basins, and construction and operation of facilities for treatment of nitrate-affected groundwater in areas of former agricultural land use.

Efforts have been made for nearly a century to optimise recharge of storm flows. Increasingly the groundwater basins are being managed to optimise acquisition and replenishment of imported water at times (seasonal and wet years) when greater quantities are available or costs are lower. Water transfer and water banking arrangements are being established to support storage of groundwater where capacity is available and then the export of the water to other basins when needed. Because California's water resources already have been heavily developed and linked by major engineering projects, groundwater management outside the immediate Los Angeles area plays a major role in supply for the area. For example, water originating in northern California or the Colorado River could be recharged and stored in a water bank area in central California or the Mojave Desert, then later extracted and transferred to the Los Angeles area, where it would be used, treated, and then recharged again within the same basin or a downstream basin. Groundwater producers and managers are developing and implementing cyclic and long-term storage programmes, and executing transfers and leases of production rights and stored groundwater. Treated wastewater is used for replenishment, sometimes applied through spreading basins and sometimes used for injection in seawater intrusion barrier wells. Management of groundwater quantity and quality are increasingly integrated. For example, the Optimum Basin Management Programme recently developed and currently being implemented by the Chino Basin addresses water quality issues extensively. This programme even addresses the links between groundwater extraction patterns, groundwater quality, and land subsidence (Wildermuth, 1999).

Increasingly, management of both water supply and groundwater quality has merged. By the late 1990s, the scope of impact from industrial solvents in the inland groundwater basins was relatively well understood and efforts were underway to resume pumping of affected groundwater, with treatment of the water and subsequent use. Historical agricultural practices in the San Fernando, San Gabriel, and Chino basins, and a high concentration of dairy farms in the Chino Basin, contributed to groundwater degradation by nitrate and other salts. In addition, the salinity of imported Colorado River water used for basin replenishment, and the use of treated wastewater, also with elevated salt concentrations, has exacerbated the increase of salt concentrations in the basins. Water treatment facilities for removal of nitrate and other salts have been constructed or designed for several of the basins, and treated water from these facilities is put to use.

8.7 APPROACHES FOR AN INTEGRATED GROUNDWATER MANAGEMENT STRATEGY

The integrated water resources model for Latin America proposed by the World Bank (1998) is used below to evaluate groundwater management strategies in the three cities being discussed. The World Bank (1998) proposes the following strategies:
- Stakeholder involvement and participation.
- Market-driven mechanisms should be used to allocate water resources.
- Ensure water availability to the poorest users.
- Comprehensive and balanced water resources sectorial laws and regulations.
- Watershed or basin management approach.
- Institutional strengthening.

With some exceptions and area-specific differences, these approaches are being followed in the three urban areas discussed here.

Agadir

The 1995 Water Law authorised the creation of river basin agencies to improve and decentralize water management in Morocco. The mandate for each agency was to prepare master plans and oversee their implementation, providing appropriate and necessary technical support. Each agency was also to issue permits, monitor water quality and quantity, enforce relevant laws, and respond to drought. Responsibility for implementation remained with the traditional local and national authorities.

A 5-year plan and a 20-year master plan have been prepared for guiding water resource management in Morocco, including the Souss-Massa Basin. Both plans are pending approval at the national level. The long-range water resource objectives for the Souss-Massa Basin are:
- Ensure that the optimal worth of water resources is defined by the maximum worth of every cubic meter of water.
- Preserve the existing hydro-agricultural infrastructure by drawing upon the dynamic nature of the private sector.
- Mobilise surface water resources through the construction of small, medium, and large dams to satisfy the new water demands and reduce water deficits at the level of groundwater aquifers.
- Preserve the inherited hydraulic infrastructure against the effects of erosion and, in particular, the silting up of reservoirs.
- Meet growing water demands by using efficient water management techniques.
- Encourage nonconventional water management methods, in particular for saline and reclaimed water, following treatment and use for the irrigation of green spaces.

- Develop and access water sources throughout the region by promoting small and medium-sized hydraulic projects.
- Define regulatory measures and financial initiatives for the efficient use and conservation of clean water resources in the Souss-Massa Basin.
- Move from traditional water-supply management to water-demand management.
- Develop new policies, introduce institutional reforms, disseminate best practices, and strengthen the ability of water user associations and end users to participate in decision-making processes related to water.
- Accelerate the implementation of new policies to limit or reverse the overexploitation of aquifers.

Lima

SEDAPAL's current strategy is to maintain an adequate supply of water to the city using a conjunctive management approach. More water from wells is used during the low flow months of the Rímac River (May through November), but more river water is used during the high flow months of December through April. In addition, pilot recharge programmes along the middle portion of the Rímac River have been implemented to capture excess water during the high flow months. One of the concerns during the high flow months is that the turbidity of the river increases to more than 20 000 NTU, which forces SEDAPAL to close its intakes on the Rímac River and depend more on water stored in its reservoirs. Another concern is the uncontrolled use of water by private parties. Approximately 75% of industries in Peru are located in Lima, and approximately 90% of them use groundwater as their water supply (UNESCO, 1993). Although industries and other private parties are required to obtain permits from SEDAPAL before drilling a well, there are dozens and perhaps hundreds of wells throughout the city that are not registered yet continue to extract and use water at little cost and with little incentive to ration its use.

Los Angeles

Historical and current groundwater management practices in the Los Angeles area are generally consistent with the World Bank guidelines (1998). Stakeholder involvement and participation occurred to a high degree in the negotiations and adjudications of the basins, and continue through the diverse management structures in place. Market-driven mechanisms are used to allocate water resources, but have become much more complex as the value of water for environmental uses has been given higher priority than in the past. High-quality water for domestic use is readily available to nearly all users, and basin or watershed management approaches have been adopted. Groundwater in the Los Angeles area is intensively managed, probably as intensively as anywhere in the world. The management structure is complex but has been effective in achieving the goals of local groundwater producers and consumers.

Even with intensive management of its groundwater supplies, the Los Angeles area is heavily reliant on imported water to meet its total water demand. To evaluate the results of groundwater management in the Los Angeles area will require consideration of the impacts of water management throughout the state and perhaps regionally. In the past few decades, environmental benefits of water have received a higher social, political, and legal priority within the United States and particularly California. Water diversions to the Los Angeles aqueduct have been reduced in an attempt to restore habitat in the Mono Basin and Owens Valley, and a significant effort has been devoted to resolving water supply and water quality issues associated with the delta system of the San Joaquin and Sacramento rivers. In addition, other western states have developed infrastructure to allow them to utilise their full entitlement of Colorado River water, which will reduce the amount available to southern California. Thus, management of groundwater in the Los Angeles basins will become even more important in the future.

8.8   CONCLUSIONS

Many similarities exist between the hydrogeologic and development characteristics of the three cities discussed in this chapter. They have similar climates and hydrogeologic settings; they are underlain by aquifer systems having substantial storage capacity but limited natural recharge; they are water deficient in context of current water demands; they are regional population centres; and they support industrial or agricultural economies of national importance. However, striking differences between the areas do exist, including groundwater rights and management authority, availability of imported water, and historical and current degree of groundwater management. Of the three areas, Los Angeles has the most complex structure of groundwater management and manages its groundwater basins most intensively. Some of the groundwater management practices developed or utilised in the Los Angeles area may be of value to other cities. These include optimisation of storm flow recharge, replenishment using treated wastewater, cyclic or long-term storage, water banking, and others. Many of the practices utilised in Los Angeles, however, are largely dependent on the availability of imported water, and similar import may not be feasible or desirable in other areas.

Renewed or further depletion of groundwater resources in each of the cities would have adverse but possibly dissimilar consequences in each city. The inability to supply sufficient water for irrigation of the existing irrigated crops might result in a significant decline in foreign currency to the government of Morocco because the Souss-Massa area contributes over 70% of the exported fruits and vegetables. In addition, the loss of agricultural employment in the Souss-Massa Basin would contribute to a population migration from rural areas into Greater Agadir. In Los Angeles, as with most other areas in the USA, the major rural to city migration has already occurred but the population is still increasing. Relatively little irrigated agriculture remains in the immediate Los Angeles area, and the loss of irrigation water supply to the local area could pose an economic hardship on a few individuals but not on the regional economy as a whole. However, the loss of water supply for irrigated agriculture in other parts of California would have an enormous impact both on those parts of the state and on the state as a whole. Because Lima represents 30% of Peru's population, continued degradation of groundwater quality and depletion of groundwater resources would also have a significant impact on the country's economy.

Morocco's concept of maximising use of all water supplies is similar to that historically followed both in California and elsewhere in the western United States. Currently, social and political factors, including litigation by non-governmental organizations (NGOs), have reversed that trend in California. The value of water for habitat, wildlife, and recreational beneficial uses is now being given higher priority. This re-evaluation of priorities has been possible largely because economic conditions and the general standard of living in the United States are high enough to allow people to view these beneficial uses as important, rather than as non-essential luxuries. As noted previously, increased import of water from Andean watersheds to Lima is planned. In Agadir, interbasin transfers may not be feasible on a large scale because there are not significant areas in the region with surplus water. Placing a higher priority on ecological concerns in the Agadir region would place additional burdens on the area because it is currently using water at an unsustainable rate.

In each of these arid-climate urban areas, development and corresponding changes in water use and management have produced changes in the hydrogeologic system. The sources, distribution, and magnitude of groundwater recharge have changed greatly, resulting in a net increase in areal recharge in the urbanised areas. Groundwater quality degradation also has occurred. The type, extent, and impact of degradation has been influenced by the nature of development and the hydrogeologic conditions over which it has occurred.

In summary, these three major urban areas share a number of characteristics, but each area has sufficiently distinct geographic, historical, political, and economic characteristics to suggest there is no universal approach to water resource management. Donor nations should recognise this when attempting to export their own water management policies as part of aid packages.

CHAPTER 9

Shallow porous aquifers in Mediterranean climates

Oliver T.N. Sililo and Steve Appleyard

9.1   INTRODUCTION

Hydrogeological environment

In this chapter, the effect of urban development on groundwater in shallow porous aquifers in Mediterranean climates is described. The main characteristics of this environment include:
- A Mediterranean type of climate with hot, dry summers and mild to cold wet winters.
- Unconfined shallow porous aquifers.
- Aquifer recharge mainly from precipitation occurring in winter.
- High vulnerability of the shallow aquifers to pollution.

Two urban areas are used as examples: Perth in Western Australia and Cape Town in South Africa. Perth is the only large urban centre within Western Australia. It is built on top of the Swan Coastal Plain and has a population of 1.3 million people. The Metropolitan region covers an area equivalent to many large European cities (Fig. 9.1). Perth is very much a garden city, with most of the population living in detached houses with large gardens of lawn and exotic shrubs on 500–1 000 m^2 blocks.

The Cape Metropolitan area is built on top of the Cape Flats (Fig. 9.2) and has a population of about 2.6 million people. 95% of the population has formal housing. The type of housing varies from high cost, low density with sophisticated infrastructure to low cost, high density dwellings with less sophisticated infrastructure. 5% of the population live in informal settlements with a limited number of communal water supply systems and bucket latrines.

Climate

The climate in Cape Town and Perth is typically Mediterranean with hot, dry summers and mild to cold wet winters. The hot, dry summers are caused by belts of anticyclones that pass through the regions between October and March. Little rainfall is associated with these anticyclones because the air descending within the high pressure zones has a greater capacity to retain moisture. Such weather conditions are usually accompanied by clear skies. During the cool winter months, rainfall results from low-pressure cells that cross the regions as cold fronts. The average annual rainfall ranges between 700 and 1 300 mm for Perth and from about 500 to 800 mm for the Cape Flats. Approximately 90% of the rain falls between April

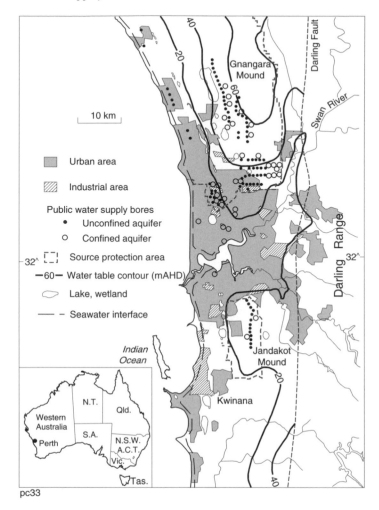

Figure 9.1.   Location map of the Perth Region showing the extent of the urban area and ground-water features.

and October. The average annual potential evaporation is about 1 800 mm for Perth and 1 070 mm for Cape Town. The mean maximum temperatures for Perth ranges between 17 and 30 °C while the mean minimum temperature ranges between 9 and 18 °C. For Cape Town, the average daily temperatures range from 22 °C in summer to 11 °C in winter.

Geology

The geology of the Swan Coastal Plain and the Cape Flats is similar. They both have a young sedimentary cover of Quaternary age underlain by older formations. The Swan Coastal Plain is a sub-region of the Perth Basin. This basin comprises deep layers of sand, clay and minor limestone sediments, up to 15 km thick and up to 286 million years old. The younger formations in the Basin vary between 10 and 100 m in thickness, and are less than 25 million years old (late Tertiary and Quaternary). These are known as the superficial formations,

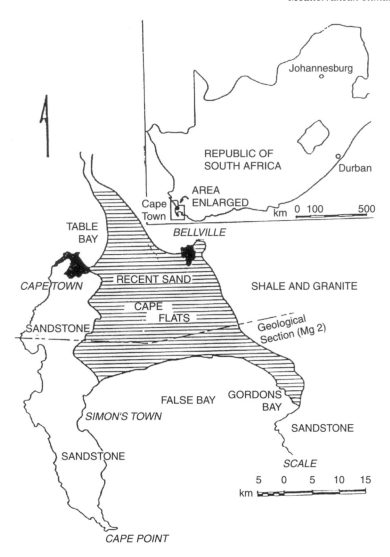

Figure 9.2.   Location map of the Cape Flats area.

and consist of sand and limestone and discrete beds of silt and clay. The geology of the area is shown in Figures 9.3 and 9.4. The superficial deposits are underlain by the Leederville Formation and in the Gnangara mound by the Osborne Formation. Beneath the Leederville Formation is the South Perth Shale which overlies the Yarragadee formation.

   The Cape Flats, which covers an area of about $630\,km^2$, consists of a young sedimentary cover of Quaternary age underlain by Malmesbury Group metasediments and Cape Suite granites (Fig. 9.5). The unconsolidated young sediments consist essentially of sand deposits locally interbedded with peat, clay layers and varying amounts of calcareous material. These sediments were deposited partly under fluvio-marine conditions and partly under aeolian conditions.

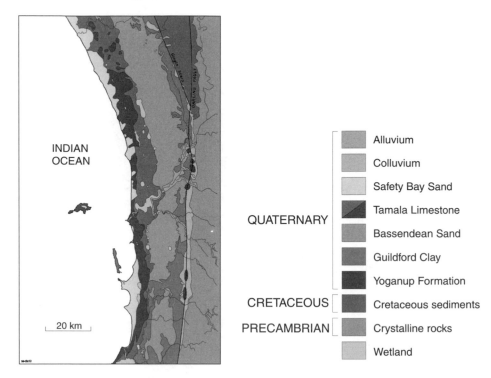

Figure 9.3.   Geological map of the Perth region.

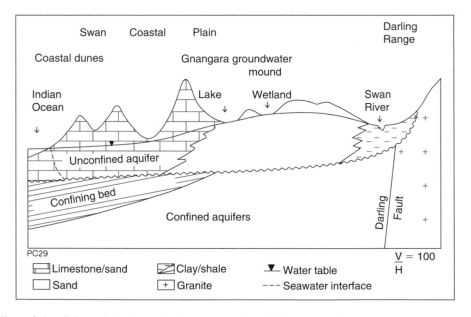

Figure 9.4.   Schematic hydrogeological cross section of the Perth region.

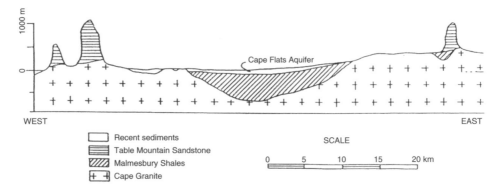

Figure 9.5.   East-west cross section of the Cape Flats.

Hydrogeology

Both the Swan Coastal Plain and the Cape Flats have an upper shallow unconfined aquifer which occur in the Quaternary age "superficial" deposits. The maximum aquifer thickness is 45–50 m in the Cape Flats and up to 70 m in Perth. The hydraulic conductivities that have been measured in the aquifers are similar: 24.2 m/d for Perth (Benker et al., 1996) and 15–50 m/d in the Cape Flats (Vandoolaeghe, 1989). The aquifers are recharged principally from precipitation. Vandoolaeghe (1989) estimated that the net groundwater recharge through sandy soils of the Cape Flats varies between 15 and 37% of annual precipitation. Similar values have been estimated for Perth: 15% of the annual 900 mm rainfall under native heath and woodland and 30% in urban areas (Davidson, 1995).

In Perth, the water table configuration of the unconfined aquifer is characterised by two major groundwater mounds, the Gnangara groundwater mound to the north and the Jandakot groundwater mound to the south. Groundwater flows southwesterly in the unconfined aquifer to discharge into rivers, drains, lakes and the ocean. The unconfined aquifer is underlain by three confined aquifers comprising Jurassic-Cretaceous sediments. Groundwater in these aquifers flows very slowly and is more than 40 000 years old. At some localities the groundwater in the confined aquifers is fresh to a depth of at least 1 000 m. The discharge of water in the confined aquifers is mainly offshore, into the ocean via a veneer of superficial formations. The Gnangara mound is a major recharge area for both the unconfined aquifer and the confined aquifers. Part of the mound overlies the confined aquifers at shallow depth, where overlying confining beds are thin or absent (Fig. 9.4).

The Cape Flats aquifer pinches out against impermeable boundaries in the east, west and north, whilst the southern boundary is defined by the coastline (Fig. 9.2). The calcareous clay and calcrete layers, when present, act as aquitards and result in semi-confining conditions. The hydraulic parameters of the aquifer have been determined in a series of abstraction and recovery tests. Transmissivity values are 50–650 m^2/d, with typical values between 200 and 350 m^2/d (Tredoux, 1984). The effective porosity is typically 0.10–0.12 with a maximum value of 0.25. The long-term yield of prototype production wells is about 17 L/s. The pumping tests also showed that the deposits were markedly anisotropic, the vertical permeability being smaller by a factor of 10–20 or even up to 100 when compared with the horizontal permeability. The bedrock is generally regarded as impervious although high yielding boreholes located in these units have been reported.

Table 9.1.   Indigenous groundwater quality in the Cape Flats aquifer (after Tredoux et al., 1980).

Chemical	Concentration (mg/L unless shown)		
	Median	Minimum	Maximum
Sodium	57	20	760
Potassium	1.5	0.9	12
Calcium	95	36	150
Magnesium	11	3.9	93
Ammonia (as N)	<0.5	<0.5	1.2
Sulphate	30	4	166
Chloride	99	35	1 317
Total alkalinity (as $CaCO_3$)	248	80	391
Nitrate (as N)	<0.1	<0.1	2.6
Nitrite (as N)	<0.05	<0.5	0.1
Total phosphate	<0.1	<0.1	0.35
pH	7.7	7	8.2
Electrical conductivity (mS/m)	78	43	499

A large number of groundwater dependent wetlands occur on both the Cape Flats and the Swan coastal plain. Most of the wetlands dry out seasonally under natural conditions, and this is important for maintaining the environmental health of these ecosystems.

Groundwater chemistry

The chemical characteristics of the indigenous groundwater of the Cape Flats aquifer are shown in Table 9.1. Generally, the water in most parts of the aquifer has a fairly low salinity and relatively high temporal hardness. Saline water generally occurs in the peripheral areas of the aquifer. In addition, the occurrence of argillaceous units and marine deposits at different stratigraphic levels results in zones of saline water within the aquifer. The abundance of shelly material throughout the aquifer results in a groundwater that is saturated with respect to calcium carbonate.

In Perth, groundwater salinity (total dissolved solids – TDS) is less than 250 mg/L at the crest of the Gnangara and Jandakot groundwater mounds, and typically rises to around 1 000 mg/L in the unconfined aquifer close to the coast and estuary. The low salinity groundwater also extends to at least 1 000 m below the Gnangara mound in both the Leederville and Yarragadee aquifers, but is flanked by higher salinity groundwater (up to 3 000 mg/L) derived from groundwater flowing from the north. Groundwater pH ranges from 4.5 to 6.5 at the centre of Gnangara Mound to 6.5 to 8.0 in the limestone coastal areas. Limestone near the coast also causes water to be very hard. Elsewhere, the hardness varies from soft to slightly hard. Dissolved iron varies from <1 to 10 mg/L, with higher concentrations occurring in groundwater discharge areas around the coast and stream boundaries (Gomboso, 1997).

9.2   BENEFICIAL USES OF GROUNDWATER

Water supply

In Perth, groundwater forms an important component of the city's water supply, providing 70% of water used, and 40% of the municipal water supply. The groundwater in storage represents some 500 years current annual abstraction. Bores as deep as 1 000 m supply water with a TDS of 180 mg/L. The shallow water table and unconsolidated sand aquifer makes groundwater easily available to most private properties and as a consequence, there are about 120 000 privately owned narrow diameter bores, or dug wells with spear points, which are used for garden irrigation. The gardens require watering for about six months of the year due to the dry Mediterranean type climate, and are responsible for 80% of domestic water use.

Unconfined and confined groundwater is used conjunctively with surface water resources for municipal supply. The confined groundwater under the north and western part of the city allows peak loads to be maintained from deep sources, while the replenishable shallow resources are used all year round. Artesian bores are commonly located at reservoir sites and treatment plants, and unlike the shallow groundwater which requires treatment for pH, turbidity and dissolved iron, the confined groundwater can often be put straight into the reservoirs. Shallow groundwater is also drawn from borefields located within newer northern residential areas although it is recognised that changes in groundwater quality will eventually restrict its use for drinking water.

Under present land use conditions, the maximum total and sustainable abstraction from the unconfined and confined aquifers is estimated to be about $600 \times 10^6 \, \text{m}^3/\text{y}$ (Davidson, 1995). Current groundwater abstraction within the Perth area is about $300 \times 10^6 \, \text{m}^3/\text{y}$, of which 75% is abstracted from the unconfined aquifer. Current rates of abstraction are therefore well within sustainable limits, and significant additional abstraction can occur.

Water supply for Cape Town is mainly obtained from surface water. Although it has been estimated that the Cape Flats Aquifer is capable of providing about 15% of the present annual water consumption in the Greater Cape Town area (Tredoux, 1984), this resource is still under-utilized. Currently groundwater is being abstracted in bulk for irrigation in two areas. The Philippi agricultural area is responsible for pumping in excess of $13 \times 10^6 \, \text{m}^3/\text{y}$ while the Cape Town Municipality pumps more than $5 \times 10^6 \, \text{m}^3/\text{y}$ at Mitchells Plain and Strandfontein. In certain areas, irrigation requirements are met by means of shallow ponds sunk in the upper 5 m of the aquifer. Minor quantities of groundwater are also abstracted for private and industrial use.

Wetlands

Wetlands are areas of high biodiversity and high biological productivity, and support some of the wildlife in the regions either directly or indirectly. Many of these wetlands are also visited each year by migratory wading birds from the northern hemisphere and, in Perth, these habitats are protected under Australia's obligation to an international treaty on wetland protection. Wetlands in Perth are also important for recreation and aesthetic purposes. In some areas where local private groundwater abstraction and transpiration by large trees in established gardens and parks has lowered the water table, water levels within wetlands are artificially maintained by groundwater pumping.

In South Africa, wetlands are favourite haunts for bird-watchers but have received scant regard until very recently (Davies & Day, 1986).

9.3   EFFECT OF URBAN DEVELOPMENT ON WATER BALANCE

Very little work has been done in Cape Town to determine the effect of urban development on water balance. In Perth, shallow wells were first used for water supply, but the recognition in the 1880s of the health risk by contamination from cesspits led to the development of artesian water. The confined groundwater beneath the city was the major component of the city water supply until 1940, after which surface water reservoirs were built in the Darling Range 30–50 km away. Attention refocused on the shallow groundwater resources of the Gnangara and Jandakot groundwater mounds (Fig. 9.1) in the 1960s, and these resources have been progressively developed since then. In addition to the public water supply, it is estimated that one in four suburban plots has a shallow bore for garden watering, making groundwater a de facto distribution system, and per capita use for those households with a bore is estimated to be about 400 m^3/y.

Recharge, which under native heath and woodland was about 15% of the annual 900 mm rainfall, has been doubled by urban development to around 30% (Davidson, 1995). Removal of native vegetation, runoff from roads and roofs and import of water, together with local recharge into storm water sumps in the dune swales, has resulted in a rising water table, creating lakes out of swamps. Rising water levels in newly urbanised areas have also necessitated drainage, either by gravity to the estuary or ocean, or by pumping to infiltration basins where the water table is deeper. It is some years before a suburb establishes a new equilibrium with mature gardens and a higher density of private bores.

The increasing use of private bores, especially during the 1977 drought year in which restrictions were placed on garden watering from the public supply, led to concern about the drop in water table in some areas, and intrusion of saline water from the Swan River estuary. This prompted the Perth Urban Water Balance Study (Cargeeg et al., 1987; Smith & Allen, 1987). As a result of the findings, environmental constraints were placed on the use of the shallow groundwater because of the need to maintain water levels in lakes and wetlands, and to maintain the condition of phreatophytic vegetation. The water table around public supply borefields is now maintained at a minimum level which is generally at the lower end of the natural seasonal range. In some circumstances this involves artificially maintaining the levels of lakes by pumping in groundwater.

The maximum decline in the water table level has been about 0.1 m/y since 1975, and the potentiometric levels in the confined aquifers below the city are declining at maximum rates of 0.7 m/y in the Leederville aquifer, and 0.5 m/y in the Yarragadee aquifer (Davidson, 1995). This is inducing increased leakage from the overlying unconfined aquifer in areas where land use controls protect the groundwater quality. Drilling of private bores is being encouraged in some areas to reduce the use of high quality, treated scheme water for garden irrigation, and also to control rising shallow water table levels.

9.4   EFFECT OF URBAN DEVELOPMENT ON GROUNDWATER QUALITY

The vulnerability to pollution of the shallow aquifers underlying the Cape Flats and Perth is quite high. This is mainly due to the fact that the aquifers are unconfined and that recharge takes place practically over the whole area.

In Perth, numerous occurrences of groundwater contamination have been reported (Table 9.2). These include contamination from commercial and industrial point sources within the industrial areas of Perth, and widespread low-level diffuse contamination from fertiliser use on gardens and from septic tank leachate. Point sources of contamination include a range of light industries, petrol service stations, and pest control depots (Appleyard, 1995a) which have either accidentally or deliberately disposed of wastes, and about 100 former

Table 9.2.   Point source contaminants detected in private bores.

Point sources	Number of sources	Contaminants detected
General industry Metal finishing, foundries, power stations, mechanical workshops, cement production, laboratories, photo processors, laundries, drycleaners, chemical production	250	Trichloromethane, chlorinated benzenes, BTEX hydrocarbons, PAHs, surfactants, metals, boron
Animal based industry Wool scourers, tanneries, piggeries, meat processing	176	Ammonia, high BOD, hydrogen sulphide, bacteria
Food production Production of starch, bakeries, dairies, canning and packaging, wineries, breweries, soft drink	47	Ammonia, high BOD, hydrogen sulphide, bacteria
Waste disposal Domestic waste disposal; liquid waste disposal, sewage treatment plants, large septic systems	153	Ammonia, high BOD, hydrogen sulphide, bacteria, phenols
Burial sites, human and animal	20	Ammonia, high BOD, hydrogen sulphide, bacteria, iron
Pest control depots	~100	Atrazine, fenamiphos, diazinon
Fuel storage	~100	Aliphatic hydrocarbons, BTEX (particularly benzene)

waste disposal sites (Hirschberg, 1993a,b). There are about 2 000 known or suspected sources of groundwater contamination within the region, and contamination plumes from 100–1 000 m or more in length have extended from some of these sites through residential areas where private bores are used (e.g. Benker et al., 1996). A wide range of contaminants have been detected in private bores wells near many of these sites, commonly at levels which exceed national drinking water criteria. Although this water is generally not used for drinking, other routes of exposure, such as droplet inhalation or eating irrigated produce have not been widely assessed. Groundwater contamination in at least one private bore was sufficiently severe to be toxic on prolonged skin contact and to kill plants irrigated with the water (Appleyard, 1995a).

There is also a widespread addition of nitrate from fertiliser use on gardens, and of nitrate, ammonia, and bacteria from septic tanks. Concentrations of nitrogen (mostly present as nitrate) beneath urban areas in Perth generally exceed 1 mg/L, and often 10 mg/L compared to concentrations of less than 1 mg/L commonly measured beneath areas of native vegetation (Appleyard & Bawden, 1987; Appleyard, 1995b; Davidson, 1995). The concentration generally increases with the age of urban development (Barber et al., 1996; Appleyard, 1995b). Currently, about 1 600 t/y of nitrogen and 480 t/y of phosphorus is applied annually to lawn areas in Perth. Although much of the phosphorus is bound up in soil profiles, up to 80% of the applied nitrogen may leach to the water table (Sharma et al., 1996). Gerritse et al. (1990) estimated that 80–260 kg/ha/y of nitrogen is applied in fertilisers to urban areas of Perth and, on this basis nitrate concentrations in groundwater in these areas should be at least 40 mg/L as N. The fact that concentrations are generally much lower than this suggests that denitrification is taking place in the aquifer. However, the large scatter of nitrate concentrations in groundwater in urban areas of Perth indicate that denitrification is not occurring uniformly, and that nitrogen inputs are not uniformly distributed.

Fertiliser use is particularly high in horticultural areas, which are typically located at the fringes of urban areas in Perth. Typically, 500–1 500 kg/ha/y of nitrogen is applied in fertilisers (mostly as poultry manure) to crops (Lantzke, 1997), and this exceeds plant uptake rates by a factor of 4 to 7 (Pionke et al., 1990). Nitrate concentrations up to 100 mg/L as N occur beneath horticultural areas, although concentrations generally decrease to background levels within a few hundred metres of irrigated areas due to dispersion and denitrification.

Another source of nitrogen contamination in groundwater in Perth is the use of septic tanks. Currently, about 30% of the urban area, mainly developed in the 1950s and 1960s, is unsewered. There is an ongoing programme to replace septic tanks with sewer connections, but existing groundwater contamination will take many years to dissipate.

In the Cape Flats, although pollution incidences are known to occur, these have not been reported widely. The major sources of pollution include solid waste disposal sites, sewage treatment works, informal settlements and agricultural areas. Groundwater quality monitoring at two existing solid waste sites and a sewage treatment works revealed that groundwater pollution was occurring (Tredoux, 1984). At one of the landfills, it was found that the pollution plume had moved more than 30 m from the disposal site. Underneath the landfill, the groundwater had high concentration of ammonium and the maximum values exceeded 600 mg/L.

A study of agricultural contamination in vegetable farming areas of the Cape Flats where inorganic fertilisers, sewage sludge and manure were used intensively found elevated levels of nitrates (38 mg/L) and DOC (30 mg/L) in shallow groundwater (Conrad et al., 1999).

In a study in one of the townships, Wright et al. (1993) found that stormwater runoff in the area was polluted throughout the year predominantly with microbiological contaminants with corresponding high concentrations of nutrients and organics. The major source of pollution was found to be litter and faecal waste.

9.5   MANAGEMENT OF GROUNDWATER RESOURCES

The management of groundwater contamination problems in Western Australia is carried out jointly by the Department of Environmental Protection, the Water and Rivers Commission, and the Health Department. Although the Department of Environmental Protection is generally the lead agency and administers the Environmental Protection Act, water quality complaints are normally referred to the Water and Rivers Commission to determine whether there is a significant contamination problem.

The investigation and remediation of groundwater contamination in industrial areas in Perth is generally carried out by the responsible polluter under the "polluter pays" principle, and industrial processes are licensed under the Environmental Protection Act to minimise the risk of groundwater contamination taking place. However, when a polluter cannot be identified, or if contamination took place legally prior to the introduction of the Act (the Western Australian Environmental Protection Act has limited retrospective powers), then the responsibility for investigating and possibly remediating occurrences of groundwater contamination at so-called "orphan sites" falls on the government. This is generally the case in residential areas, many of which were former industrial areas, or are located adjacent to old industrial areas.

The identification and management of contamination problems in residential areas of Perth is particularly difficult due to a lack of a groundwater monitoring programme in these areas, and the absence of well licensing. Most of the contamination problems are reported by members of the public when they detect unpleasant odours or discolouration of the water pumped from wells, or when groundwater is having an adverse impact on plants in gardens (Appleyard & Manning, 1997). However, only certain contaminants produce these responses, generally at very high concentrations, and it is likely that many contamination problems remain undetected in residential areas.

In South Africa, a new water law has been passed which empowers the government to regulate the use, flow and control of all water in the republic (National Water Act, 1998). A major implication of the new law is that groundwater, which has not been subject to regulation before with the exception of a number of special control areas, will now need countrywide government attention to ensure that it is managed according to laid down criteria. The government is responsible for defining a national water resource strategy for management of water resources. The national strategy also provides the framework within which water will be managed in defined water management areas. The law provides for catchment management agencies, whose main task is to progressively develop a catchment strategy for the water resources within its water management areas.

The protection strategy in the new water law is based on the "polluter pays" principle. The person who owns, controls or uses the land in question is responsible for taking measures to prevent pollution of groundwater. If these measures are not taken, the water management institution concerned may itself do whatever is necessary to prevent the pollution, or to remedy its effects, and to recover all reasonable costs proportionally from people responsible for the pollution.

9.6 GROUNDWATER AS AN ENVIRONMENTAL AGENT

In Perth, the discharge of groundwater contaminated by nutrients into waterways and the coastal marine environment is contributing to the degradation of water quality in these environments (Hillman, 1981; Appleyard, 1990, 1992). There are frequent algal blooms in the Swan Estuary and in some coastal waters, and large areas of seagrass beds have been lost in the nearshore marine environment due to nutrient enrichment.

Excessive fertiliser use on gardens and public open space is a major source of nutrient contamination in the Perth metropolitan area, as discussed above. About 160 t/y of nitrogen is discharged by groundwater to the Swan River, and up to 10 t/y of nitrogen for each kilometre of coastline is discharging into the marine environment (Appleyard & Powell, 1998). The issue of nutrient discharge is particularly important in Cockburn Sound, a sheltered marine embayment in the southern part of the Perth metropolitan area. Discharge from the numerous industries in the Kwinana industrial strip has contributed to the loss of more than 80% of the seagrass meadows in the Sound, and there have been frequent algal blooms in this water body. Most of the surface sources of nitrogen discharge have been shut down, and now groundwater is the single largest source of nitrogen discharge, contributing about 300 t/y of nitrogen from a number of industrial contamination plumes, although there is considerable uncertainty in this estimate due to limited monitoring data

(Appleyard & Haselgrove, 1995). The discharge varies with tides and with variations in sea level caused by the El-Niño Southern Oscillation, and is greatest when the sea level is low, particularly in spring with large spring tides. Groundwater discharge probably also varies diurnally, and is likely to be greatest in the morning with offshore breezes and decrease with the afternoon sea breezes which typically occur in Perth in summer months. A number of industries in the area have instigated groundwater remediation programmes to reduce the impact of groundwater discharge on the marine environment in this area.

9.7   GROUNDWATER PROTECTION

In Perth, the quality of groundwater in the unconfined aquifer which is used for municipal supply is protected by controlling the land use in designated groundwater protection areas (Fig. 9.1). There are three types of protection areas:

- Priority 1 (P1) source protection areas are defined to ensure that there is no degradation of water quality used for public supply. P1 areas are declared over land where the provision of the highest quality public drinking water is the prime beneficial land use. P1 areas would typically include land under government ownership where there is no development, or where forestry or silviculture takes place.
- Priority 2 (P2) source protection areas are defined to ensure that there is no increased risk of pollution to groundwater quality by regulating previously existing land uses. P2 areas are declared over land where low intensity development (such as rural) already exists. Provision of public water supply is a high priority in these areas, but there may be some degradation of water quality.
- Priority 3 (P3) are declared over land where water supply needs co-exist with other land uses such as residential, commercial and light industrial developments. Protection of groundwater quality in P3 areas is achieved through management guidelines rather than restrictions on land use. If the water source does become contaminated, then water may need to be treated or an alternative water source be found.

9.8   GROUNDWATER POLLUTION CASE STUDIES

Three case studies are discussed below:

- Pesticide contamination in the Kwinana industrial area, Perth. This is an example of a groundwater investigation at an industrial site where a polluter has escaped liability.
- Solvent (TCE) contamination in a residential area, Perth; an example of an investigation in a residential area.
- Groundwater contamination by industrial liquid waste, Cape Town; an example of an investigation at an industrial site in Cape Town.

Pesticide contamination in the Kwinana industrial area (Perth)

Introduction
A plant for the production of agricultural chemicals was established in the Kwinana industrial area in Perth in the early 1960s, about 1.5 km inland from a sheltered marine embayment. Effluent from the plant was disposed of in an unlined pond on the site, and groundwater nearby became contaminated with the herbicides 2,4-D and 2,4,5-T, and with chlorinated

phenols. This disposal practice was discontinued in 1984, when disposal by deep-well injection into a saline aquifer commenced. The agricultural chemical plant changed ownership in 1985, prior to the enactment of the current Environmental Protection Act. The new owners cannot be held responsible for the results of previous activities on the site, so that the responsibility for the management of the groundwater contamination problem has passed to the Western Australian Government.

Site hydrogeology

The Kwinana industrial area is underlain by sediments of Quaternary age up to 30 m thick, which contain fresh groundwater. Near the coast, two aquifers occur: an upper unconfined aquifer of variably cemented sand and silt of moderate permeability; and a deeper semi-confined aquifer of highly permeable, locally cavernous limestone and sand. These aquifers are separated by a 0.5–1 m thick silty or clayey shelly layer which forms a local aquitard, and are underlain by Cretaceous sediments of low permeability.

The presence or absence of the aquitard separating the two aquifers is important in determining whether or not particular cases of groundwater contamination may be easily remediated. In areas where the aquitard is present, and groundwater contamination is restricted to the sand aquifer, it may be possible to use pump-and-treat or other technologies to improve groundwater quality. However, where the aquitard is absent, and contamination can pass into the limestone aquifer, it may be extremely difficult to treat contaminated groundwater due to the high permeability of the limestone, and the uncertain nature of solute flow-paths in this cavernous aquifer.

Investigation programme

Investigations in 1982 indicated that an extensive groundwater contamination plume had developed near the agricultural chemical plant. The plume contained up to 180 mg/L of 2,4-D, and peak concentrations of 2,4,5-T up to 60 mg/L, but no dioxins were detected in groundwater (Appleyard, 1993). Solute transport modelling suggested that the plume could reach the ocean in about 70 years assuming negligible retardation and biodegradation. The modelling suggested that concentrations of herbicides and phenols were sufficiently high to be potentially harmful to marine life on discharge.

A more extensive drilling and groundwater sampling programme was carried out in 1989 and 1990 (Appleyard, 1993) to determine whether contamination had penetrated the aquitard and had entered the limestone aquifer, which is utilised downgradient of the agricultural chemical plant for industrial water supply. Sampling for 2,4-D, 2,4,5-T and total phenols indicated that contamination was mostly restricted to the sand aquifer (Fig. 9.6). The contamination in the limestone aquifer was found to be limited in extent and not severe. The most likely source of the contamination is leakage from the overlying sand aquifer where the aquitard has been perforated by badly constructed bores sunk into the limestone. Two water supply bores in the agricultural chemical plant were plugged to prevent further leakage.

The data suggest that the leading edge of the contamination plume in the sand aquifer has not moved since 1982, but that the centre of mass of the plume may have moved about 200 m to the west. The western margin of the plume also shows a sharp transition from contam-inated to uncontaminated groundwater (Fig. 9.6) characteristic of biodegradation at the margins of the plume. This is supported by more recent data provided by ongoing monitoring programmes that peak concentrations of herbicides are decreasing. However, further work is required to confirm whether natural biodegradation of the plume is actually taking place.

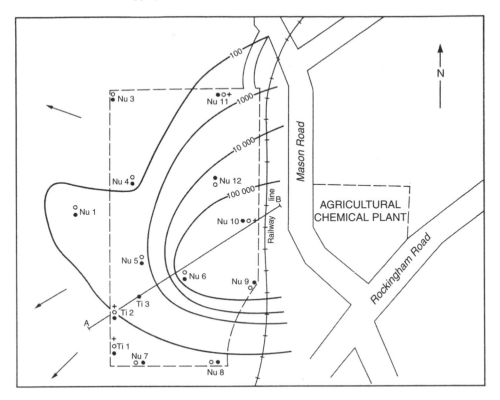

Figure 9.6   Extent of groundwater contamination near the agricultural chemical plant, Kwinana, February 1990.

Solvent (TCE) contamination in a residential area (Perth)

Preliminary investigations

In October 1992, a resident complained to government agencies about the appearance of a strong smell in water from a bore he used for garden irrigation. The smell was sufficiently bad to upset residents in neighbouring houses and flats. The odour was found to be due to hydrogen sulphide, but at levels much higher than typically detected in Perth groundwater. Field testing also indicated high concentrations of ammonium, and a leak from a nearby sewer main was suspected. However, as the bore was located downgradient of an industrial area, a full suite of inorganic analyses, and a GC-MS scan for organic constituents, were also carried out. As a result, the solvent trichloroethene (TCE) was detected (Benker et al., 1996). The concentration of TCE was 2000 µg/L, well above national and international drinking water criteria.

Health-risk investigation programme

Western Australian government agencies coordinated the sampling of additional private bores to determine the extent and severity of contamination, and a soil-gas survey was also undertaken. These investigations indicated that the TCE contamination plume had a lateral extent of about 900 m (Fig. 9.7), but little was known about the vertical distribution of contamination due to limited information about the screened depth of many bores. The aim of this phase of

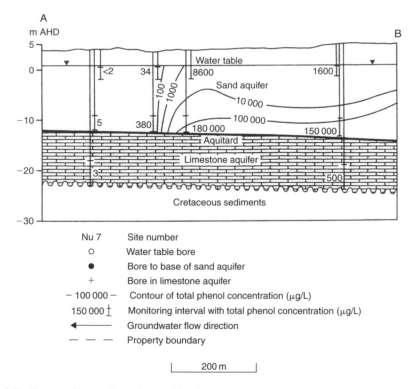

Figure 9.7   Extent of TCE plume in a residential area of Perth.

the investigation programme was to assess whether or not groundwater contamination posed a threat to public health, rather than trying to identify and prosecute a polluter.

The major routes of exposure of the public to TCE contamination were likely to be through drinking contaminated groundwater, through inhalation of TCE vapour during pumping, and possibly through eating garden-grown fruit and vegetables irrigated with contaminated groundwater. To ensure exposure was minimised, residents were warned by letter and media releases not to drink groundwater or eat garden-grown produce until cleared by further testing. The owner of the most severely contaminated bore was asked not to use the bore until testing was completed. Residents were kept informed of subsequent investigations by further letters, by informal meetings, and by making government officers available for advice by telephone contact. The management of the public response to the contamination problem was complicated by:

- Public confusion about the nature of the contaminant. Residents who lived near the most severely contaminated bore mistook the hydrogen sulphide smell for TCE; for example, a pregnant woman living next door to the most contaminated bore was concerned about the effects that the strong odour was having on her unborn child.
- The owner of the most contaminated bore grew most of his family's fruit and vegetables in his garden, and supplied others with produce.

Additional sampling was carried out to ensure that the routes of exposure were not a health risk. Air sampling was carried out and indicated that both TCE and hydrogen sulphide levels near pumping bores were well below levels of health concern. Chemical analysis of garden-grown fruit and vegetables also indicated that they were safe to eat. Consequently, residents were informed that groundwater was safe for garden irrigation, but it was reiterated that untreated groundwater in Perth should not be used for drinking (largely because of the risk of microbial contamination).

Additional investigations
A drilling programme was carried out in the area in order to determine the vertical distribution of contamination within the superficial aquifer, to locate sources of contamination, and to evaluate the physical and chemical factors that influence the migration of chlorinated solvents and their breakdown products. The drilling indicated that TCE contamination was restricted to a 6 to 8 m thick interval in the aquifer (Benker et al., 1996), and was associated with high concentrations of ammonium, possibly due the co-disposal of effluent in large infiltration basins in the nearby industrial area.

Sampling indicated that there were no degradation products like dichloroethene or vinyl chloride in the aquifer, suggesting that anaerobic dechlorination of TCE is not occurring, despite the presence of suitable chemical conditions in the aquifer. The drilling suggested that there were probably multiple sources of TCE. However, negotiations with a company that admitted to using TCE as a degreaser and disposing residue in an infiltration basin, resulted in the company voluntarily undertaking a soil cleanup programme on its site.

Resampling of private bores in 1997 indicated that TCE concentrations in groundwater had declined significantly, despite a lack of evidence that natural degradation of this chemical was taking place in the superficial aquifer.

Groundwater contamination by industrial liquid waste (Cape Town)

Background
In January 1996, the Environmental Unit of a metal plating plant discovered that one of the settling tanks used for treating effluent from the metal plating process was leaking. This tank was buried below surface and there was concern that groundwater may have been impacted. The management of the plant approached the CSIR to determine whether groundwater pollution had occurred and if so whether or not the pollution plume had travelled beyond the property boundaries. The plant is located on top of an unconfined aquifer, which consists predominantly of sands with intercalations of peat units.

Investigation programme and results
Twelve well points were installed and water samples collected for chemical analysis. A full cation–anion analysis was conducted for all the samples. In addition, analysis was conducted for CN, Ni, Cu, Zn, DOC and EC. The results showed that chemical deterioration of groundwater had occurred at the site as a result of leakage from the faulty tank. The degree of spreading was found to vary among the contaminant groups. Significant, though, was the fact that the pollution plume had not migrated beyond the property boundaries. Salinity as shown by electrical conductivity and represented by major ions sodium, calcium, chloride and sulphate caused the most significant contamination at the site (Fig. 9.8). It was found that the heavy metals, Ni, Cu and Zn, as well as cyanide pollution, was highly localised. The pollution was concentrated around the leaking tank area and showed a marked decline a few metres from the source.

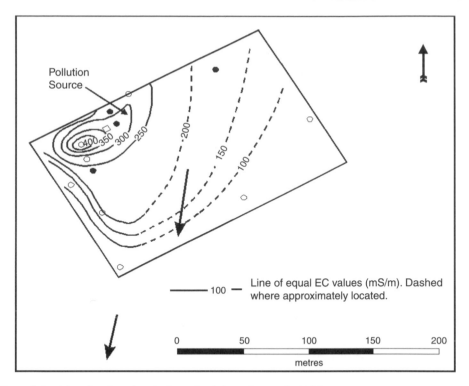

Figure 9.8   Map showing electrical conductivity values, July 1996.

Remedial strategy

Two remedial strategies were recommended: source removal and contaminant withdrawal. The leaking unit which had caused most of the contamination at the site was removed. The contaminated soil around the source area was excavated and disposed of in a landfill. A pumping programme was then conducted from the excavated hole for 3 months. The pumped water was treated before it was discharged into an industrial sewer. Generally, a marked decline in concentration levels occurred as a result of pumping. For example, the DOC measured in water samples from one well point declined from 62 mg/L at the start of the remediation to 11 mg/L at the end while the EC declined from 435 mS/m to 250 mS/m at the end of the remediation period.

9.9   THE WAY FORWARD

Perth

The ready availability of groundwater has contributed to the garden culture, and there are currently more than 130 000 mostly unlicensed private bores which are used for garden irrigation. The gardens are major sources of groundwater contamination from excessive fertiliser use. The widespread use of garden bores has meant that there are potential health risks for householders who may be exposed to contaminants including solvents, pesticides and petroleum hydrocarbons from up to 2 000 potential contaminated sites in the city

(Appleyard & Manning, 1997; Appleyard et al., 1999). The use of garden bores is deeply entrenched in Perth's culture, and has become an alternative water distribution system. It would be politically very difficult to license and regulate their use, and they have a positive benefit of reducing the usage of high quality treated water for garden use.

The challenges for groundwater management in urban Perth arise because the issues cannot be controlled by "command and control" legislative processes, but require the whole community to become better informed about the potential environmental or health impacts of their lifestyle choices and to voluntarily make changes where necessary. Water resource management agencies in this context are required to be information providers, educators and facilitators of change, rather than prescribers of community actions. Although the community in Perth have a good understanding of the importance of groundwater for water supply, there is a poor understanding of the interaction between groundwater and surface water bodies and of the potential environmental effects of excessive fertiliser use. To address this, the Water and Rivers Commission have instigated a number of school and community based education programmes to increase awareness of the role of groundwater as an environmental agent (Appleyard, 2000). The Commission is also promoting wise use of garden fertilisers and the use of local native plants in gardens that do not need fertiliser for effective growth (Appleyard & Powell, 1998).

A number of community action groups are also becoming interested in the issue of measuring and managing groundwater discharge to wetlands and waterways. Several groups are working in partnership with local and State government agencies, some with the assistance of Commonwealth (Federal) government funding. These may be the beginnings of an Integrated Catchment Management (ICM) approach to dealing with groundwater-related environmental issues in Perth. This approach is being successfully used in rural areas to manage environmental problems caused by land salinisation and the eutrophication of waterways, and should also help manage environmental problems in urban areas.

The public need to have access to high quality information about the distribution of groundwater contamination to make informed choices about where groundwater use is appropriate, particularly in Perth where groundwater contamination plumes may extend up to 2 km from large contamination sources. The Water and Rivers Commission has published a "Groundwater Atlas" showing groundwater features and the distribution of contaminated sites in Perth, and this is widely used in the community. As part of its commitment to proposed contaminated sites legislation, the Commission is also developing a groundwater contamination database (Nelson et al., 1999) to capture and quality assure contamination data. It is intended that the database will be accessible by the public with the enactment of the new legislation.

Cape Town

In line with the New Water Act, the Department of Water Affairs and Forestry has released a national policy and strategy document for groundwater quality management in South Africa (DWAF, 2000). The department proposes to achieve groundwater protection goals through a combination of source-directed control measures, resource-directed measures and rehabilitation measures. Of these, the source-directed controls, like discharge controls and activity controls were available under the previous legislation. In the new Act, however, there is greater focus on addressing surface water, groundwater, quantity and quality in an integrated way.

Through the proposed resource-directed measures, it is envisaged that a systematic way of setting resource classes and resource quality objectives for all significant water resources will be implemented. Furthermore, a Reserve will be determined for these water resources as soon as reasonably practicable after the class has been set. The Reserve consists of two parts: a basic human needs reserve which provides for the essential needs of individuals served by the water resource in question, and an ecological reserve which relates to the water required to protect aquatic ecosystems of the water resource. Resource quality objectives will be used for the regulation or prohibition of those land-based activities which may affect groundwater quality. Groundwater protection zones will also be introduced through this measure.

One of the major tasks for the implementation of the new Act is establishment of suitable institutions for appropriate and participative management of water resources. The Act provides for three levels of management, namely national government, catchment management agencies at the regional management level and water user associations acting cooperatively at the local level. The catchment management level with devolved management responsibilities will be crucial for addressing groundwater in a planned and integrated way in the public interest. It will also have to provide the essential support for local level management, which had been completely lacking in South Africa for groundwater management.

The biggest challenge for water management in South Africa will be the bridging period of the next ten to twenty years in which transformation from a highly centralised situation to new functional and highly participative institutions at different levels will have to take place (Braune, 2000). Information will be the cornerstone on which implementation of the new water policy and legislation will rest and on which new institutions will function effectively. New instruments in the Act, like catchment management plans, will mean the reserve and controlled activities will be very information-intensive. The need for integrated water resource management will create the need for much greater integration of the relevant information systems, e.g. hydrology, hydrogeology, water quality and land type, use and cover. The many new participants in a devolved management system will need to share in these information systems which are presently still very centralised. Widespread public participation will require information that is transparent and user-friendly, together with systems that can be used inter-actively to facilitate multi-user decision-making. The basis for these increased information needs will have to be a more intensive and highly systematic monitoring of water resources status and trends as well as a compliance with policy and regulation at all levels (Braune, 2000).

References

Abeyasekere, S. (1987). *Jakarta – A History*. Oxford University Press.

ACE (Associated Consulting Engineers & Burns & McDonnell International) (1993). Overview of existing waste management systems. *Fourth Urban Development Project, Municipal Environmental Component. Phase I, Final report*. MHUP. Yemen.

Acton, D.W. & Barker, J.F. (1992). In-situ biodegradation potential of aromatic hydrocarbons in anaerobic groundwaters. *Journal of Contaminant Hydrology*, 9, 325–352.

Acworth, R.I. (1987). The development of crystalline basement aquifers in a tropical environment. *Quarterly Journal of Engineering Geology*, 20, 265–272.

Adanu, E.A. (1991). Source and recharge of groundwater in the basement terrain in the Zaria-Kaduna area, Nigeria: applying stable isotopes. *Journal of African Earth Sciences*, 13, 229–234.

Alderwish, A.M. (1992). *Hydrochemical trends in the Sana'a Basin Hydrogeological system*. Unpublished M.Sc. thesis, University of London.

Alderwish, A.M. (1993). Hydrochemical trends in the Sana'a basin hydrogeological systems. In *Applied Sciences Researches, Proceedings of Conference, Book II*. Sana'a, Yemen (Arabic).

Alderwish, A.M. (1996). *Estimation of groundwater recharge to aquifers of Sana'a basin, Yemen*. Unpublished Ph.D. thesis, University of London.

Alderwish, A.M. (2000). *Environmental Aspects of Sana'a and Taiz, Republic of Yemen*. World Bank & Ministry of Construction, Housing and Urban Planning.

Alderwish, A.M. & Al Eryani, M. (1999). Water right aspects of the proposed sources for Sana'a water supply. In *Proceedings of IAH 31st Conference, Bratislava, Slovakia*, 3–7.

Alderwish, A.M. & Dottridge, J. (1998). Recharge components in a semi-arid area: the Sana'a Basin, Yemen. In N.S. Robins (Ed.), *Groundwater Pollution, Aquifer Recharge and Vulnerability*. Geological Society Publication, 130, 169–177.

Alderwish, A.M. & Dottridge, J. (1999). Urban recharge and its influence on groundwater quality in Sana'a, Yemen. In J. Chilton (Ed.), *Groundwater in the Urban Environment: Selected City Profiles*, Volume 21 of the International Contribution to Hydrogeology. Rotterdam: Balkema, 85–90.

Al-Hamdi, M. (1994). Groundwater Pollution due to domestic waste water in the Sana'a City area. M.Sc. thesis EE 138, IHE, Delft, the Netherlands.

Aller, L., Bennet, T., Lehr, J.H., Petty, R.J. & Hackett, G. (1987). DRASTIC: a standardised system for evaluating ground water pollution potential using hydrogeologic settings. *U.S. Environmental Protection Agency Report* EPA/600/2-87-036.

Aller, L., Bennett, T.W., Hackett, G., Petty, R.J., Lehr, J.H., Sedoris, H., Nielsen, D.M. & Denne, J.E. (1989). Handbook of suggested practices for the design and installation of ground-water monitoring wells. *U.S. Environmental Protection Agency Report* EPA 600/4-89/034.

Alvarez, M.E. & Pillai, S. (1997). Reversible inactivation of viruses in groundwater. *El Paso, Texas A&M Agricultural Research and Extension Centre Report.*

Anderson, M.P. & Woessner, W.W. (1992). *Applied Groundwater Modeling.* Academic Press, CA, USA.

Anon (1985). Leaking sewer causes typhoid in Israel. *World Water*, 8.

Appleyard, S.J. (1990). The flux of nitrogen and phosphorus from groundwater to the ocean in the Perth Metropolitan Region. *Geological Survey of Western Australia Hydrogeology Report*, 1990/64.

Appleyard, S.J. (1992). Estimated nutrient loads discharged into the Swan-Canning Estuary from groundwater. *Geological Survey of Western Australia Hydrogeology Report*, 1992/20.

Appleyard, S.J. (1993). Impact of stormwater infiltration basins on groundwater quality, Perth metropolitan region, Western Australia. *Environmental Geology*, 21, 227–236.

Appleyard, S.J. (1993). Resourcing the investigation and management of groundwater contamination in Western Australia – a case study from herbicide manufacture near Perth. *Journal of Australian – Geology and Geophysics*, 14, 177–181.

Appleyard, S.J. (1994). The impact of stormwater infiltration basins on groundwater quality, Perth metropolitan region. *Geological Survey of Western Australia*, 37, 19–36.

Appleyard, S.J. (1995a). Investigation of groundwater contamination by fenamiphos and atrazine in a residential area: source and distribution of contamination. *Ground Water Monitoring and Remediation*, 15, 110–113.

Appleyard, S.J. (1995b). The impact of urban development on recharge and groundwater quality in a coastal aquifer near Perth, Western Australia. *Hydrogeology Journal*, 3(2), 65–75.

Appleyard, S.J. (1996). Impact of liquid waste disposal on potable groundwater resources near Perth, Western Australia. *Environmental Geology*, 28(2), 106–110.

Appleyard, S.J. (2000). Fertilisers and the hidden sea: increasing awareness of the impacts of groundwater discharge on the marine environment in Perth. In *Proceedings of the Marine Education Society of Australia (MESA) Conference "Beyond 2000 – Future Directions in Marine Education"*, Fremantle, April 2000.

Appleyard, S.J. & Bawden, J. (1987). The effects of urbanisation on nutrient levels in the unconfined aquifer underlying Perth, Western Australia. In *Proceedings of an International Conference on Groundwater Systems under Stress.* Australian Water Resources Council, Conference Series No 13, Brisbane 1986, 587–594.

Appleyard, S.J. & Haselgrove, K.D. (1995). Groundwater discharge of nutrients to a sheltered marine system: data uncertainty and constraints on management. In *Proceedings of the Australian Systems Conference*, Perth, September 1995, 144–151.

Appleyard, S.J. & Manning, P. (1997). Managing the impact of contamination on domestic groundwater use in an urban setting. In *Proceedings of the 24th Australian–New Zealand Water Resource Conference*, Auckland, New Zealand, 241–245.

Appleyard, S.J. & Powell, R.J. (1998). Preserving urban bushland and growing local plants to protect water resources. In *Proceedings of the "Managing our Bushland" Conference*, Perth, October 1998, 72–75.

Appleyard, S.J., Davidson, W.A. & Commander, D.P. (1999). The effects of urban development on the utilisation of groundwater resources in Perth, Western Australia. In J. Chilton

(Ed.), *Groundwater in the Urban Environment: Selected City Profiles.* Volume 2, City Case Studies. Rotterdam: Balkema, 97–102.

Argue, J.R. (1997). On-Site Retention and Use (OSRU): The new kid on the stormwater block. *Stormwater and Soil Erosion 97: Future Directions for Australian Soil and Water Management*, Brisbane.

Asomaning, G. (1992). Groundwater resources of the Birmin basin in Ghana. *Journal of African Earth Sciences*, 15, 375–384.

Aust, H. & Sustrac, G. (1992). Impact of development on the geologic environment. In G.I. Lumsden (Ed.), *Geology and the Environment in Western Europe.* Oxford: Clarendon Press, 202–280.

Baedecker, M.J. & Cozzzarelli, M. (1992). The determination and fate of unstable constituents of contaminated groundwater. In S. Lesage & R.E. Jackson (Eds.), *Groundwater contamination and analysis at hazardous waste sites.* Marcel Dekker, Inc., 425–461.

Bagchi, A. (1990). *Design, construction & monitoring of sanitary landfills.* New York: John Wiley & Sons.

Bair, E.S., Springer, E. & Roadcap, G.S. (1991). Delineation of traveltime-related capture areas of wells using analytical flow models and particle-tracking analysis. *Groundwater*, 29, 387–397.

Bair, E.S. & Roadcap, G.S. (1992). Comparison of flow models used to delineate capture zones of wells: 1. Leaky-Confined Fractured-Carbonate Aquifer. *Groundwater*, 30, 199–211.

Bakker, M. & Strack, O. (1996). Capture zone delineation in two-dimensional groundwater flow models. *Water Resources Research*, 32, 1309–1315.

Banerjee, S. (1984). Solubility of organic mixtures in water. *Environmental Science and Technology*, 18(8), 587–591.

Bannerman, R.R. (1973). Problems associated with development of ground water in igneous and metamorphic rocks: A case study in Ghana. *Ground Water*, 11, 31–34.

Bannerman, R.R. (1975). The role of groundwater in rural supplies in Ghana. *Hydrological Sciences Bulletin*, 20, 191–201.

Barbash, J.E. & Resek, E.A. (1996). *Pesticides in ground water: distribution, trends and governing factors.* Chelsea, Michigan: Ann Arbor Press.

Barber, C. (1997). Perspectives on innovations and solutions for groundwater management. In J. Chilton (Ed.), *Groundwater in the Urban Environment.* Rotterdam: Balkema 1, 3–10.

Barber, C., Briegel, D.J., Power, T.R. & Hosking, J.K. (1992). Pollution of groundwater by organic compounds leached from domestic solid wastes: A case study from Morley, Western Australia. In S. Lesage & R.E. Jackson (Eds.), *Groundwater contamination and analysis at hazardous waste sites.* Marcel Dekker, Inc., 357–380.

Barber, C., Otto, C.J., Bates, L.E. & Taylor, K.J. (1996). Evaluation between land-use changes and groundwater quality in a water-supply catchment, using GIS technology: the Gwelup wellfield. *Hydrogeology Journal*, 4(1), 20–29.

Barber, L.B. (1992). Hierarchical analytical approach to evaluating the transport and biogeochemical fate of organic compounds in sewage-contaminated groundwater, Cape Cod, Masachusetts. In S. Lesage & R.E. Jackson (Eds.), *Groundwater contamination and analysis at hazardous waste sites.* Marcel Dekker, Inc., 73–120.

Barcelona, M.J., Wehrmann, H.A. & Varljen, M.D. (1994). Reproducible well purging procedures and VOC stabilization criteria for ground-water sampling. *Ground Water*, 32, 12–22.

Barker, R.D., White, C.C. & Houston, J.F.T. (1992). Borehole siting in an African accelerated drought relief project. In E.P. Wright & W.G. Burgess (Eds.), *Hydrogeology of crystalline basement aquifers in Africa*. Geological Society Special Publication 66, 183–201.

Barrell, R.A.E. & Rowland, M.G.M. (1979). The relationship between rainfall and well water pollution in a West African (Gambian) village. *Journal of Hygiene*, 83, 143–150.

Barrett, M.H., Hiscock, K.M., Pedley, S., Lerner, D.N., Tellam, J.H. & French, M.J. (1999). Marker species for identifying urban groundwater recharge sources: A review and case study in Nottingham, UK. *Water Research*, 33(14), 3083–3097.

Barrett, M.H. (1999). Unpublished data, University of Surrey, U.K.

Barrett, M.H., Hiscock, K.M., Pedley, S.J., Lerner, D.N. & Tellam, J.H. (1997). The use of marker species to establish the impact of the city of Nottingham, UK on the quantity and quality of its underlying groundwater. In J. Chilton (Ed.), *Groundwater in the urban environment: Problems, Processes and Management*. Rotterdam: Balkema, 85–90.

Barrett, M.H., Howard, G., Pedley, S., Taylor, R.G. & Nalubega, M. (1998b). A comparison of the extent and impacts of sewage contamination on urban groundwater in developed and developing countries. In *Proceedings of the WHO Conference: Water, Sanitation and Health*, 24–28 November, Bad Elster, Germany.

Barrett, M.H., Johal, K., Howard, G., Pedley, S. & Nalubega, M. (2000). Sources of faecal contamination of shallow groundwater in Kampala. Accepted for publication at IAH Congress, Cape Town.

Barrett, M.H., Yang, Y., Lerner, D.N., French, M.J., Tellam, J.H. (1997). The impact of cities on the quantity and quality of their underlying groundwater, incorporating the importance of non-agricultural sources of nitrate in UK groundwaters. Unpublished report, University of Bradford.

Bear, J. & Verruijt, A. (1987). *Modelling groundwater flow and pollution*. Dordrecht, The Netherlands: Reidel.

Beckett, J. (Ed.) (1997). *A centenary history of Nottingham*. Manchester University Press.

Bedient, P.B. (1999). *Ground Water Contamination: Transport and Remediation*. 2nd edition. New Jersey: Prentice-Hall.

Bedient, P.B., Rifai, H.S. Newell, C.J. (1994). *Groundwater Contamination: Transport and Remediation*. New Jersey: PTR Prentice-Hall Inc.

Beller, H.R., Grbic-Galic, D. & Reinhard, M. (1992). Microbial degradation of toluene under sulphate reducing conditions and the influence of iron on the process. *Applied Environmental Microbiology*, 58, 786–793.

Benker, E., Davis, G.B., Appleyard, S., Barry, D.A. & Power, T.R. (1996). Tricloroethene (TCE) contamination in an unconfined sand aquifer underlying a residential area of Perth, Western Australia. *Hydrogeology Journal*, 4(1), 20–29.

Bennett, J.R.P. (1976). Hydrological study of the Lagan Valley. *Geological Survey of Northern Ireland Open Report* 57.

Berg, L.R., Johnson, G.B. & Raven, P.H. (1997). *Environment*. Saunders College Publishing, U.S.A.

Bernard, C., Carbiener, R., Cloots, A.R., Froehlicher, R., Schenek, C. & Zilliox, L. (1992). Nitrate pollution of groundwater in the Alsatian Plain (France) – A multidisciplinary study of an agricultural area: The Central Ried of the Ill River. *Environmental Geology Water Sciences*, 20(2), 125–137.

British Geological Survey (1994). Impact of urbanisation on groundwater in Mérida, Mexico: final report. *British Geological Survey Technical Report*, Overseas Geology Series, WC/94/38.

Biberhofer, J. & Stevens, R.J.J. (1987). Organochlorine contaminants in ambient waters of Lake Ontario. *Environment Canada*, Scientific Series 159, 1–11.

Bishop, P.K., Lerner, D.N., Jakobsen, R., Gosk, E. & Burston, M.B. (1993a). Investigation of a solvent polluted industrial site on a deep sandstone/mudstone sequence in the UK. 1: Site description and groundwater flow. *Journal of Hydrology*, 149, 209–229.

Bishop, P.K., Lerner, D.N., Jakobsen, R., Gosk, E., Burston, M.B. & Chen, T. (1993b). Investigation of a solvent polluted industrial site on a deep sandstone/mudstone sequence in the UK. 2: Contaminant sources, distributions, transport and retardation. *Journal of Hydrology*, 149, 231–256.

Blomquist, W. (1992). *Dividing the Waters*. ICS Press, San Francisco, U.S.A.

Bodhankar, N. & Chatterjee, B. (1994). Pollution of limestone aquifer due to urban waste disposal around Raipur, Madhya Pradesh, India. *Environmental Geology*, 23, 209–213.

Boeckh, E. (1992). An exploration strategy for high-yield boreholes in the West African crystalline basement. In E.P. Wright & W.G. Burgess (Eds.), *The hydrogeology of crystalline basement aquifers in Africa*. Geological Society (London), 87–100.

Borden, R.C. & Yanoschak, T.M. (1990). Ground and surface water quality impacts of North Carolina Sanitary landfills. *Water Resources Bulletin*, 26(2), 269–277.

Boutaeb, S., Bouchaou, L., Mudry, J., Hsissou, Y., Mania, J. & Chauve, P. (2000). Hydrogeologic effects on the quality of water in the Qued Issen Watershed, Western Upper Atlas Mountains, Morocco, *Hydrogeology Journal*, 8(2), 230–238.

Braune, E. (2000). Towards comprehensive groundwater resource management in South Africa. In O. Sililo (Ed.), *Groundwater: past achievements and future challenges*. Rotterdam: Balkema.

Brimhall, G.H., Chadwick, O.A., Lewis, C.J., Compston, W., Williams, I.S., Danti, K.J., Dietrich, W.E., Power, M.E., Hendricks, D. & Bratt, J. (1991). Deformational mass transport and invasive processes in soil evolution. *Science*, 255, 695–702.

Brink C. van den, Heimovaara, T. & Marinussen, M.P.J.C. (2000). Conditions for the Use of Innovative Remediation Techniques set by new Policy. In: *Remediation of Hazardous Waste Contaminant Soil, 2nd edition.*

British Geological Survey (BGS), the Academic Hydrology Section of the Autonomous University of Yucatan (FIUADY) and the National Water Commission of Mexico (CNA), 1995.

Brunn, S.D. & Williams, J.E. (1983). *Cities of the World: Worlds Regional Urban Development*. New York: Harper & Row. ISBN 0-06-381225-8.

Buckley, D.K. & MacDonald, D.J. (1994). Geophysical logging of a karstic limestone aquifer for hydrogeological purposes at Mérida, Yucatan, Mexico. *British Geological Survey Technical Report*, WD/94/4C.

Buckley, D.K. & Zeil, P. (1984). The character of fractured rock aquifers in eastern Botswana. In *Challenges in African Hydrology and Water Resources, Proceedings of the Harare Symposium*, July 1984, IAH Series Publication 144, 25–36.

Burgess, W.G., Dottridge, J. & Symington, R.M. (1998). Methyl tertiary butyl ether (MTBE): a groundwater contaminant of growing concern, In J. Mather, D. Banks, S. Dumpleton,

M. Fermor (Eds.), *Groundwater Contaminants and their migration* Geological Society, Special Publication 128, 29–34.

Burke, J.J., Sauveplane, C. & Moench, M. (1999). *Groundwater management and socio-economic responses.* National Resources Forum, 23, 303–313.

Burston, M.W., Nazari, M.M., Bishop, P.K. & Lerner, D.N. (1993). Pollution of groundwater in the Coventry region by chlorinated solvents. *Journal of Hydrology*, 149, 137–161.

Calderon, R.L. (2000). The epidemiology of chemical contaminants of drinking water. *Food and Chemical Toxicology*, 38(1), S13–20.

California Department of Water Resources (1961). Planned utilization of the ground water basins of the coastal plain of Los Angeles County – Appendix A, ground water geology. *California Department of Water Resources*, Bulletin 104.

California Department of Water Resources (1996). Adjudicated Basins and Watermasters in California, *Water Facts*, Number 3.

Cannell, R.Q. & Burford, J.R. (1976). *The Use of Lysimeters at the Agricultural Research Centre.* Lysimeter Conference, Institute of Geological Sciences, London, October 1976.

Cargeeg, G.C., Townley, L.R., Smith, G.R., Appleyard, S.J. & Smith, R.A. (1987). *Perth urban water balance study – volumes 1 and 2.* Western Australia Water Authority, WP29.

Carmon, N. Shamir, U. & Meiron-Pistner, S. (1997). Water Sensitive Urban Planning, *Journal of Environmental Planning and Management*, 40(4), 413–434.

Chanda, D.S., Kirk, S. & Watkins, J. (1997). Groundwater pollution threats to public water supplies from urbanisation. In J. Chilton (Ed.), *Groundwater in the Urban Environment, Problems, Processes and Management, Proceedings of the 27th IAH groundwater congress, Nottingham*, ISBN 90-5410 837 1, 297–301.

Charsley, T.J., Rathbone, P.A. & Lowe, D.J. (1990). Nottingham: A geological background for planning and development. *BGS Technical Report*, WA/90/1.

Chilton, P.J. & Foster, S.S.D. (1995). Hydrogeological characterisation and water-supply potential of basement aquifers in tropical Africa. *Hydrogeology Journal*, 3, 36–49.

Chilton, P.J. & Smith-Carington, A.K. (1984). Characteristics of the weathered basement aquifer in Malawi in relation to rural water-supplies. In *Challenges in African Hydrology and Water Resources, Proceedings of the Harare Symposium*, July 1984, IAHS Publication 144, 57–72.

Chiras, D. (1998). *Environmental Science: a systems approach to sustainable development.* 5th Edition. Belmont, CA: Wadsworth Publishing Company, U.S.A.

Christensen, T.H., Kjeldsen, P., Albrechtsen, H.J., Perron, G., Nielsen, P.H., Bjerg, P.L. & Holm, P.E. (1994). Attenuation of landfill leachate pollutants in aquifers. *Critical Reviews in Environmental Science and Technology*, 24(2), 119–202.

Church, P.E. & Friesz, P.J. (1993). Effectiveness of highway drainage systems in preventing road-salt contamination of groundwater: preliminary findings. *Transportation Research record 425*, HRB, National Research Council, Washington, D.C., 56–64.

CIRIA (1989). The engineering implications of rising groundwater levels in the deep aquifer beneath London. *Construction Industry Research and Information Association Special Publication*, 69, ISBN 0 86017 303 8.

Clark, D. (1998). Interdependent urbanization in an urban world: an historical overview. *The Geographical Journal*, 164, 85–95.

Clark, L. (1985). Groundwater abstraction from Basement Complex areas of Africa. *Quarterly Journal of engineering Geology (London)*, 18, 25–34.

Clesceri, L.S., Greenburg, A.E. & Trussel, R.R. (1989). Standard Methods for the Examination of Water and Wastewater, 17th Edition. *American Public Health Association Publication.*

Colvin, C.A. & Conrad, J. (1998). Contamination of groundwater by agricultural activities in South Africa.

Comley, H.H. (1945). Cyanosis in infants caused by nitrates in well water. *Journal of the American Medical Association*, 129 (in press).

Conrad, J.E., Colvin, C., Sililo, O.T.N., Gorgens, A., Weaver, J. & Reinhardt, C. (1999). Assessment of the Impact of Agricultural Practices on the Quality of Groundwater Resources in South Africa. *WRC Report*, Pretoria, 641/1/99.

Conservation Foundation (1987). Groundwater: Saving the Unseen Resource – A Guide to Groundwater Pollution: Problems, Causes, and Government Responses. Conservation Foundation, Washington, D.C.

Craun, G.F. (1981). Outbreaks of waterborne disease in the United States 1971 to 1978. *Journal of American Waterworks Works Association*, 73(7), 360–369.

Craun, G.F. (1984). Health aspects of groundwater pollution. In G. Bitton & C.P. Gerba (Eds.), *Groundwater pollution microbiology*. John Wiley and Sons, 135–179.

Crawford, N.C. & Ulmer, C.S. (1994). Hydrogeologic investigations of contaminant movement in karst aquifers in the vicinity of a train derailment near Lewisburg, Tennessee. *Environmental Geology*, 23, 41–52.

Crewe, L. & Beaverstock, J. (1998). Fashioning the city: Cultures of consumption in contemporary urban spaces. *Geoform*, 29(3), 287–308.

Cronin, A.A. (2000). *Groundwater Flow and Isotope Geochemical Modelling of the Triassic Sandstone Aquifer, Northern Ireland*. Unpublished Ph.D. Thesis, Queens University, Belfast.

Cronin, A.A. (2001). Improved regional groundwater flow modelling using geochemical and isotopic constraints, submitted for publication at the International Association of Hydrogeologists Annual Conference, Munich, September 2001.

Crowe, A.S. & Mutch, J.P. (1991). EXPRES: An expert system for assessing the fate of pesticides in the subsurface: Progress Report – Incorporation of the LP/LI screening model into EXPRES. *National Water Research Institute Contribution*, Environment Canada, Burlington, Ontario. 90–68.

Custodio, E. (1997). Groundwater quantity and quality changes related to land and water management around urban areas: Blessings and misfortunes. In J. Chilton (Ed.), *Groundwater in the Urban Environment 1 Problems, Processes and Management, Proceedings of the 27th IAH groundwater congress, Nottingham.* ISBN 90 5410 837 1, 11–22.

Dar Al-Handasah (1999). Sana'a 1 environmental impact assessment. Sana'a water supply and sanitation project. *NWSA*, ROY, March 1999.

Davidson, J.M. (1995). These contaminants know no boundaries: monitoring well position, plume movement among factors influencing NAPL detection. *International Ground Water Technology*, 1(3), 19–20.

Davidson, W.A. (1995). Hydrogeology and groundwater resources of the Perth Region, Western Australia. *Geological Survey of Western Australia*, Bulletin 142.

Davies, B.R. & Day, J.A. (1986). *The biology and conservation of South Africa's vanishing waters*. Centre for extra-mural studies, University of Cape Town, Rondebosch, South Africa.

Davis, G.B. (1997). Site clean-up: the pros and cons of disposal and in situ and ex situ remediation. *Land Contamination & Remediation*, 5(4), 287–290.

Davis, S.N. & Turk, L.J. (1964). Optimum depth of wells in crystalline rocks. *Ground Water*, 2, 6–11.

De Borde, D.C., Woessner, W., Lauerman, B. & Ball, P.N. (1998). Virus Occurrence and Transport in a school septic tank system and unconfined aquifer. *Ground Water*, 36(5), 825–834.

De Roche, J.T. & Breen, K.J. (1988). Hydrogeology and Water Quality near a Solid- and Hazardous-Waste Landfill, Northwood, Ohio. *United States Geological Survey Water-Resources Investigations Report* 88-4093.

DENR (South Australian Department of Environment and Natural Resources) (1994). *Blue Lake Management Plan.* DENR Report, ISBN 0 7308 0620 0.

Department of Environmental Affairs (1992). *Hazardous waste in South Africa.* Volume 1: Situation analysis. R.G. Noble, (Ed.) CSIR, Pretoria, South Africa.

DGH, Du Maroc, R. (1999). Resources en Eau Dans Les Bassins Souss-Massa et Cotiers Nord D'Agadir Etat Actuel et Perspectives D'Avenir, Handouts at the US AID-FORWARD DEH Team debriefing March 17, Rabat, Morocco.

Dillon, P.D. (1989). An evaluation of the sources of nitrate in groundwater near Mount Gambier, South Australia. *Australian CSIRO Water Resource Series Report*, 1.

Dillon, P. (1997). Groundwater pollution by sanitation on tropical islands. IHP-V Project 6-1. UNESCO, Paris.

Document of the World Bank (1999). Project appraisal document on a proposed credit in the amount of SDR 18.4 Million to the ROY for a Sana'a Water Supply & Sanitation Project. Infrastructure Development Group, Middle East and North Africa Region.

DoE NI (1994). *Groundwater vulnerability map of Northern Ireland*, Department of the Environment Northern Ireland, scale 1:250,000.

Domenico, P.A. & Schwartz, F.W. (1990). *Physical & Chemical Hydrogeology.* Brisbane: John Wiley and Sons.

Dragisic, V., Miladinovic, B. & Milenic, D. (1997). Pollution of groundwaters on Donji Milanovac. In Chilton *et al.* (Eds.), *Groundwater in the urban environment: Problems, Processes and Management.* Rotterdam: Balkema, 395–398.

Dreher, J.E. & Gunatilaka, A. (1999). Management of urban groundwater: solutions for water quality and quantity problems; evaluation of a groundwater management system for Vienna after three years of operation. D.L. Danielpol, C. Griebler, J. Gibert, H.P. Nachtnebel & J. Notenboom (Eds.), *Groundwater ecology – A tool for management of water resources – EC Advanced study Course 1999*, 253–275.

Drever, J.I. (1997). *The Geochemistry of Natural Waters.* New Jersey: Prentice-Hall.

Driscoll, F.G. (1989). *Groundwater and Wells.* 2nd edition. Johnson Filtration System Inc., St. Paul, Minnesota.

Duda, A. & Nawar, M. (1996). Implementing the world bank's water resources management policy: A priority on toxic substances from nonpoint sources. *Water Science and Technology*, 33(4–5), 45–51.

Duncan, H.P. (1995). *A Bibliography of Urban Storm Water Quality.* Cooperative Research Centre for Catchment Hydrology, Melbourne.

Department of Water Affairs and Forestry (2000). *Groundwater Quality Management Policies and Strategies.* DWAF, Pretoria.

Eastwood, P.R., Lerner, D.N., Bishop, P.K. & Burston, M.W. (1991). Identifying land contaminated by chlorinated hydrocarbon solvents. *Journal of IWEM*, 5, 163–171.

Eaton, T.T. & Zaporozec, A. (1997). Evaluation of groundwater vulnerability in an urbanizing area. In J. Chilton (Ed.), Groundwater in the Urban Environment 1 Problems, Processes and Management, *Proceedings of the 27th IAH groundwater congress*, Nottingham, ISBN 90 5410 837 1, 577–582.

Ebohon, O.J. (1996). The scope and limits of sustainable development in Africa built environment sector. *International Journal of Sustainable Development and World Ecology*, 3, 1–12.

Eckenfelder, W.W. (1989). *Industrial water pollution control*. New York: McGraw-Hill.

Edwards, E.A., Wills, L.E., Reinhard, M. & Grbic-Galic, D. (1992). Anaerobic degradation of toluene and xylene by aquifer microorganisms under sulphate-reducing conditions. *Applied Environmental Microbiology*, 58, 794–800.

Edwards, K.C. (1966). Nottingham and its region. Prepared for the meeting of the British Association for the Advancement of Science, Nottingham.

Edworthy, K.J. (1991). *Report on Environmental Aspects of Water Management in Northern Yemen*. Technical Secretariat of the High Council for Water., UNDP/DTCD Project Yem/88/001.

EEA (1999). *Groundwater quality and quantity in Europe* – Environmental assessment report No 3, In S. Nixon (Ed.), European Environmental Agency, Luxembourg. ISBN 92-9167-146-0.

Ehlers, J. & Grieger, J. (1983). Groundwater chemistry in the Hamburg region. In F.X. Dunin, G. Matthess, & R.A. Gras (Eds.), *Relation of groundwater quantity and quality, Proceedings of the Hamburg Symposium, Aug. 1983*. IAHS Publication 146, 285–293.

Eisen, C. & Anderson, M.P. (1979). The effect of urbanisation on groundwater quality – a case study. *Ground Water*, 17(5), 456–462.

Eiswirth, M. & Hotzl, H. (1997). The impact of leaking sewers on urban groundwater. In J. Chilton (Ed.), *Groundwater in the urban environment: Problems, Processes and Management*. Rotterdam: Balkema, 399–404.

Eiswirth, M. & Hotzl, H. (1994). Groundwater Contamination by leaky sewer systems. *Water Down Under*, National Conference of Institution of Engineers, Adelaide, Australia, November 1994, 111–114.

El Arabi, N.E. & Fatma, A. (1997). Impacts of sewage-based irrigation on groundwater: an Egyptian case. In Towards Sustainable Irrigation in the Mediterranean Region, *Proceedings of International Conference on Water Management, Salinity and Pollution Control*, Bari, Italy, September 1997, 215–228.

El Arabi, N.E. (1999). Problems of groundwater quality related to the urban environment in Greater Cairo. In *Proceedings of the IUGG 99 Symposium HS5 "Impacts of Urban Growth on Surface Water and Groundwater Quality"*, Birmingham, UK, July 1999, 29–37.

El Arabi, N.E., Rashed, M. & Vermeulen, A. (1996). Environmental impacts of sewage water irrigation on groundwater. In *Proceedings of the 16th ICID Congress on Irrigation and Drainage "Managing Environmental Changes Due to Irrigation and Drainage"*, Cairo, September 1996, 102–114.

El Gohary, F. (1994). Comparative Environmental Risks in Cairo: Water Pollution Problems. In *Comparing Environmental Risks in Cairo, Egypt*, Volume 2. US AID report.

El Sammani, M.O., El Hadi, A.S., Talha, M., El Hassan, B.M. & Haywood, I. (1989). Management problems of Greater Khartoum. In R.E. Stren & R.R. White (Eds.), *African cities in crisis*, Westview Press, 247–275.

Ellis, J.B. (1997). Groundwater pollution from infiltration of urban stormwater runoff. In *Proceedings of XXVII IAH Congress on Groundwater in the Urban Environment, Nottingham*, Balkema.

Ellis, J.B. & Hvitved-Jacobsen, T. (1996). Urban drainage impacts on receiving waters. *Journal Hydrological Research*, 36(6), 771–783.

Emmett, A.J. & Telfer, A.L. (1994). Influence of karst hydrology on water quality management in southeast South Australia. *Environmental Geology*, 23, 149–155.

Engelbrecht, J.F.P. (1993). An assessment of health aspects of the impact of domestic and industrial waste disposal activities on groundwater resources – A literature review. Water Research Commission, South Africa No 371/1/93.

Engelen, G.B. & Kloosterman, F.H. (1996). *Hydrological systems analysis; methods and applications*. Kluwer Academic, Dordrecht, the Netherlands, ISBN 0-7923-3986-X.

Esrey, S.A., Potash, J.B., Roberts, L. & Shiff, C. (1991). Effects of improved water supply and sanitation on ascaris, diarrhoea dracunculiarsis, hookworm infection, schistosomiasis, and trachoma. *Bulletin of the World Health Organisation*, 69, 609–621.

Eswaran, H. & Bin, W.C. (1978). A study of a deep weathering profile on granite in peninsular Malaysia: I. Physico-chemical and micromorphological properties. *Journal of the Soil Science Society of America*, 42, 144–149.

European Commission (1995). Karst Groundwater Protection Final Report. *COST Action 65 Report EUR 16547EN*. European Commission.

EWS (South Australian Engineering and Water Supply Department) (1973). Report of the committee on water pollution control in the southeast. *EWS Report*, 73/15.

Fabricius, T. & Whyte, A. (1980). *Evaluation of Ontario Municipal experiments to reduce the use of road salt*. Institute for Environmental Studies Snow and Ice Control Working Group Working Paper SIC-7.

Faillat, J.P. (1990). Origine des nitrates dans les nappes de fissures de la zone tropicale humide – exemple de la Côte D'Ivoire. *Journal of Hydrology*, 113, 231–264.

Faillat, J.P. & Rambaud, A. (1991). Deforestation and leaching of nitrogen as nitrates into underground water in intertropical zones: the example of Côte D'Ivoire. *Environmental Geology and Water Sciences*, 17, 133–140.

Falkland, A. (Ed.) (1991). *Hydrology and water resources of small islands: a practical guide*. Studies and reports in hydrology 49. UNESCO, Paris.

Fellman, E. & Barker, J.A. (1997). Scoping the impact of highway drainage in groundwater quality. In *Proceedings of XXVII IAH Congress on Groundwater in the Urban Environment, Nottingham*, Balkema.

Fernandez, J.W. (1998). Providing Water in Latin America, *New York Law Journal*, April 27.

Fetter, C.W. (1993). *Contaminant Hydrogeology*. New York: Macmillan Publishing Company.

Fetter, C.W. (1988). *Applied Hydrogeology* (2nd edition). Columbus, Ohio: Merrill Publishing Company.

Fetter, C.W. (1999). *Contaminant Hydrogeology* (2nd edition). New York: Macmillan Publishing Company.

Filipovic, S. (1988). Heavy metals as indicators of water relations in karst and protection problems. In *Karst Hydrogeology and Karst Environment Protection, Proceedings of the 21st Congress of the International Association of Hydrogeologists*, Guilin, China, 1037–1044.

Finnish National Road Administration (1993). *Winter Maintenance Program in Finland, 1995–6*. Helsinki.

Finnish National Road Administration (1996). *Groundwater Protection along Roads*. Helsinki.

Firman, T. (1998). The restructuring of Jakarta Metropolitan Area: a "global city" in Asia. *Cities*, 15(4), 229–243.

Flipse, J.R., Katz, B.G., Lindner, J.B. & Markel, R. (1984). Sources of nitrate in ground-water in a sewered housing development, Central Long Island, New York. *Ground Water*, 22(4), 418–426.

Foppen, J.W., Hoogeveen, R., Wiesenekker, H., Al Haimy, I. & Ali, K.Z. (1996). *Hydro-chemistry of the Sana'a Basin and microbiology of the groundwater below Sana'a.* SAWAS Technical report 13, TNO/NWSA.

Ford, D.C. & Ewers, R.O. (1978). The development of limestone cave systems in the dimensions of depth and length. *Canadian Journal of Earth Sciences*, 15, 1783–1798.

Ford, M. & Tellam, J.H. (1994). Source, type and extent of inorganic contamination within the Birmingham urban aquifer system. *Journal of Hydrology*, 156, 101–135.

Ford, M., Tellam, J.H. & Hughes, M. (1992). Pollution-related acidification in the urban aquifer, Birmingham. *Journal of Hydrology*, 140, 297–312.

FORWARD (1999). *Supporting the Creation of a Watershed Authority in the Souss-Massa River Basin of Morocco.* Prepared for the US Agency for International Development, Contract HNE-C-00-96-90027-00.

Foster, S., Lawrence, A. & Morris, B. (1998). Groundwater in urban development: assessing management needs and formulating policy strategies. *World Bank Technical Paper No 390.*

Foster, S.S.D. (1984). African groundwater development – the challenges for hydrogeo-logical science. In *Challenges in African Hydrology and Water Resources, Proceedings of the Harare Symposium, July 1984*, IAH Series Publication 144, 3–12.

Foster, S.S.D. (1985). Groundwater pollution protection in developing countries. In G. Matthess, S.S.D. Foster & A.C. Skinner (Eds.), Theoretical Background, Hydro-geology and Practice of Groundwater Protection Zones, *IAH Special Volume* 6, 167–200.

Foster, S.S.D. (1990). Impacts of urbanization on groundwater. In *Hydrologic Processes and Water Management in Urban Areas, Proceedings of the Duisburg Symposium,* April 1988. International Association of Hydrological Sciences (IAHS) Publ. No. 198, 187–207.

Foster, S.S.D. & Skinner, A.C. (Eds.), Theoretical Background, Hydrogeology and Practice of Groundwater Protection Zones, *IAH Special Volume* 6, 167–200.

Foster, S.S.D. & Morris, B.L. (1994). Effects of urbanisation on groundwater recharge. In W.B. Williamson (Ed.), *Groundwater Problems in Urban Areas, Proceedings of ICE conference London,* June 1993, Thomas Telford, 43–63.

Foster, S.S.D., Lawrence, A.R. & Morris, B.L. (1996). Groundwater Resources beneath rapidly urbanizing cities-implications and priorities for water supply management. In *Report of the Habitat II Conference, Beijing, China, March 1996*, 356–365.

Frapporti, G., Vriend, S.P. & Gaans van, P.F.M. (1993). Hydrogeochemistry of the shallow Dutch groundwater: Interpretation of the national groundwater quality monitoring Network. *Water Resources Research*, 29, 2993–3004.

Fraser, B., Howard, K.W.F. & Williams, D.D. (1996). Monitoring biotic and abiotic processes across the hyporheic/groundwater interface. *Hydrogeology Journal* 4(2), 36–50.

Freeman, T.W. (1966). *The conurbations of Great Britain.* Manchester: Manchester University Press.

Freeze, R.A. & Cherry, J.A. (1979). *Groundwater.* New Jersey: Prentice-Hall.

Gardner-Outlaw, T. & Engelman, R. (1997). *Sustaining water, easing scarcity.* Population Action International (Washington). http://www.populationaction.org

Garlake, P. (1982). *Great Zimbabwe described and explained.* Harare: Zimbabwe Publishing House.

Geirnaert, W., Groen, M., Van Der Sommen, J. & Leusink, A. (1984). Isotope studies as a final stage in groundwater investigations on the Africa shield. In *Challenges in African Hydrology and Water Resources, Proceedings of the Harare Symposium, July 1984*, IAH Series Publication 144, 141–153.

Gelinas, Y., Randall, H., Robidoux, L. & Schmit, J.P. (1996). Well water survey in two districts of Conakry (Republic of Guinea), and comparison with the piped city water. *Water Research*, 30, 2017–2026.

Geomatrix Consultants, Inc. (1997a). Phase III Groundwater Assessment and Remediation Planning Report, California Institute for Men, California. *Unpublished consultant's report.*

Geomatrix Consultants, Inc. (1997b). Conceptual Hydrogeologic Model, Charnock Wellfield Regional Assessment. *Unpublished consultant's report.*

Gerba, C.P. & Bitton, G. (1984). Microbial pollutants: Their survival and transport pattern in groundwater. In G. Bitton & C.P. Gerba (Eds.), *Groundwater pollution microbiology.* John Wiley and Sons, 65–88.

Gerba, C.P., Yates, M.V. & Yates, S.R. (1991). Quantification of factors controlling viral and bacterial transport in the subsurface. In C.J. Hurst (Ed.), *Modeling the Environmental fate of microorganisms.* American Society for Microbiology, Washington, D.C. 77–88.

Gerritse, R.G., Barber C. & Adeney, J.A. (1990). The impact of residential urban areas on groundwater quality: Swan Coastal Plain, Western Australia. *CSIRO Water Resources*, Series 3.

Gibert, J. (1990). Behaviour of aquifers concerning contaminants: Differential permeability and importance of the different purification processes. *Water Science and Technology*, 22(6), 101–108.

Gikunju, J.K., Githui, K. & Maitho, T.E. (1992). Fluoride levels in borehole water around Nairobi. *Flouride*, 25, 111–114.

Gilboa, Y. (1971). Replenishment sources of the alluvial aquifers of the Peruvian coast. *Ground Water*, 9(4), 39–46.

Girard, P. & Hillaire-Marcel, C. (1997). Determining the source of nitrate pollution in the Niger discontinuous aquifers using the natural $^{15}N/^{14}N$ ratios. *Journal of Hydrology*, 199, 239–251.

Goerlitz, D.F. (1992). A review of studies of contaminated groundwater conducted by the US Geological Survey Organics Project, Menlo Park, California, 1961–1990. In S. Lesage & R.E. Jackson (Eds.), *Groundwater contamination and analysis at hazardous waste sites.* Marcel Dekker, Inc. 295–355.

Gomboso, J. (1997). Economic and land-use planning tools for groundwater protection in Perth, Western Australia. In Chilton *et al.* (Eds.), *Groundwater in the Urban Environment. Proceedings of XXVII IAH Congress, Nottingham.* Rotterdam: Balkema, 587–596.

Gong, H., Menlou, L. & Xinli, H. (2000). Management of groundwater in Zhengzhou city, China. *Water Resources*, 34(1), 57–62.

Gonzales-Herrera, R.A. (1992). Evolution of groundwater contamination in the Yucatan karstic aquifer. Unpublished M.Sc. Thesis, University of Waterloo, Ontario, Canada.

Gooddy, D.C., Morris, B.L., Vasquez, J. & Pachecho, J. (1993). Organic contamination of the karstic limestone aquifer underlying the city of Mérida, Yucatan, Mexico. *British Geological Survey Technical Report WD/93/8*, Wallingford, UK.

Government of Canada (1991). *The State of Canada's Environment.* Government Publication Office, Ottawa.

Greenbaum, D. (1992). Structural influences on the occurrence of groundwater in SE Zimbabwe. In E.P. Wright & W.G. Burgess (Eds.), *The hydrogeology of crystalline basement aquifers in Africa.* Geological Society (London), 77–86.

Greswell, R.B., Lloyd, J.W., Lerner, D.N. & Knipe, C.V. (1994). Rising groundwater in the Birmingham area. In W.B Wilkinson (Ed.), *Groundwater Problems in Urban Areas.* London: Thomas Telford, 355–368.

Grischek, T., Nestler, W., Piechniczek, D. & Fischer, T. (1996). Urban groundwater in Dresden, Germany. *Hydrogeology Journal*, 4(1), 48–63.

GSI (1999). *Groundwater Protection Schemes.* Geological Survey of Ireland in association with the Department of the Environment and Local Government and the Irish Environmental Protection Agency, ISBN 1-899702-22-9.

Gugler, J. & Flanagan, W.G. (1978). *Urbanization and social change in West Africa.* London: Cambridge University Press.

Hadiwinoto, S. & Leitmann, J. (1994). Urban environmental profile – Jakarta. *Cities*, 11(3), 153–157.

H&ES Ltd. (1997). *White Mountain to Black Mountain, Groundwater Survey: Landfill Impact Study*, prepared for Department of the Environment, Northern Ireland by Hydrogeological and Environmental Services Ltd., Belfast, July 1997.

Hagedorn, C. (1984). Microbiological aspects of groundwater pollution due to septic tanks. In G. Bitton & C.P. Gerba (Eds.), *Groundwater pollution microbiology.* Brisbane: John Wiley and Sons, 181–195.

Haitjema, H.M. (1995). *Analytic Element Modeling of Groundwater Flow.* San Diego CA, USA: Academic Press.

Hanley, K. (1979). *The physics of snow and ice control on roads and highways.* Institute for Environmental Studies, Snow and Ice Control Working Group Working Paper, SIC-2.

Harris, N. (1990). Urbanisation, economic development and policy in developing countries. *Habitat International*, 14, 3–42.

Harris, G.P. (1994). *Nutrient loadings and algal blooms in Australian waters – a discussion paper.* Land and Water Resources Research and Development Corporation Occasional Paper 12/94. Canberra, ACT, Australia.

Haskoning Royal Dutch Engineers and Architects (1991). *Hazardous Waste in The Republic of Yemen.* Support to the Secretariat of the Environmental Protection Council.

Hassall, K.A. (1982). *The Chemistry of Pesticides; their metabolism, mode of action and uses in crop protection.* Verlag Chemie, Weinheim.

Hazell, J.R.T., Cratchley, C.R. & Jones, C.R.C. (1992). The hydrogeology of crystalline in northern Nigeria and geophysical techniques used in their exploration. In E.P. Wright & Burgess, W.G. (Eds.), *Hydrogeology of crystalline basement aquifers in Africa.* Geological Society Special Publication 66, 155–182.

Headworth, H.G. & Fox, G.B. (1986). The South Downs Chalk aquifer: its development and management. *Journal of the Institution of Water Engineers and Scientists*, 40, 345–361.

Hem, J.D. (1985). Study and Interpretation of the chemical characteristics of natural water. U.S. Geological Survey, Water Supply Paper 1473.

Hershaft, A. (1976). The plight and promise of on-site wastewater treatment. *Compost Science*, 17(5), 6–13.

Hiebert, F.K. & Bennett, P.C. (1992). Microbial control of silicate weathering in organic-rich groundwater. *Science*, 258, 278–281.

Hill, A.R. (1982). Nitrate distribution in the ground water of the Alliston region of Ontario, Canada. *Ground Water*, 20, 696–702.

Hill, M.J., Hawksworth, G. & Tattersall, G. (1973). Bacteria, nitrosamines and cancer of the stomach. *British Journal Cancer*, 28, 562–567.

Hillman, M.O. (1981). *Groundwater as a source of nutrients to the coastal lagoon system, Perth, Western Australia*. Masters of Engineering Science thesis, University of Western Australia.

Hirschberg, K.-J.B. (1986). *Liquid-waste disposal in Perth, A hydrogeological assessment*. Western Australia Geological Survey, Report 19, 55–61.

Hirschberg, K.-J.B. (1993a). *Municipal waste disposal in Perth and its impact on ground-water quality*. Western Australia Geological Survey, Report 34, 97–109.

Hirschberg, K.-J.B. (1993b). The location and significance of point sources of groundwater contamination in the Perth Basin. Western Australia Geological Survey, Report 34, 37–46.

Holman, I. & Palmer, R. (1999). Groundwater protection incorporated into land use planning: a study from Cebu City, the Philippines. In *Impacts of Urban Growth on Surface and Groundwater Quality, Proceedings of IUGG Symposium, Birmingham, July 1999*. IAHS 259.

Houston, J. (1990). Rainfall–Runoff–Recharge Relationships in the Basement Rocks of Zimbabwe. *Groundwater Recharge; a guide to understanding and estimating natural recharge*. IAH 8: Chapter 21. Heise, Hannover.

Houston, J. (1995). Exploring Africa's ground water resources. *International Groundwater Technology*, 1, 29–32.

Houston, J.F.T. (1982). Rainfall and recharge to dolomite aquifer in a semi-arid climate at Kabwe, Zambia. *Journal of Hydrology*, 59, 173–187.

Houston, J.F.T. & Lewis, R.T. (1988). The Victoria Province drought relief project: Borehole yield relationships. *Ground Water*, 26, 418–426.

Howard, K.W.F. (1985). Denitrification in a limestone aquifer. *Journal of Hydrology*, 76, 265–280.

Howard, K.W.F. (1993). Impacts of Urban and Industrial Development on Groundwater – Mexico City. In *Perceptions, Reality and Resolution, Proceedings of an International Seminar on the Environment, Toluca, Mexico*, February 1993. Universidad Autonoma del Estado de Mexico.

Howard, K.W.F. (1997). Impacts of Urban Development on Groundwater. In E. Eyles (Ed.), *Environmental Geology of Urban Areas, Special Publication of the Geological Association of Canada*. Geotext #3, 93–104.

Howard, K.W.F. (1997). Incorporating Policies for Groundwater Protection Into the Urban Planning Process. In J. Chilton (Ed.), *Groundwater in the Urban Environment, 1 Problems, Processes and Management, Proceedings of the 27th IAH groundwater congress, Nottingham*, ISBN 90 5410 837 1, 31–40.

Howard, K.W.F. (1998). Monitoring the impact of road de-icing chemicals on groundwater. In T. Nystén & T. Suokko (Eds.), *De-icing and dustbinding – risk to aquifers, Proceedings of International Symposium, Helsinki, Finland*. October 1998, 51–62.

Howard, K.W.F. & Beck, P.J. (1993). Hydrogeochemical implications of groundwater contamination by road de-icing chemicals. *Journal of Contaminant Hydrology*, 12(3), 245–268.

Howard, K.W.F. & Haynes, J. (1993). Groundwater contamination due to road de-icing chemicals – salt balance implications. *Geoscience Canada*, 20, 1–8.

Howard, K.W.F. & Livingstone, S. (1997). Contaminant source audits and ground-water quality assessment. In E. Eyles (Ed.), *Environmental Geology of Urban Areas*. Special Publication of the Geological Association of Canada. Geotext #3, 105–118.

Howard, K.W.F. & Karundu, J. (1992). Constraints on the development of basement aquifers in east Africa – water balance implications and the role of the regolith. *Journal of Hydrology*, 139, 183–196.

Howard, K.W.F. & Taylor, L.C. (1998). Hydrogeochemistry of Springs in Urban Toronto. In *Gambling with Groundwater, Proceedings of the International Association of Hydrogeologists XXVIII Congress*, Las Vegas, November 1998.

Howard, K.W.F., Boyce, J.I., Livingstone, S.J. & Salvatori, S.L. (1993). Road salt impacts on groundwater quality – the worst is yet to come! *GSA Today*, 3(12), 301–321.

Howard, K.W.F., Eyles, N. & Livingstone, S. (1996). Municipal landfilling practice and its impact on groundwater resources in and around urban Toronto, Canada. *Hydrogeology Journal*, 4(1), 64–79.

Howard, K.W.F., Hughes, M., Charlesworth, D.L. & Ngobi, G. (1992). Hydrogeologic evaluation of fracture permeability in crystalline basement aquifers of Uganda. *Journal of Applied Hydrogeology*, 1, 55–65.

Hoxley, G. & Dudding, M. (1994). Groundwater contamination by septic tank effluent: two case studies in Victoria, Australia. *XXVth Congress of the International Association of Hydrogeologists/International Hydrology and Water Resources Symposium of the Institution of Engineers*, Australia. Adelaide, 21–25 November. Preprints of Papers NCPN 94/10, 145–152.

Hughes, A.J., Tellam, J.H., Lloyd, J.W., Stagg, K.A., Bottrell, S.H., Barker, A.P. & Barrett, M.H. (1999). Sulphate isotope signatures from three urban Triassic Sandstone aquifers. In B. Ellis (Ed.), *Impacts of urban growth on surface and ground waters, Proceedings of IAHS symposium HS5, Birmingham, July 1999*. IAHS publication 259.

Hunt, J.R., Sitar, N. & Udell, K.S. (1988). Nonaqueous phase transport: a cleanup. 1. Analysis of mechanisms. *Water Resources Research*, 24, 1247–1258.

Hydroscience Inc. (1977). *The Paraiba do Sul river water quality study*. WHO-UNDP BRA-73/003 Technical Report 6.

IDRC (1993). *Farming Logic in Kampala*. IDRC Reports, October 1993, 10–11.

Ilahiane, H. (1996). Small-scale irrigation in a multiethnic oasis environment: the case of Zaouit Amelkis village, Southeast Morocco. *Journal of Political Ecology*, 3, 89–105.

Ion, N.J. (1996). The causes and effects of rising groundwater in Merseyside and Manchester. Unpublished Ph.D. thesis, University of Liverpool.

Italconsult (1973). Water Supply for Sana'a and Hodeidah. *Sana'a Basin groundwater studies, volume 1*, UNDP. Yem 507 WHO. Yemen 3202. Rome.

Iturregui, P. (1996). *Problemas Ambientales de Lima*, Lima. Fundación Friedrich Ebert.

Jackson, D. & Lloyd, J.W. (1983). Groundwater chemistry of the Birmingham Triassic Sandstone aquifer and its relation to structure. *Quarterly Journal of Engineering Geology*, 16, 135–142.

Jackson, R.E. (Ed.) (1980). *Aquifer contamination and protection*. Studies and reports in hydrology, 30. UNESCO.

Jacobsen, G. (1983). Pollution of a fractured rock aquifer by petrol – a case study. *Journal of Australian Geology and Geophysics*, 8, 313–322.

Jankowski, J. & Acworth, R.I. (1997). Development of a contaminant plume from a municipal landfill: Redox reactions and plume variability. In J. Chilton (Ed.), *Groundwater in the Urban Environment, 1 Problems, Processes and Management'*, Proceedings of 27th IAH groundwater congress, Nottingham, 439–444. ISBN 90 5410 837 1.

Johnson, R.L., Johnson, P.C., McWhorter, D.B., Hinchee, R.E. & Goodman, I. (1993). An overview of in situ air sparging. *Ground Water Monitoring and Remediation*, 13(4), 127–135.

Joliffe, I. & Songberg, G. (1998). Development of a Stormwater Management Strategy with Groundwater Constraints. *Waterfall*, 9, 25–30.

Jones, A.L. & Sroka, B.N. (1997). *Effects of highway de-icing chemicals on shallow unconsolidated aquifers in Ohio, Interim Report 1988–93*. US Geological Survey, Water Resources Investigations Report 97-4027. Columbus, Ohio.

Jones, M.J. (1985). The weathered zone aquifers of the basement complex areas of Africa. *Quarterly Journal of Engineering Geology*, 15, 47–54.

Jones, P.H., Jeffrey, B.A., Watler, P.K. & Hutchon, H. (1986). *Environmental Impact of Road Salting – State of the Art*. Ontario Ministry of Transport, Research report RR237.

Kafundu, R.C. (1986). A general outline of groundwater supplies in Malawi. In *Geohydrology of drought-prone areas in Africa* (Commonwealth Technical Series No. 202), 68–79.

Karundu, J. (1992). Hydrogeology of some fractured aquifers in Southeast Uganda – Nyabisheki catchment. *Hydrogéologie*, 1, 37–46.

Kalin, R.M. & Roberts, C. (1997). Groundwater Resources in the Lagan Valley Sandstone Aquifer. *Journal of Chartered Institution of Water and Environmental Management*, 11, 133–139.

Kamrin, M., Nugent, M., *et al.* (n.d.). *Nitrate – A Drinking Water Concern*. Michigan State University: Centre for Environmental Toxicology and Institute of Water Research.

Karakouzian, M., Candia, M.A., Wyman, R.V., Watkins, M.D. & Hudyma, N. (1997). Geology of Lima, Peru. *Environmental & Engineering Geoscience*, 3, 55–88.

Keetelaar, C. (1995). Discussion paper on trends in water demand and options for source development. SAWAS Technical Note 12. NWSA/TNO, Sana'a/Delft.

Kerndoff, H., Schleyer, R., Milde, G. & Plumb, R.H. (1992). Geochemistry of groundwater pollutants at German Waste Disposal sites. In S. Lesage & R.E. Jackson (Eds.), *Groundwater contamination and analysis at hazardous waste sites*. Marcel Dekker, Inc. 245–271.

Keswick, B.H. (1984). Sources of groundwater pollution. In G. Bitton & C.P. Gerba (Eds.), *Groundwater pollution microbiology*. Brisbane: John Wiley and Sons, 39–64.

Keuper, H. (1965). *Urbanization and migration in West Africa*. London: Cambridge University Press.

Key, R.M. (1992). An introduction to the crystalline basement of Africa. In E.P. Wright & Burgess, W.G. (Eds.), *Hydrogeology of crystalline basement aquifers in Africa*. Geological Society Special Publication 66, 29–57.

Kiely, G. (1997). *Environmental Engineering*. McGraw-Hill Publishers. ISBN 0-07-709127-2.

Kimmel, G.E. (1984). Non-point contamination of groundwater on Long Island, New York. In *Studies in geophysics: Groundwater Contamination*. Washington, D.C.: National Academy Press, 120–126.

Kinsella, J. (1995). LNAPL behaviour complicates monitoring. *International Ground Water Technology*, 1(4), 17–18.

Kinyamario, J.I. & Macharia, J.N.M. (1992). Aboveground standing crop, protein-content and dry-matter digestibility of a tropical grassland range in the Nairobi National Park, Kenya. *African Journal of Ecology*, 30(1), 33–41.

Kishi, Y., Inouchi, K. & Kakinuma, T. (1988). Seawater Intrusion into Confined Coastal Aquifers. In *Water for World Development, Proceedings of the IWRA World Congress on Water Resources, Ottawa*, II, 390–399.

Kjeldsen, P. (1993). Groundwater pollution source characterization of an old landfill. *Journal of Hydrology*, 142, 349–371.

Knight, M.J. (1993). Organic chemical contamination of groundwater in Australia – an overview to 1993. *AGSO Journal of Australian Geology and Geophysics*, 14(2/3), 107–122.

Kolpin, D.W., Barbash, J.E. & Gilliom, R.J. (1998). Occurrence of pesticides in shallow ground water of the United States: initial results from the National Water-Quality Assessment Program. *Environmental Science and Technology*, 325, 558–566.

Kolpin, D.W., Kalkhoff, S.J., Goolsby, D.A., Sneck-Fahrer, D.A. & Thurman, E.M. (1997). Occurence of selected herbicides and herbicide degradation products in Iowa's ground water, 1995. *Ground Water*, 35(4), 679–68.

Kolpin, D.W., Squillace, P.J., Zogorski, J.S. & Barbash, J.E. (1997). Pesticides and volatile organic compounds in shallow groundwater of the United States. In J. Chilton (Ed.), *Groundwater in the Urban Environment, 1 Problems, Processes and Management', Proceedings of 27th IAH groundwater congress, Nottingham*, ISBN 90 5410 837 1. 469–474.

Kresic, N., Papic, P. & Golubovic, R. (1992). Elements of groundwater protection in a karst environment. *Environmental Geology and Water Sciences*, 20(3), 157–164.

Kruseman, G.P. (1996). Sources for Sana'a Water Supply (SAWAS) Final Technical report and Executive Summary. NWSA/TNO, Netherlands/Yemen.

Ku, H.F.H., Hagelin, N.W. & Buxton, H.T. (1992). Effects of Urban Storm-Runoff Control on Groundwater Recharge in Nassau County, New York. *Ground Water*, 30(4), 507–514.

La Touche, M.C.D. (1997). The Water Resources of Lima, Peru. *Water and Environmental Management Journal*, 11(6), 437–439.

Lamba, D. (1994). The forgotten half – environmental health in Nairobi's poverty areas. *Environ Urban*, 6(1), 164–173.

Lamplugh, G.W., Smith, B. & Mill, H.R. (1914). Quality of Nottinghamshire waters. In *Water supply of Nottinghamshire from underground sources*. Memoirs of the Geological Survey of England and Wales, 128–165.

Lance, J.C. (1984). Land disposal of sewage effluents and residues. In G. Bitton & C.P. Gerba (Eds.), *Groundwater pollution microbiology*. John Wiley and Sons, 197–224.

Land, D.H. (1966). *Hydrogeology of Bunter in Nottinghamshire*. Water supply papers of the Geological Survey of Great Britain, Hydrogeological Report 1.

Lanen, H. (1998). *Monitoring for groundwater management in (Semi-) Arid Regions*. UNESCO Publishing.

Langenegger, D. (1981). High nitrate concentrations in shallow aquifers in a rural area of central Nigeria caused by random deposits of domestic refuse and excrement. In W. van Duijvenbooden, P. Glasbergen & H. van Lelyveld (Eds.), *Quality of Groundwater*, *Studies in Environmental Science*, 17, 135–140.

Langmuir, D. (1997). *Aqueous Environmental Chemistry*. Prentice-Hall, Inc.

Lantzke, N. (1997). Phosphorus and nitrogen loss from horticulture on the Swan Coastal Plain. Agriculture WA Miscellaneous Publication 16/97.

Lawrence, A.R., Stuart, M.E., Barker, J.A. & Tester, D.J. (1996). Contamination of chalk groundwater by chlorinated solvents: A case study of deep penetration by non-aqueous phase liquids. *Journal CIWEM*, 10, 263–272.

Leaf, M.J. (1994). The Suburbanisation of Jakarta: A Concurrence of Economics and Ideology. *Third World Planning Review*, 16(4), 341–356.

Lerner, D.N. (1986). Leaking pipes recharge ground water. *Ground Water*, 24(5), 654–662.

Lerner, D.N. (1989). Groundwater recharge in urban areas. *Atmospheric Environment* 24(1), 29–33.

Lerner, D.N. (1992). Borehole catchments and time-of-travel zones in aquifers with recharge. *Water Resources Research*, 28, 2621–2628.

Lerner, D.N. (1994). Urban groundwater issues in the UK. In *Proceedings of "Water Down Under", XXV Congress of the IAH and the 22nd Hydrology and Water Resources Symposium of the Institution of Engineers*, Australia, Adelaide, November 1994, 289–293.

Lerner, D.N. (1996). Urban groundwater: An asset for a sustainable city. *European Water Pollution Control*, 6(5), 43–51.

Lerner, D.N. (1997). Too much or too little: Recharge in urban areas. In Chilton *et al.* (Eds.), *Groundwater in the urban environment: Problems, Processes and Management*. Rotterdam: Balkema, 41–47.

Lerner, D.N. (1999). Loadings of non-agricultural nitrogen in urban groundwater. *Impact of urban growth on surface water and groundwater quality*. IAHS publication 259.

Lerner, D.N. (2002). Identifying and quantifying urban recharge: a review. *Hydrology Journal*, 10, 143–152.

Lerner, D.N. & Barrett, M.H. (1996). Urban groundwater issues in the United Kingdom. *Hydrogeology Journal*, 4, 80–89.

Lerner, D.N. & Tellam, J.H. (1992). The protection of urban groundwater from pollution. *Journal of the Institution of Water and Environmental Management*, 6(1), 28–36.

Lerner, D.N. & Wakida, F.T. (1999). Identifying, quantifying and managing sources of non-agricultural nitrates in UK groundwater. In *Proceeedings of IAWQ, International conference on Diffuse Pollution, Perth, Australia, May 1999*, CSIRO, Perth, 384–394. ISBN 0643063544.

Lerner, D.N., Mansell-Moullin, M., Dellow, D.J. & Lloyd, J.W. (1982). Groundwater Studies for Lima, Peru. In M.J. Lowing (Eds.), *Optimal Allocation of Water Resources, Proceedings of the Exeter Symposium*, IAHS Publication 135, 17–30.

Lerner, D.N., Issar, S.I. & Simmers, I. (1990). Recharge due to Urbanisation. *Groundwater Recharge, A guide to understanding and Estimating Natural Recharge – International Contributions to Hydrogeology*, 8, IAH publication. Verlag Heinz Heise Publishers, 201–214.

Lerner, D.N., Yang, Y., Barrett, M.H. & Tellam, J.H. (1999). Loadings of non-agricultural nitrogen in urban groundwater. In *Impacts of Urban Growth on Surface Water and Groundwater Quality, Proceedings of IUGG 99 symposium, Birmingham, July 1999*. IAHS 259.

Lerner, D.N., Burston, M.W. & Bishop, P.K. (1993a). Hydrogeology of the Coventry region: an urbanised, multi-layer dual-porosity aquifer system. *Journal of Hydrology*, 149, 111–135.

Lerner, D.N., Gosk, E., Bourg, A.C.M., Bishop, P.K., Burston, M.W., Mouvet, C., DeGranges P. & Jakobsen, R. (1993b). Postscript: summary of Coventry groundwater investigation and implications for the future investigations. *Journal of Hydrology*, 149, 257–292.

Lewis, W.J., Foster, S.S.D. & Drasar, B.S. (1980). *The risk of groundwater pollution by on-site sanitation in developing countries*. International Reference Centre for Wastes Disposal Report No. 01/82.

Li, L., Barry, D.A., Stagnitti, F. & Parlange, J. (1999). Submarine groundwater discharge and associated chemical input to a coastal sea. *Water Resources Research*, 35(11), 3253–3259.

Livingston, E.H. (1997a). Successful Stormwater Management: Selecting and Putting the Puzzle Pieces Together. *Stormwater and Soil Erosion, 1997*, Future Directions for Australian Soil and Water Management, Brisbane.

Livingston, E.H. (1997b). *Stormwater BMP's: The Good, the Bad and the Ugly*. Florida Department of Environmental Protection and Watershed Management Institute, Inc.

Lloyd, B., Bartram, J.K., Rojas, R., Pardon, M., Wheeler, D. & Wedgwood, K. (1991). *Surveillance and improvement of Peruvian drinking water supplies*. Robens Institute, 1–54.

Longe, E.O., Malomo, S. & Olorunniwo, M.A. (1987). Hydrogeology of Lagos metropolis. *Journal of African Earth Sciences*, 6, 163–174.

Lopes, T.J. & Bender, D.A. (1998). Non-point sources of volatile organic compounds in urban areas: relative importance of land surfaces and air. *Environmental Pollution*, 101, 221–230.

Los Angeles Department of Water and Power (2000). [Online]. Available: http://www.LADWP.com

Lovelock, P.E.R. (1977). Aquifer properties of the Permo-Triassic sandstone in the UK. *Bulletin of the Geological Survey of Great Britain*, 56.

Lu, J.C.S., Eichenberger, B. & Stearns, R.J. (1985). *Leachate from Municipal Landfills, Production and Management*. Noyes, Park Ridge, New Jersey.

Lyngkilde, J. & Christensen, T.H. (1992a). Redox zones of a landfill leachate pollution plume (Vejen, Denmark). *Journal of Contaminant Hydrology*, 10, 273–289.

Lyngkilde, J. & Christensen, T.H. (1992b). Fate of organic contaminants in the redox zones of a landfill leachate pollution plume (Vejen, Denmark). *Journal of Contaminant Hydrology*, 10, 291–307.

MacDonald, J.A. & Kavanaugh, M.C. (1994). Restoring contaminated groundwater: an achievable goal? *Environmental Science and Technology*, 28(8), 362A–368A.

Mackay, D.M. (1991). *Geochemical behaviour of organics in aquifer systems*. In Organic Chemical Contaminants in Groundwater, a Short Course. University of New South Wales Centre for Groundwater Management and Hydrogeology.

Macilwain, C. (1995). US report raises fears over nitrate levels in water. *Nature*, 377, 4.

MacLaughlan, R.G. & Walsh, K.P. (1998). Using biogeochemical indicators for decision making at sites contaminated with hydrocarbons. In *Proceedings of "Groundwater: Sustainable Solutions", IAH Congress, Melbourne, February 1998*, 503–508.

MacRitchie, S.M., Pupp, C., Grove, G., Howard, K.W.F. & Lapcevic, P. (1994). *Groundwater in Ontario; hydrogeology, quality concerns, management*. National Hydrology Research Institute NHRI Contribution CS-94011, Environment Canada.

Maguire, D., Goodchild, M.F. & Rind, D.W. (1993). *Geographical Information Systems (Volume 1 – Principles)*. London: Longman.

Malmquist, P.-A. & Hard, S. (1981). Groundwater quality changes caused by stormwater infiltration, *2nd International Conference on Urban Storm Drainage, Urbana.* 89–97.

Malomo, S., Okufarasin, V.A., Olorunniwo, M.A. & Omode, A.A. (1990). Groundwater chemistry of weathered zone aquifers of an area underlain by basement complex rocks. *Journal of African Earth Sciences*, 11, 357–371.

Manning, P.I., Robbie, J.A. & Wilson, H.E. (1970). *Geology of Belfast and the Lagan Valley Memoir.* Geological Survey of Northern Ireland.

Manning, P.I. (1972). *The Development of Water Resources in Northern Ireland.* Institution of Civil Engineers, Northern Ireland Association.

Marcoux, A. (1996). *Population change-natural resources-environment linkages in the Arab States Region.* Population Division of the United Nations and the Food and Agriculture Organisation (FAO) of the United Nations. [Online]. Available: http://www.undp.org/popin/fao/arabstat.htm

Marsily, G. de (1986). *Quantitative Hydrology.* Orlando, Florida: Academic Press.

Marter, A. & Gordon, A. (1996). Emerging issues confronting the renewable natural resources sector in sub-Saharan Africa. *Food Policy*, 21, 229–241.

Mbonu, M. & Travi, Y. (1994). Labelling of precipitation by stable isotopes (18O, 2H) over the Jos Plateau and the surrounding plains (north-central Nigeria). *Journal of African Earth Sciences*, 19, 91–98.

McCarthy, J. & Shevenell, L. (1998). Obtaining representative ground water samples in fractured and karstic formation. *Ground Water*, 36(2), 251–260.

McConville, C. (1999). *Using Stable Isotopes to estimate Groundwater Recharge in a Temperate Zone.* Unpublished Ph.D. Thesis, Civil Engineering Department, Queens University, Belfast.

McFarlane, M.J. (1991a). Some sedimentary aspects of lateritic weathering profile development in the major bioclimatic zones of tropical Africa. *Journal of African Earth Sciences*, 12, 267–282.

McFarlane, M.J. (1991b). Aluminium menace in tropical wells. *New Scientist*, 131, 38–40.

McFarlane, M.J. (1992). Groundwater movement and water chemistry associated with weathering profiles of the African surface in Malawi. In E.P. Wright & W.G. Burgess (Eds.), *Hydrogeology of crystalline basement aquifers in Africa.* Geological Society Special Publication, 66, 101–129.

McFarlane, M.J. & Bowden, D.J. (1992). Mobilisation of aluminium in the weathering profiles of the African surface in Malawi. *Earth Surface Processes and Landforms*, 17, 789–805.

McFarlane, M.J. & Bowden, D.J. (1994). The behaviour of chromium in weathering profiles associated with the African surface in parts of Malawi. In D.A. Robinson & Williams, R.B.G. (Eds.), *Rock Weathering and Landform Evolution.* John Wiley, 321–338.

McFeters, G.A. & Stuart, D.G. (1972). Survival of coliform bacteria in natural waters: field and laboratory studies with membrane-filter chambers. *Applied Microbiology*, 24(5), 805–811.

Memon, P.A. (1982). The growth of low-income settlements: planning response in the peri-urban zone of Nairobi. *Third World Planning Review*, 4, 145–158.

Mendenhall, W.C. (1905a). *Development of Underground Waters in the Eastern Coastal Plain Region of Southern California.* U.S. Geological Survey, Water-Supply and Irrigation Paper 137.

Mendenhall, W.C. (1905b). *Development of Underground Waters in the Central Coastal Plain Region of Southern California.* U.S. Geological Survey, Water-Supply and Irrigation Paper 138.

Mendenhall, W.C. (1905c). *Development of Underground Waters in the Western Coastal Plain Region of Southern California.* U.S. Geological Survey, Water-Supply and Irrigation Paper 139.

Mendonça, A.F., Pires, A.C.B. & Barros, J.G.C. (1994). Pseudo sinkhole occurrences in Brasilia, Brazil. *Environmental Geology*, 23, 36–40.

Mendoza, C.A. & Frind, E.O. (1990a). Advective-dispersive transport of dense organic vapours in the unsaturated zone: 1. Model development. *Water Resources Research*, 26(3), 379–387.

Mendoza, C.A. & Frind, E.O. (1990a). Advective-dispersive transport of dense organic vapours in the unsaturated zone: 2. Sensitivity analysis. *Water Resources Research*, 26(3), 388–398.

Mercer, J.W. & Cohen, R.M. (1990). A review of immiscible fluids in the subsurface: properties, models, characterisation and remediation. *Journal of Contaminant Hydrology*, 6, 107–163.

Mercer, J.W., Skipp, D.C. & Giffin, D. (1990). *The basics of pump-and-treat technology.* U.S. Environment Protection Agency Report, EPA/600/8-90/003.

Metropolitan Water District of Southern California (2000) [Online]. Available: http://www.mwd.dst.ca.us

Milukas, M.Y. (1993). Energy for secondary cities – the case of Nakuru, Kenya. *Energy Policy*, 21(5), 543–558.

Misstear, B.D., White, M., Bishop, P.K. & Anderson, G. (1996). *Reliability of Sewers in environmentally vulnerable areas.* CIRIA Project Report 44.

Misstear, B.D. & Bishop, P.K. (1997). Groundwater contamination from sewers: Experience from Britain and Ireland. In Chilton *et al.* (Eds.), *Groundwater in the urban environment: Problems, Processes and Management.* Rotterdam: Balkema, 491–496.

Mocanu, V.D. & Mirca, V.D. (1997). Risk assessment of groundwater contamination from the south-eastern Bucharest landfill. In Chilton *et al.* (Eds.), *Groundwater in the urban environment: Problems, Processes and Management.* Rotterdam: Balkema.

Mohamed Ait Kadi (undated) Water Challenges for Low Income Countries with High Water Stress: the Need for a Holistic Response, Morocco's Example. Ministere de l'Agriculture, Rabat, Maroc.

Moller, H.M.F. & Markussen, L.M. (1994). Groundwater pollution in urban areas. In Williamson, W.B. (Ed.), *Groundwater problems in urban areas. Proceedings of ICE conference*, London, June 1993. Thomas Telford, 159–171.

Moody, D., Carr, J., Chase, E.B. & Paulson, R.W. (1988). *Water Supply Paper 2325.* United States Geological Survey.

Moore, J.W. & Ramamoorthy, S. (1984). *Organic chemicals in natural waters: applied monitoring and impact assessment.* Berlin Springer-Verlag, New York.

Morris, B.L. & Graniel, E. (1992). Effects of urbanization on groundwater resources of Mérida, Yucatan. In *Proceedings of the 1st Latin American Hydrogeological Congress, Venezuela, October 1992.*

Morris, B.L., Lawrence, A.R. & Stuart, M.E. (1994). *The impact of urbanisation on groundwater quality (Project summary report).* BGS Technical Report, WC/94/56.

Morris, B.L., Lawrence, A.R. & Foster, S.S.D. (1997). Sustainable groundwater management for fast-growing cities: mission achievable or mission impossible? In J. Chilton (Ed.), *Groundwater in the Urban Environment, 1 Problems, Processes and Management, Proceedings of the 27th IAH groundwater congress, Nottingham*, 55–66. ISBN 90 5410 837 1.

Morris, B.L., Lawrence, A.R. & Stuart, M. (1994). *Impact of Urbanisation on Groundwater Quality – Project Summary Report*. British Geological Survey Report WC/94/56, prepared for the Overseas Development Administration Project 91/13.

Morton, T.G., Gold, A.J. & Sullivan, W.M. (1988). Influence of over watering and fertilization on nitrogen losses from home lawns. *Journal of Environment*, 17, 124–130.

Moses, C. (1980). *Sodium in Medicine and Health*. Reese Press, Inc. ISBN 0 9605214 1 0.

Mosgiprovodkhoz (1986). Sana'a Basin Water Resource Scheme: volume 2, Geology and Hydrogeology. *Final report for Ministry of Agriculture, Sana'a*. Yemen.

Motyka, J., Witczak, S. & Zuber, A. (1994). Migration of lignosulfonates in a karstic-fractured-porous aquifer: history and prognosis for a Zn-Pb mine, Pomorzany, southern Poland. *Environmental Geology*, 24, 144–149.

Mouritz, M. & Newman, P. (1997). *Water Sensitive Urban Design: A Tool for Urban Integrated Catchment Management – a Case Study of Bayswater, Perth*. Urban Water Research Association of Australia.

Mueller, D.K., Hamilton, P.A., Helsel, D.R., Hitt, K.J. & Ruddy, B.C. (1995). Nutrients in ground water and surface water of the United States – an analysis of data through 1992. *US Geological Survey Water Resources Investigations Report 95/4031*.

Mull, R., Harig, F. & Pielke, M. (1992). Groundwater management in the urban areas of Hanover, Germany. *Journal of the Institute of Water and Environmental Management*, 6, 199–206.

Muszkat, L., Raucher, D., Magaritz, M., Ronen, D. & Amiel, A.J. (1993). Unsaturated zone and ground-water contamination by organic pollutants in a sewage-effluent-irrigated site. *Ground Water* 31(4), 556–565.

Myllylä, S. (1995). Cairo – a Mega-City and its water resources. Paper Presented at the 3rd Nordic conference on Middle Eastern Studies: Ethnic Encounter and Culture Change, Joensuu, Finland, June 1995 [Online]. Available: http://www.hf-fak.uib.no/institutter/smi/paj/Myllyla.html

Nahon, D. & Tardy, Y. (1992). The ferruginous laterites. In C.R.M. Butt & H. Zeegers (Eds.), *Regolith exploration geochemistry in tropical and sub-tropical terrains: Handbook of exploration geochemistry*, 4, 41–55.

Naidu, R., Kookana, R.S., Oliver, D.P., Rogers, S. & McLaughlin, M.J. (Eds.) (1996). *Contaminants and the Soil Environment in the Australasia-Pacific Region*. Kluwer Academic Press.

Nazari, M.N., Burston, M.W., Bishop, P.K. & Lerner, D.N. (1993). Urban Groundwater Pollution: A Case Study from Coventry, UK. *Ground Water,* 31(3), 417–424.

Nelson, S.W., Wallace-Bell, P. & Yu, X. (1999). Site LEGACI: A systematic approach to groundwater contamination investigations in Western Australia. In *Proceedings of conference Contaminated Site Remediation: Challenges Posed by Urban and Industrial Contaminants, Fremantle, March 1999*. 251–258.

Ngecu, W.M. & Gaciri, S.J. (1998). Urbanisation impact on the water resources with major third world cities: a case study for Nairobi and its environs. *Episodes*, 21(4), 225–228.

NHMRC/AWRC (National Health and Medical Research Council/Australian Water Resources Council) (1987). *Guidelines for drinking water quality in Australia*. Australian Government Publishing Service, Canberra.

Nkotagu, H. (1996). Application of environmental isotopes to groundwater recharge studies in a semi-arid fractured crystalline basement area of Dodoma, Tanzania. *Journal of African Earth Sciences*, 22, 443–457.

Nkotagu, H. (1996). Origins of high nitrate in groundwater in Tanzania. *Journal of African Earth Sciences*, 21, 471–478.

Norra, S. (1995). *2nd Intermediate Work Report, Sana'a Urban Ecology*. MCHUP/GTZ Planning assistance for urban development.

Notenboom, J. (1999). Managing ecological risks of groundwater pollution – A tool for management of water resources. D.L. Danielpol, C. Griebler, J. Gibert, H.P. Nachtnebel & N. Notenboom (Eds.), EC Advanced study course 1999. 63–75.

Novotny, V. (Ed.) (1995). *Nonpoint Pollution and Urban Stormwater Management*. Technomic Publishing Co. Inc.

NRA (1992). *Policy and Practice for the Protection of Groundwater.* National Rivers Authority of England and Wales.

Nwankwoala, A.U. & Osibanjo, O. (1992). Baseline levels of selected organochlorine pesticide residues in surface waters in Ibadan (Nigeria) by electron capture gas chromatography. *The Science of theTotal Environment*, 119, 179–190.

Nystén, T. & Suokko, T. (1998). De-icing and dustbinding: the risk to aquifers. In *Proceedings of an International Symposium, Helsinki, Finland. October 14–16, 1998.*

O'Connor, A. (1983). *The African city*. Hutchinson, 359.

Ollier, C.D. (1984). *Weathering* (2nd edition). Essex: Longman, 270.

Olson, A.C. & Ohno, T. (1989). Determination of free cyanide levels in surface and ground waters affected by highway salt storage facilities in Maine. In *Proceedings of Focus Conference on Eastern Groundwater Issues, National Water Well Association, Kitchener, Ontario, Oct. 17–19, 1989.* 63–77.

Onibokun, A.G. (Ed.) (1999). *Managing the monster: urban waste and governance in Africa*. International Development Research Centre, Ottawa, 269.

Ontario Ministry of the Environment (1991). *Parameters System Listing (PALIS)*. Drinking Water Section of the Water Resources Branch Report, February 1991.

Orange County Water District (1999). *Master Plan Report.*

Organisation for Economic Co-operation and Development (OECD) (1989). *Curtailing usage of de-icing agents in winter maintenance*. Paris.

Osibanjo, O. (1982). Third world science at the crossroads: basic or applied research? *Chemistry in Britain*, 18, 270–271.

Owoade, A. (1995). The potential for minimizing drawdowns in groundwater wells in tropical aquifers. *Journal of African Earth Sciences*, 20, 289–293.

Padco Inc. (1998). *Morocco Urban Development Assessment*. U.S. Agency for International Development, Contract No PCE-1-00-96-00002-00, Task Order No 813.

Palmer Development Group (1995). *Evaluation of solid waste practice in developing urban areas in South Africa*. Water Research Commission Report.

Palmer, A.N. (1984). Recent trends in karst geomorphology. *Journal of Geological Education*, 32, 247–253.

Panno, S.V., Krapac, I.G., Weibel, C.P. & Bade, J.D. (1996). *Groundwater contamination in karst terrain of southwestern Illinios*. Illinois Geological Survey Environmental Geology Report 151.

Patterson, B.M., Power, T.R. & Barber, C. (1993). Comparison of two integrated methods for the collection and analysis of volatile organic compounds in ground water. *Ground Water Monitoring and Remediation*, 13, 118–123.

Pedley, S. & Howard, G. (1997). The public health implications of microbiological contamination of groundwater. *Quarterly Journal of Engineering Geology*, 30, 179–188.

Perchanok, M.S., Manning, D.G. & Armstrong, J.J. (1991). *Highway de-icers: standards, practice and research in the Province of Ontario.* Ontario Ministry of Transportation, Research and Development Branch, MAT-91-13.

Pionke, H.B., Sharma, M.L. & Hirschberg, K.J.B. (1990). Impact of irrigated horticulture on nitrate concentrations in groundwater. *Agriculture, Ecosystems and Environment,* 32, 119–132.

Pitt, R., Clark, S. & Parmer, K. (1994). *Project Summary: Potential groundwater contamination from intentional and non-intentional stormwater infiltration.* US Environmental Protection Agency.

Pitt, R., Field, R., Clark, S., Parmer, K. & Charbeneau, R.J. (1995). Potential groundwater contamination from stormwater infiltration Groundwater management. In *First International Symposium on Water Resources Engineering, San Antonio, Texas, United States.* 127–132.

Platenburg, R.J., Hoencamp, T.E. & El Arabi, N.E. (1997). Soil quality management at industrial sites: Dutch experience relevant for Egyptian context. In *Proceedings of the First Conference and Trade Fair on Environmental Management and Technologies "Environment '97", Cairo, Egypt,* Volume 2.

Poland, J.F., Garrett, A.A. & Sinnott, A. (1959). *Geology, Hydrology, and Chemical Character of Ground Waters in the Torrance-Santa Monica Area, California.* U.S. Geological Survey Water Supply Paper 1461.

Powell, K.L., Barrett, M.H., Pedley, S., Tellam, J.H., Stagg, K.A., Greswell, R.B. & Rivett, M.O. (2000a). Enteric virus detection in groundwater using a glasswool trap. In O. Sililo *et al.* (Eds.), *Groundwater: Past Achievements and Future Challenges,* Rotterdam: Balkema, 813–816.

Powell, K.L., Barrett, M.H., Pedley, S., Cronin, A.A., Taylor, R.G., Tellam, J.H., Greswell, R.B. & Trowsdale, S.A. (2001). *Virus-colloid interactions: implications for groundwater monitoring.* Submitted for publication at the International Association of Hydrogeologists Annual Conference, Munich, September 2001.

Powell, K.L., Tellam, J.H., Barrett, M.H., Pedley, S., Stagg, K., Greswell, R.B. & Rivett, M.O. (2000b). *Optimisation of a new method for detection of viruses in groundwater.* Report to the Environment Agency, UK, National Groundwater and Contaminated Land Centre Project NC/99/40.

Powelson, D.K. & Gerba, C.P. (1995). Fate and transport of microorganisms in the vadose zone. In Wilson, L.G., Everett, L.G. & Cullen, S.J. (Eds.), *Handbook of vadose zone characterization & monitoring.* Geraghty & Miller Environmental Science and Engineering series. Lewis Publishers, 123–135.

Powelson, D.K., Simpson, J.R. & Gerba, C.P. (1990). Virus transport and survival in saturated and unsaturated flow through soil columns. *Journal of Environment Quality,* 19, 396–401.

Price, M. (1994). *Introducing Groundwater.* Chapman & Hall, London. ISBN 0-412-48500-1.

Price, M. & Reed, D.W. (1989). The influence of mains leakage and urban drainage on groundwater levels beneath conurbations in the UK. *Proceedings of Institution of Civil Engineers,* 86(1), 31–39.

Priddle, M.W., Jackson, R.E. & Mutch, J.P. (1989). Contamination of the sandstone aquifer of Prince Edward Island, Canada by aldicarb and nitrogen residues. *Ground Water Monitoring Review,* 134–140.

Puls, R.W. & Paul, C.J. (1995). Low-flow purging and sampling of ground water monitoring wells with dedicated systems. *Ground Water Monitoring and Remediation*, 15(1), 116–123.

Pupp, C. (1985). *An Assessment of ground water contamination in Canada, Part 1.* Environmental Interpretation Division, EPS, Environment Canada, Ottawa.

Quiazon, H.P. (1972). *Hydrogeology of Cebu.* Bureau of Mines and Geosciences. Manila, Philippines.

Quinlan, J.F. & Ewers, R.O. (1985). Ground water flow in limestone terranes: strategy, rationale and procedure for reliable, efficient monitoring of ground water quality in karst areas. In *Proceedings of the 5th National Symposium and Exposition on Aquifer Restoration and Ground Water Monitoring, U.S. National Water Well Association, Dublin, Ohio*, 197–234.

Quinlan, J.F. (1988). Protocol for reliable monitoring of groundwater quality in karst terranes. In Karst Hydrogeology and Karst Environment Protection, *Proceedings of the 21st Congress of the International Association of Hydrogeologists*, Guilin China. 888–893.

Ramamurthy, L.M. (1983). *Environmental isotope and hydrogeochemical studies of selected catchments in South Australia.* Unpublished Ph.D. Thesis, Flinders University, South Australia.

Ramesh, R. (1999). Groundwater quality management: pollution perspectives. In *Impacts of Urban Growth on Surface and Groundwater Quality, Proceedings of IUGG '99 Symposium HS5,* Birmingham, July 1999. IAHS Publication 259, 47–58.

Rao, P.S.C., Davis, G.B. & Johnston, C.D. (1996). Technologies for enhanced remediation of contaminated soils and aquifers: overview, analysis and case studies. In R. Naidu, R.S. Kookana, D.P. Oliver, S. Rogers & M.J. McLaughlin (Eds.), *Contaminants and the Soil Environment in the Australasia-Pacific Region.* Kluwer Academic Press, 361–410.

Regenstein, L. (1998). The Myth of the "Banned" Pesticides. In D.D. Chiras, *Environmental Science: a systems approach to sustainable development*, 5th Edition. Wadsworth Publishing Company, U.S.A., 446.

Reitzel, S., Faquhar, G. & Mcbean, E. (1991). Temporal characterization of municipal solid waste leachate. *Canadian Journal of Civil Engineering*, 19(4), 668–679.

Rivers, C.N., Barrett, M.H., Hiscock, K.M., Dennis, P.F., Feast, N.A. & Lerner, D.N. (1996). Application of nitrogen isotopes to identify nitrogen contamination of the Sherwood Sandstone aquifer beneath the city of Nottingham, UK. *Hydrogeology Journal*, 4(1), 90–102.

Rivett, M.O., Lerner, D.N., Lloyd, J.W. & Clark, L. (1990). Organic contamination of the Birmingham aquifer. *Journal of Hydrology*, 113, 307–323.

RIVM/RIZA (1991). *Sustainable use of groundwater: problems and threats in the European Community.* Report prepared for Ministerial Seminar on Groundwater, The Hague, 26/27 November 1991. RIVM-report 600025001.

Robertson, W.D., Cherry, J.A. & Sudicky, E.A. (1991). Groundwater contamination from two small septic systems on sand aquifers. *Ground Water*, 29, 82–92.

Robins, N.S. (1995). Groundwater in Scotland and Northern Ireland: Similarities and differences. *Quarterly Journal of Engineering Geology*, 28, 163–169.

Robins, N.S. (1996). *Hydrogeology in Northern Ireland.* British Geological Survey, HMSO, ISBN 0 11 8845241.

Rogers, P. (1995). *La Grande Secheresse: Strategic Considerations*, prepared for US Agency for International Development, Morocco.

Rojas, R., Howard, G. & Bartram, J.K. (1995). Groundwater quality and water supply in Lima, Peru. In H. Nash & G.J.H. McCall (Eds.) *Groundwater Quality*. Chapman and Hall, 159–167.

Rose, J. & Gerba, C. (1991). Assessing Potential Health Risks from Viruses and Parasites in Reclaimed Water in Arizona and Florida, USA. *Water Science & Tecnology*, 23, 2091–2098.

Roy, W.R. & Griffin, R.A. (1990). Vapour-phase interaction and diffusion of organic solvents in the unsaturated zone. *Environmental Geology and Water Sciences*, 15, 101–110.

Rushton, K.R. & Weller, J. (1985). Response to pumping of a weathered-fractured granite aquifer. *Journal of Hydrology*, 80, 299–309.

Rushton, K.R. & Bishop, T.J. (1993). *Water resource study of the Nottinghampshire Sherwood Sandstone aquifer, Eastern England: mathematical model of the Sherwood Sandstone aquifer*. Unpublished report for the National Rivers Authority, University of Birmingham.

Ryan, M. & Meiman, J. (1996). An examination of short-term variations in water quality at a karst spring in Kentucky. *Ground Water*, 34(1), 23–30.

Sahgal, U.K., Sahgal, R.K. & Kakar, R. (1989). Nitrate pollution of ground water in the Lucknow area. *Proceedings of International Workshop on Appropriate Methodologies for Development and Management of Groundwater Resources in Developing Countries*. National Geophysical Research Institute, Hyderabad, India. 871–892.

Sanchez-Diaz, L.F. & Gutierrez-Ojeda, C. (1997). Overexploitation effects of the aquifer system of Mexico City. In Chilton *et al.* (Eds.), *Groundwater in the Urban Environment, 1 Problems, Processes and Management, Proceedings of the 27th IAH groundwater congress*, Nottingham. Rotterdam: Balkema, 353–357.

Sandström, K. (1995). Modeling the effects of rainfall variability on groundwater recharge in semi-arid Tanzania. *Nordic Hydrology*, 26, 313–330.

Sartor, J.D. & Boyd, G.B. (1972). *Water Pollution Aspects of Street Surface Contaminants. Washington, D.C.* United States Environmental Protection Agency.

Schwille, F. (1984). Migration of organic fluids immiscible with water in the unsaturated zone. In B. Yaron, G. Dagan & J. Goldschmid (Eds.), *Pollutants in the Unsaturated Zone Between Soil Surface and Groundwater*. Springer Verlag Ecological Studies 47, 27–48.

Schwille, F. (1988). Dense Chlorinated Solvents in Porous and Fractured Media. Lewis Publishers.

Scott, W.S. & Wylie, N.P. (1980). The Environmental Effects of Snow Dumping: A Literature Review. *Journal of Environmental Management*, 10, 219–240.

SEDAPAL (1997). *Historia del Abastecimiento de Agua Potable de Lima 1535–1996.* SEDAPAL, Lima.

Sekhar, M., Mohan Kumar, M.S. & Sridharan, K. (1994). A leaky aquifer model for hard rock aquifers. *Journal of Applied Hydrogeology*, 3, 32–39.

Sharma, M.L., Herne, D.E., Byrne, J.D. & Kin, P.G. (1996). Nutrient discharge beneath urban lawns to a sandy coastal aquifer, Perth, Western Australia. *Hydrogeology Journal*, 4(1), 103–117.

Siddiq, Lutfiam, G. & Foppen, J.W. (1996). *Well Inventory and assessment of trends in the use of aquifers below Sana'a*. SAWA Technical Note, 18. NWSA/TNO, Sana'a/Delft.

Simpson, R.W. (1994). Quantification of processes. In W.B. Williamson (Ed.), *Groundwater Problems in Urban Areas, Proceedings of ICE conference London, June 1993*, published by Thomas Telford (1994), 105–120.

Singh, J., Wapakala, W.W. & Chebosi, P.K. (1984). Estimating groundwater recharge based on the infiltration characteristics of layered soil. In *Challenges in African Hydrology and Water Resources, Proceedings of the Harare Symposium*, July 1984, IAHS Publication 144, 37–45.

Slater, S.S. (1997). *Common Law Methods to Redress Groundwater Contamination.* Perth, University of Western Australia.

Smaill, A. (1994). Water resource management conflicts is a shallow fractured rock aquifer underlying Auckland city, New Zealand. In *Management to Sustain Shallow Groundwater Systems, Proceedings of the XXV Congress of the International Association of Hydrogeologists, Adelaide, Australia*, 325–328.

Smedley, P. (1996). Arsenic in rural groundwater in Ghana. *Journal of African Earth Sciences*, 22, 459–470.

Smith, A.E. (1988). Transformations in soil. In R. Grover (Ed.), *Environmental Chemistry of Herbicides*. Boca Raton, Florida: CRC Press. 171–200.

Smith, R.A. & Allen, A.D. (1987). The unconfined aquifer and effects of urbanisation – Perth W.A. In *Proceedings of the International Conference on Groundwater Systems under Stress*, Australian Water Resources Council Conference Series 13, Brisbane, 1986, 575–585.

Snow, J. (1854). *Snow on Cholera.* In a reprint of two papers together with a bibliographical memoir by B.W. Richardson and an introduction by W.H. Frost, 1936. The Commonwealth Fund, London, Humphrey Milford, Oxford University Press.

Somasundaram, M.V., Ravindran, G. & Tellam, J.H. (1993). Ground-water pollution of the Madras urban aquifer. *Ground Water*, 31, 4–11.

Sotornikova, R. & Vrba, J. (1987). Some remarks on the concepts of Vulnerability Maps. In *Proceedings of international conference on the Vulnerability of Soil and Groundwater to Pollutants*, TNO Committee on Hydrogeologic Research, The Hague, 471–475.

Spalding, R.F. & Exner, M.E. (1993). Occurrence of nitrate in groundwater – a review. *Journal of Environmental Quality*, 22, 392–402.

Spears, D.A. (1987). An investigation of metal enrichment in Triassic Sandstones and porewaters below an effluent spreading site, West Midlands. *Quarterly Journal of Engineering Geology*, 20, 117–129.

Squillace, P.J., Moran, M.J. & Lapham, W.W. (1999). Volatile organic compounds in untreated ambient groundwater of the United States, 1985–1995. *Environmental Science & Technology*, 33, 4176–4187.

Stauffer, J. (1998). *The Water Crisis: Constructing Solutions to Freshwater Protection.* London: Earthscan Publications.

Steffy, D.A., Johnston, C.D. & Barry, D.A. (1995). A field study of the vertical immiscible displacement of LNAPL associated with a fluctuating water table. In *Proceedings of the Groundwater Quality, Protection and Remediation Conference, Prague, Czech Republic.* IAHS Publication 225, 49–57.

Stren, R.E. (1989). The administration of urban services. In R.E. Stren & R.R. White (Eds.), *African cities in crisis*. Westview Press, 37–67.

Stuart, M.E. & Milne, C.J. (1997). Groundwater quality implications of wastewater irrigation in Leon, Mexico. In Chilton et al. (Eds.), *Groundwater in the urban environment: Problems, Processes and Management*. Rotterdam: Balkema, 193–198.

Stumm, W. & Morgan, J.J. (1981). *Aquatic Chemistry.* Brisbane: John Wiley and Sons.

Stuyfzand, P.J. (1999). Patterns in groundwater chemistry resulting from groundwater flow, *Hydrogeology Journal*, 7, 15–27.

Subba Rao, C. & Subba Rao, N.V. (1999). The impact of urban industrial growth on groundwater quality in Visakhapatnam, India. In *Impacts of Urban Growth on Surface Water and Groundwater Quality, Proceedings of IUGG 99 symposium, Birmingham, July 1999*. IAHS 259.

Sukhija, B.S., Reddy, D.V. & Saxena, V.K. (1989). Study of groundwater pollution at Tirumala, Tirupathi, India. In *Proceedings of International Workshop on Appropriate Methodologies for Development and Management of Groundwater Resources in Developing Countries*, Hyderabad, India. 893–898.

Swallow, K.C. (1992). Nonpriority pollutant analysis and interpretation. In S. Lesage & R.E. Jackson (Eds.), *Groundwater contamination and analysis at hazardous waste sites*. Marcel Dekker Inc., 37–51.

Swearingen, W.D. (1987). *Moroccan Mirages, Agrarian Dreams and Deceptions, 1912–1986*. Princeton, New Jersey: Princeton University Press.

Swearingen, W.D. & Bencherifa, A. (1996). *The North African Environment at Risk*. Boulder, Colorado: Westview Press.

Tardy, Y. (1992). Diversity and terminology of lateritic profiles. In I.P. Martini & W. Chesworth (Eds.), *Weathering, soils and paleosols*, 379–406.

Taylor, R.G., Barrett, M.H., Baines, O.P., Trowsdale, S.A., Lerner, D.N. & Thornton, S.F. (2000). Depth variations in aquifer hydrochemistry using a low-cost, multilevel piezometer. In Sililo *et al.* (Eds.), *Groundwater: Past Achievements and Future Challenges*. Rotterdam: Balkema, 651–654.

Taylor, R.G. (1998). *Tectonically controlled landscape evolution and its relation to the lithology, hydrology and hydrogeology of weathered crystalline rock in Uganda*. Doctoral Thesis, University of Toronto.

Taylor, R.G. & Howard, K.W.F. (1995). Averting shallow well contamination in Uganda. In *Sustainability of water and sanitation systems, Proceedings of the 21st WEDC Conference, Kampala*, 62–65.

Taylor, R.G. & Howard, K.W.F. (1996). Groundwater recharge in the Victoria Nile basin of east Africa: support for the soil moisture balance approach using stable isotope tracers and flow modelling. *Journal of Hydrology*, 180, 31–53.

Taylor, R.G. & Howard, K.W.F. (1998a). The dynamics of groundwater flow in the regolith of Uganda. In P.J. Dillon & I. Simmers (Eds.), *Shallow groundwater systems*. International Contributions to Hydrogeology, 18, 97–113.

Taylor, R.G. & Howard, K.W.F. (1998b). Post-Palaeozoic evolution of weathered land surfaces in Uganda by tectonically controlled cycles of deep weathering and stripping. *Geomorphology*, 25, 173–192.

Taylor, R.G. & Howard, K.W.F. (1998c). Lithological evidence for then evolution of deeply weathered land surfaces in Uganda by tectonically controlled cycles of deep weathering and stripping. *Catena* (in press).

Taylor, R.G. & Howard, K.W.F. (1999a). Geomorphic control on the hydrogeological characteristics of deeply weathered crystalline rock. *Water Resources Research* (in prep.)

Taylor, R.G. & Howard, K.W.F. (1999b). The influence of tectonic setting on the hydrological characteristics of deeply weathered terrains: evidence from Uganda. *Journal of Hydrology* (submitted, July 1998)

Taylor, R.G. & Howard, K.W.F. (1994). Groundwater quality in rural Uganda: hydrochemical considerations for the development of aquifers within the basement complex of Africa. In: J. McCall & H. Nash (Eds.), *Groundwater Quality*, 31–44.

Tebbutt, T.H.Y. (1998). *Principles of Water Quality Control*. Butterworth Heinemann.

Telfer, A.L. & Emmett, A.J. (1994). The artificial recharge of stormwater into a dual porosity aquifer, and the fate of selected pollutants. In *Proceedings of the XXV Congress of the International Association of Hydrogeologists "Management to Sustain Shallow Groundwater Systems"*, Adelaide, Australia, 33–38.

Telfer, A.L. (1993). Groundwater resource management and community consultation – Blue Lake, South Australia. *Journal of Australian Geology and Geophysics*, 14, 201–206.

Tellam, J.H. (1994). The groundwater chemistry of the Lower Mersey Basin Permo-Triassic sandstone system, (UK) and pre-industrialisation-urbanisation. *Journal of Hydrology*, 161, 287–325.

Tellam, J.H. (1995). Urban groundwater pollution in the Birmingham Triassic Sandstone aquifer. In *Preprints of 4th Annual IBC Conference on Groundwater Pollution, London*, 15–16th March, 1995.

Tellam, J.H. (1996). Interpreting the borehole water chemistry of the Permo-Triassic sandstone aquifer of the Liverpool area, UK. *Geological Journal*, 165, 45–84.

Testa, S.M. & Winegardner, D.L. (2000). *Restoration of Contaminated Aquifers: Petroleum Hydrocarbons and Organic Compounds*. CRC Press LLC.

Thomas, M.F. (1994). *Geomorphology in the tropics*. Brisbane: John Wiley and Sons.

Thompson, J.H. (1938). *Deepwell Water Supplies*. Lecture to Institution of Civil Engineers, Northern Ireland, April 1938.

Thorn, R.H. & Coxon, C.E. (1992). Hydrogeological aspects of bacterial contamination of some Western Ireland karstic limestone aquifers. *Environmental Geology and Water Sciences*, 20(1), 65–72.

Tijani, M.N. (1994). Hydrogeochemical assessment of groundwater in Moro area, Kwara state, Nigeria. *Environmental Geology*, 24, 194–202.

Tindall, J.A., Kunkell, J.R. & Anderson, D.E. (1999). *Unsaturated zone hydrology for scientists and engineers*. New Jersey: Prentice Hall.

Tindimugaya, C. (2000). *Assessment of groundwater development potential for Wobulenzi Town, Uganda*. M.Sc. Thesis, IHE Delft.

Torstensson, B.A. (1984). A new system for ground water monitoring. *Ground Water Monitoring Review*, 4(4), 131–138.

Tóth, J. (1963). Theoretical Analysis of Groundwater Flow in Small Drainage Basins. *Journal of Geophysical Research*, 68(16), 4795–4812.

Toze, S. (1997). *Microbial pathogens in wastewater*. Literature Review for Urban Water Systems Multi-divisional Research Programme, Technical Report 1/97, CSIRO Land and Water, Australia.

Train, R.E. (1979). *Quality criteria for water*. Castle House Publications Ltd., UK.

Tredoux, G. (1984). The groundwater pollution hazard in the Cape Flats. *Water Pollution Control*, 83(4), 473–483.

Tredoux, G., Ross, W.R. & Gerber, A.A. (1980). The potential of the Cape Flats Aquifer for the storage and abstraction of reclaimed effluents. *Z.dt.geol. Ges.*, 131, 22.

Turner, J.V. (1979). *The hydrologic regime of Blue Lake, southeastern Australia*. Unpublished Ph.D. Thesis, Flinders University, South Australia.

Tuthill, R.W. & Calabrese, E.J. (1979). Elevated sodium levels in the public drinking water as a contributor to elevated blood pressure levels in the community. *Archives Environmental Health*, 34, 197–203.

U.S. Environmental Protection Agency (1983). *Results of the Nation-wide Urban Runoff Program, Volume 1 – Final Report.* Report PB84-185552, Water Planning Division, Washington DC.

U.S. Environmental Protection Agency (1995). Storm Water Discharges Potentially Addressed By Phase II of the National Pollutant Discharge Elimination System Storm Water Program: Report to Congress. *Office of Water Management Report EPA* 833-K-94-002, March 1995, Washington DC.

Uganda. In Sustainability of water and sanitation systems, *Proceedings of the 21st WEDC Conference*, September 1995, Kampala, Uganda, 62–65.

Uma, K.O. (1993). Nitrates in shallow (regolith) aquifers around Sokoto Town, Nigeria. *Environmental Geology*, 21, 70–76.

UNESCO (1993). *Hydrogeologic Map of South America, Explanatory Text.* Serviço Geológico do Brasil (CPRM).

United Nations (1986). See Harris (1990).

United States Environmental Protection Agency (USEPA) (1990). *Approaches for the remediation of uncontrolled wood preserving sites.* US EPA Center for Environmental Research Information, Report EPA/625/7-90/011.

United States Environmental Protection Agency (USEPA) (1990). *Pesticides in Drinking Well Waters*, Report 20T- 1004, U.S. Environmental Protection Agency, Office of Groundwater Protection, Washington D.C.

United States Environmental Protection Agency (USEPA) (1992). *Methodologies for evaluating in-situ bioremediation of chlorinated solvents*, U.S. Environmental Protection Agency Report EPA/600/R-92/042.

United States Environmental Protection Agency (USEPA) (1992). *Another look – National survey of pesticides in drinking water wells, phase 2 report.* EPA/579/09-91/020 U.S. Environmental Protection Agency, Office of Pesticide Programs; U.S. Government Printing Office, Washington D.C.

United States Environmental Protection Agency (USEPA) (1994). *Handbook for ground water and wellhead protection.* Report EPA/625/R-94/001, USEPA, Cincinnati, Ohio, 269.

United States Environmental Protection Agency (USEPA) (1996). *Drinking Water Regulations and Health Advisories.* EPA 822-B-96-002 U.S. Environmental Protection Agency, Office of Water; U.S. Government Printing Office, Washington D.C. [Online]. Available: http://www.epa.gov/OST/pc/dwha.html

United States Environmental Protection Agency (USEPA) (1999). *Safe Drinking Water Information System factoids.* [Online]. Available: http://www.epa.gov/safewater/data/99factoids.pdf or http://www.epa.gov/safewater/data/getdata.html

United States National Research Council (US NRC) (1993). *In Situ Bioremediation; When Does it Work?* NRC publication, National Academy Press, Washington, D.C.

Upper Los Angeles River Area Watermaster (1999). *Ground Water Pumping and Spreading Plan 1998–2003* Water Years.

Upper Los Angeles River Area Watermaster (1999). Watermaster Services in the Upper Los Angeles River Area Los Angeles County 1997–98 Water Year October 1, 1997–September 30, 1998.

Urban Stormwater Pollution Task Force (1994). *Improving the Quality of Urban Stormwater in New South Wales.* Environment Protection Authority, Melbourne.

USAID (1999). FORWARD Team Briefing, March 3, Rabat, Morocco.

USDFA (1998). *Glossary of Pesticide Chemicals, April 1998* [Online]. Available: http://www.cfsan.fda.gov/~frf/pestdata.html

USGS (1999). *The Quality of Our Nation's Waters: Nutrients and Pesticides.* United States Geological Survey Circular 1225, June 1999.

USGS (2000). *Groundwater Resources for the Future – The Atlantic Coastal Zone.* United States Geological Survey Fact Sheet 085-00.

Vallius, P. & Kelloski, K. (1998). The most important aquifer in Finland, the Uti aquifer will be protected in 1995–2000. In T. Nystén & T. Suokko (Eds.), *De-icing and dust-binding – risk to aquifers, Proceedings of an International Symposium, Helsinki, Finland, October 14–16*, 259–263.

van Dam, J.C. (1983). The shape and position of the saltwater wedge in the coastal aquifers, In F.X. Dunin, G. Matthess & R.A. Gras (Eds.), *Relation of groundwater quantity and quality, Proceedings of the Hamburg Symposium*, Aug. 1983. IAHS Publication 146, 59–75.

van Hesteren, S., van de Leemkule, M.A. & Pruiksma, M.A. (1999). *Minimum soil quality: A use-based approach from ecological perspective. Part 1: Metals.* TCB Report No. R08, Technical Soil Protection Committee, The Hague, The Netherlands.

van de Leemkule, M.A., Van Hesteren, S. & Pruiksma, M.A. (1999). *Minimum soil quality: A use-based approach from ecological perspective. Part 2: Immobile organic micro pollutants.* TCB Report No. R09, Technical Soil Protection Committee, The Hague, The Netherlands.

Vandevivere, P., Welch, S.A., Ullman, W.J. & Kirchman, D.L. (1994). Enhanced dissolution of silicate minerals by bacteria at near-neutral pH. *Microbial Ecology*, 27, 241–251.

Vandoolaeghe, M.A.C. (1989). *The Cape Flats groundwater development pilot abstraction scheme.* Technical report Gh3655. Directorate Geohydrology, Department of Water Affairs and Forestry, Pretoria.

van Ryneveld, M.B. & Fourie, A.B. (1997). A strategy for evaluating the environmental impact of on-site sanitation systems. *Water SA*, 23, 279–291.

Vazquez-Suni, E. & Sanchez-Vila, X. (1999). Groundwater modelling in urban areas as a tool for local authority management: Barcelona case study (Spain) In *Impacts of Urban Growth on Surface and Groundwater Quality, Proceedings of IUGG 99 Symposium HS5, Birmingham, July 1999.* IAHS Publ. 259, 65–72.

Ventura Napa, M. (1995). *Assessment of Water Resources and Their Use by Different Productive Sectors in Peru.* Instituto Nacional de Recursos Naturales del Perú (INRENA), Lima.

Viljevac, Z. (1997a). Auckland shallow volcanic aquifers: computer simulation of recharge, Western Springs and Onehunga-Mt Wellington aquifers. In *Proceedings of "Wai Whenua", 24th Australian and New Zealand Hydrology and Water Resource Symposium, Auckland*, 37–42.

Viljevac, Z. (1997b). Western Springs Aquifer: hydrogeological characteristics and computer model. *Unpublished M.Sc. Thesis, University of Auckland, New Zealand.*

Vrba, J. & Zaporozec, A. (1994). *Guidebook on Mapping Groundwater Vulnerability.* International Association of Hydrogeologists, 16, Verlag Heinz Heise Publishers, Hannover.

Vroblesky, D.A. & Hyde, W.T. (1997). Diffusion samplers as an inexpensive approach to monitoring VOCs in ground water. *Ground Water Monitoring and Remediation*, 17(3), 177–184.

Water Replenishment District of Southern California (1996). *Annual Survey and Report on Groundwater Replenishment.*

Waterhouse, J.D. (1977). The hydrogeology of the Mount Gambier area. *South Australian Geological Survey Investigation Report 48.*

Watkins, M.D., Evans, D.A. & Lloyd, J.W. (1997). Continual Assessment of the Groundwater Resources of Lima. *Water and Environmental Management Journal,* 11(6), 440–445.

Werna, E. (1998). Urban management, the provision of public services and intra-urban differentials in Nairobi. *Habitat International,* 22(1), 15–26.

White, R.R. (1989). The influence of environmental and economic factors on the urban crisis. In R.E. Stren & R.R. White (Eds.), *African cities in crisis.* Westview Press, 1–19.

White, W.B. (1969). Conceptual models for carbonate aquifers. *Ground Water,* 7, 15–22.

Whitehead, E., Hiscock, K. & Dennis, P. (1999). Evidence for sewage contamination of the Sherwood Sandstone aquifer beneath Liverpool, UK. In *Impacts of urban growth on surface water and groundwater quality, Proceedings of IUGG 99 Symposium HS5, Birmingham 1999,* IAHS Publication 259, 179–185.

Wild, J. & Ruiz, J.C. (1987). Lima Groundwater Modelling Revisited, *First National Hydrology Symposium,* British Hydrological Society, University of Hull.

Wildermuth Environmental (1999). *Optimum Basin Management Program.* Phase I Report.

Wilhelm, S.R., Schiff, S.L. & Cherry, J.A. (1994). Biogeochemical evolution of domestic waste water in septic systems: 1. Conceptual model. *Ground Water,* 32, 905–916.

Wilson, J.L. & Conrad, S.H. (1984). Is physical displacement of residual hydrocarbons a realistic possibility in aquifer restoration? In *Proceedings of the NWWA/API Conference on Petroleum Hydrocarbons and Organic Chemicals in Ground Water – Prevention, Detection and Restoration, Dublin,* 274–298.

Wilson, J.L. (1991). Biological transformations of organic compounds in groundwater. In *Organic Chemical Contaminants in Groundwater: a Short Course.* University of New South Wales Centre for Groundwater Management and Hydrogeology.

Wilson, S.B. & Etheridge, L. (1991). Pumping and sampling contaminated groundwater: problems and techniques. In *Organic Chemical Contaminants in Groundwater: a Short Course.* University of New South Wales Centre for Groundwater Management and Hydrogeology.

Winter, T.C., Harvey, J.W., Franke, O.L. & Alley, W.M. (1998). *Ground Water and Surface Water: A Single Resource.* U.S. Geological Survey Circular.

World Bank (1995). *Kingdom of Morocco Water Sector Review,* Report 14750-MDR, June.

World Bank (1998). Integrated Water Resources Management in Latin America and the Caribbean, Technical Study Number ENV-123, December.

World Health Organisation (1984). *Guidelines for drinking-water quality, volume 1: Recommendations.* World Health Organisation, Geneva, 130.

World Health Organisation (1989). *Health guidelines for the use of wastewater in agriculture and aquaculture.* World Health Organisation Technical Report Series 778.

World Health Organisation (1991). *Urban solid waste management.* WHO Regional office for Europe; Copenhagen.

World Health Organisation (1997). Guidelines for drinking-water quality Volume 3 (Second Ed.). Surveillance and control of community supplies. World Health Organisation (Geneva).

World Health Organisation (1997). *Environmental Matters: Strategy on sanitation for high-risk communities.* Document EB101/19. WHO, Geneva, Switzerland.

Wright, A., Kloppers, W. & Fricke, A. (1993). A hydrological investigation of the stormwater runoff from the Khayelitsha Urban Catchment from the Khayelitsha urban catchment in the False Bay area, South Western Cape. *Water Research Commission Report No 323/1/93.*

Wright, E.P. (1992). The hydrogeology of crystalline basement aquifers in Africa. In *The hydrogeology of crystalline basement aquifers in Africa.* In E.P. Wright & W.G. Burgess (Eds.). The Geological Society of London, 1–28.

Wright, E.P. & Burgess, W.G. (Eds.) (1992). *The hydrogeology of crystalline basement aquifers in Africa.* Geological Society (London).

Yang, Y., Cronin, A.A. & Kalin, R.M. (2000). Optimised calibration of a regional groundwater flow model through inverse modelling. In *Groundwater: Past Achievements and Future Challenges.* O. Sililo (Ed.) Balkema, 437–442

Yang, Y., Lerner, D.N., Barrett, M.H. & Tellam, J.H. (1999a). Quantification of groundwater recharge in the city of Nottingham, UK. *Environmental Geology,* 38(3), 183–198.

Yates, M.V. (1985). Septic tank density and groundwater contamination. *Ground Water,* 23(5), 586–591.

Yates, M.V. & Yates, S.R. (1988). Modeling microbial fate in the subsurface environment. *CRC Critical Reviews in Environmental Control,* 17(4), 307–344.

Young, C.P. & Gray, E.M. (1978). *Nitrate in groundwater.* Technical Report 69, Water Research Centre, Medmenham.

Young, C.P., Hall, E.S. & Oakes, D.B. (1976). *Nitrate in groundwater.* Technical Report 31, Water Research Centre, Medmenham.

Zhu, X.Y., Xu, S.H., Zhu, J.J., Zhou, N.Q. & Wu, C.Y. (1997). Study on the contamination of fracture-karst water in Boshan District, China. *Ground Water,* 35(3), 538–545.

Zijl, W. (1999). Scale aspects of groundwater flow and transport systems. *Hydrogeology Journal,* 7, 139–150.

IAH International Contributions to Hydrogeology